Research Methods Using *R*

Research Methods Using *R*

Advanced Data Analysis in the Behavioural and Biological Sciences

Daniel H. Baker

Department of Psychology
University of York

OXFORD
UNIVERSITY PRESS

UNIVERSITY PRESS

Great Clarendon Street, Oxford, OX2 6DP,
United Kingdom

Oxford University Press is a department of the University of Oxford.
It furthers the University's objective of excellence in research, scholarship,
and education by publishing worldwide. Oxford is a registered trade mark of
Oxford University Press in the UK and in certain other countries

Published in the United States of America by Oxford University Press
198 Madison Avenue, New York, NY 10016, United States of America

British Library Cataloguing in Publication Data
Data available

Library of Congress Control Number: 2021949628

ISBN 978-0-19-289659-9

Printed in Great Britain by
Bell & Bain Ltd., Glasgow

RStudio and Shiny are trademarks of RStudio, PBC. Screenshots are taken from R Version 1.2.5033 for Mac.

Acknowledgements

First and foremost, I would like to thank all of the students who have taken my Advanced Research Methods module at the University of York over the past seven years. You have all made an invaluable contribution to this work, by asking questions, by suggesting new ways to think about methods, and through your enthusiasm and perseverance on a difficult topic. This book is for you, and for all future students who want to learn these techniques.

Many of my colleagues have had a huge influence on developing my own understanding of the topics covered in the book, especially Tim Meese, who as my PhD supervisor taught me more than I can remember about Fourier analysis, signal detection theory, and model fitting! I am further indebted to Tom Hartley, Alex Wade, Harriet Over, and Shirley-Ann Rueschemeyer, who provided invaluable feedback on the material at various stages.

I am also very grateful to Martha Bailes and the team at Oxford University Press, who have guided this project from an initial proposal through to a finished product. Martha coordinated reviews from many anonymous reviewers from diverse fields, whose comments and suggestions have strengthened the manuscript immeasurably. I am further indebted to Bronte McKeown for checking the technical content of the book, including the code and equations.

Finally, a huge thank you to my family, Gemma, Archie, and Millie, who have put up with me being buried in a laptop writing this book and tinkering with the figures for far too long!

Contents

1 Introduction

What is this book about?

Welcome to *Research Methods Using R*. The aim of this book is to take techniques that often seem inaccessible and explain the basic concepts in plain language. This is, in essence, a statistics textbook. However, it is somewhat different in flavour from other statistics texts, which in my experience fall into two camps.

Most numerous are the introductory texts, which cover a range of standard methods (descriptive statistics, t-tests, ANOVA, correlation, regression). There are many excellent examples, including *Learning Statistics with R* by Danielle Navarro (Navarro 2019), and the Discovering Statistics series by Andy Field and colleagues (Field, Miles, and Field 2012). These introductory books are appropriate for instruction at undergraduate and masters level in many disciplines within the broad umbrella of the natural and social sciences (for example biology, chemistry, neuroscience, psychology, and sociology), and knowledge of the methods they explain is a requirement of accreditation for some courses. The methods covered in this book do not necessarily require an understanding of these techniques, but I anticipate that most readers will at some point have completed an introductory statistics course that will have explained them. It would certainly not hurt to have a copy of an introductory text to hand while working through this book.

At the other extreme are specialized advanced statistics texts, filled with complex equations and aimed at trained statisticians. These books (and papers) are the absolute authority on most methods, yet they frequently assume a level of understanding and mathematical competence which frankly I do not possess! It is worth confessing from the outset that I have no advanced mathematical training, and usually aim to understand techniques at a conceptual level, rather than understand every detail of the underlying mathematics. I will attempt to do the same here, substituting diagrams and verbal explanations instead of formal equations wherever possible.

What I have aimed to produce here is a text that can be used as a starting point for the reader's use of particular techniques—a 'way in' to a method, or 'primer'. We will introduce each method by explaining the theory behind it, and then walk through some example computer code demonstrating how it can be implemented. Often, seeing a working implementation of a technique is the best way to understand how to apply it to your own research. In a sense this is a bit like a recipe book, with sections of code that can be adapted into the reader's own scripts to solve a particular problem.

Who will find this book useful?

I anticipate that advanced undergraduate and postgraduate students in the life sciences will be the most obvious audience for this book. Many of the tools are widely used in cognition, perception, neuroscience, and related topics, but there are examples from many different disciplines throughout the book. Students beginning a PhD in one of these topics will find much that is relevant—I definitely would have when I started my own PhD. It certainly doesn't hurt to already have some statistical training, e.g. from an undergraduate-level statistics course. However, at the point I started my PhD, I had forgotten most of the things I learned as an undergraduate, so this is probably not essential!

Because all of the methods are implemented in a statistical programming language called *R*, the reader should expect to develop some programming expertise as they work through the book. Chapters 2 to 4 go through some of the basics of the *R* language and demonstrate how to do some of the things I expect most readers to be familiar with already. For this reason, I think the text is suitable for a complete beginner who has never programmed before, though previous programming experience in any language should make the learning process faster.

Programming often seems a daunting prospect if you have never attempted it before, and there are many myths and stereotypes that frequently put people off. One of the most pernicious is that programming is a 'male' activity. This could not be further from the truth, and the history of computer programming—from Ada Lovelace (the first programmer) to the black female programmers who contributed to the American space programme (see the film *Hidden Figures*)—is filled with exemplary female programmers. The *R* community is explicitly and deliberately inclusive and supportive—see for example the brilliant *R-Ladies* organization (**https://rladies.org/**), which organizes meetups across the world, as well as online events, to promote gender diversity in the *R* community.

I think of learning programming as like learning a foreign language. It takes some time at first to understand the basic grammar and learn the vocabulary, but once you get the hang of the basics, programming is a very creative, and sometimes even enjoyable, skill. In the modern world it is also an extremely marketable skill, and well worth including on one's CV. So, to any readers who feel worried about the prospect of learning to program: persevere and believe in yourself. Anyone can learn to program if they put in the time and effort. It is not magic, and it is not beyond you, but it is a skill that takes practice to get the hang of.

Topics covered

The scope of this book covers a set of computational techniques that are useful in many domains of research. Because my background is in experimental psychology and perception research, most of the methods are applicable to those disciplines. However, many of the techniques are also very relevant to other fields of science. My expectation is that most readers will dip into and out of different topics, rather than read the whole book from start to finish. To this end, I have attempted to make each chapter as self-contained as possible, though most will build on (and assume knowledge of) the basic programming skills covered in Chapter 2. Most chapters begin by explaining the conceptual and theoretical uses of a technique, and end with an example implementation in the statistical programming language, *R*.

Chapter 2 introduces the *R* environment, and covers basic programming concepts including data objects, loops, conditional statements, and functions. Chapter 3 discusses how to inspect and process raw data so that it is ready for further analysis. Chapter 4 demonstrates how basic statistics that I expect will be familiar to most readers (e.g. t-tests, ANOVAs, correlations etc.) can be implemented in *R*.

Chapter 5 discusses statistical power analysis—a technique that can be used to determine how many participants should be included in an experiment. This method is becoming increasingly important in the context of the 'replication crisis', where many influential effects have failed to replicate, in part because the original studies were underpowered.

Chapter 6 introduces meta-analysis—a method for combining the results of previous studies on a single topic to estimate the overall effect size. Meta-analysis is most widely used in biomedical research, for example to synthesize the evidence regarding the effectiveness of different medicines. However, the same tools can be used in other types of research and are increasingly applied to address fundamental empirical questions.

Linear mixed-effects models are described in Chapter 7. These methods are an extension of the general linear model used in ANOVA and regression, but they are more flexible and can deal more robustly with missing data, as well as explicitly modelling variance at multiple levels.

Chapter 8 introduces a class of techniques I refer to as stochastic methods. These involve using random numbers to simulate experiments, to estimate the robustness of data by resampling, and to model systems that have random components. This is an important core skill, useful in many areas of research, but the most common application is to 'bootstrap' some statistical test by repeating it many times on resampled versions of the original data.

Chapter 9 focuses on non-linear curve fitting and function optimization; in plain language, we will see how to fit models to data that are not well described by a straight line. This is also a useful generic technique with applications in many areas of science, such as fitting models to empirical data, and predicting results in conditions that have not yet been tested.

Chapter 10 explains Fourier analysis, a mathematical technique for analysing periodic signals. This is widely used to analyse time-varying data (such as electromagnetic brain activity), and in the analysis of sound samples (e.g. of speech or music). Fourier analysis can be applied to images and even movies, and filtering operations are described that can be used to smooth noisy data or modify experimental stimuli.

Next follow four chapters on multivariate statistics—methods that can be used when you have more than one dependent measure. In Chapter 11 we discuss multivariate versions of the t-test such as Hotelling's T^2 statistic that can be used to compare conditions for which multiple dependent variables have been measured. Structural equation modelling (Chapter 12) is an extension of factor analysis and correlation, and is used to understand the structure and relationships between sets of variables. Multidimensional scaling and k-means clustering (Chapter 13) are methods for grouping complex multidimensional data and visualizing it in simpler ways. Finally, multivariate pattern analysis (MVPA; Chapter 14) is a machine learning technique that is used to categorize data into two or more groups, and is increasingly important in many research areas, including neuroimaging research and personalized medicine.

Correction for multiple comparisons is discussed in Chapter 15. We introduce traditional methods for adjusting the threshold for significance to avoid detecting effects that are not real (false positives). We then introduce two newer approaches: false discovery rate correction and cluster correction. These avoid overly stringent correction of significance criteria, which can reduce statistical power and obscure real effects.

In Chapter 16 we explain signal detection theory, a fundamental theory that underlies research in many disciplines, including perception and memory research, but is now increasingly applied in the machine learning and artificial intelligence literature. We particularly focus on forced choice methods, which are used extensively in empirical work, but often sidelined in textbooks.

Bayesian statistics are introduced in Chapter 17. The Bayesian approach uses a fundamentally different philosophy from traditional 'frequentist' statistics. In general it is better at dealing with null results and Type I errors (false positives). There are Bayesian versions of many common statistical tests.

In the final chapters, we discuss some practical considerations. In Chapter 18 we cover some techniques for graph plotting and data visualization. Unlike many works on graph plotting in *R*, we use the 'base' plotting functions instead of the popular *ggplot2* package. In Chapter 19, we discuss issues around reproducible data analysis, including methods of version control, open data formats, and automatically downloading and uploading data from public repositories.

Some words of caution

Reading a book on statistics makes you an expert statistician the same way that reading a book on juggling makes you an expert juggler. It doesn't. The only way to learn is by doing, by practising, by making mistakes, and by working things out for yourself. The same goes for programming, and most other worthwhile skills. This book aims to give you a grounding in the theory of how the methods work, and some examples of how to implement them. Applying this to your own research will be challenging and it will take time, but that is the only way to really learn.

A further caveat: this book is not meant to be the final word on any method or technique. It is intended to be an introduction (a primer) to the methods, and a starting point for your own reading and learning. So there will be lots of things I don't mention, and probably many faster or more efficient ways of programming something. Also, the *R* community is very fast moving, and new packages are being created all the time. So it is worthwhile checking online for package updates with new features, and also for more recent packages for a given method.

Implementation in *R*

Most of the methods covered in this book are not implemented in commercial statistics packages such as SPSS and SAS. Instead, we use a statistical programming language called *R* for all examples (*R* Core Team 2018). *R* has the advantage that it is an open source language, so anyone can develop their own packages (collections of code that implement statistical tests) for others to use. It is now standard practice for papers describing a new statistical technique to have an associated *R* package for readers to download. *R* is free to download, and can be installed on most operating systems. There are also some online *R* clients that can be accessed directly through a web browser without needing any installation. I mostly used another language, Matlab, throughout my PhD and postdoctoral years, and many of the methods we discuss here can also be implemented in Matlab or other programming languages such as Python. However, in the interests of consistency, we use a single language throughout, which is introduced in Chapter 2. I have also made the R code for all examples and figures available on a GitHub repository at: **https://github.com/bakerdh/ARMbookOUP**.

History

This book has grown from a lecture course that I developed in 2014 and deliver twice each year at the University of York. The course is taught to third year undergraduates and is often audited by postgraduate students on MSc and PhD courses. Many former students have gone on to have successful careers in numerate areas. I try to make the material as accessible as possible, and sometimes use props like 3D-printed surfaces to demonstrate function optimization, and a bingo machine to explain bootstrapping! The biggest concern students have is about whether they will understand the mathematical content for advanced statistical methods. For this reason I deliberately avoid equations where possible, preferring instead to explain things at a conceptual level, and I have done the same in this book.

The topics taught on the module go beyond the core undergraduate research methods syllabus, and so are rarely discussed in typical textbooks. Up until now, I have mostly used tutorial papers and package manuals as the recommended reading. But this is not ideal, and I have always felt that students on the module would benefit from a single text that presents everything in a common style. Producing a written explanation of the course content seemed the natural next step, and my publishers agreed! The original lecture content equates to about half of the topics included here, with other content added that seemed related and useful.

Introduction to the *R* environment

This chapter introduces the *R* programming language. It can be safely skipped by those already well versed in *R*. However it is worth at least skim-reading for those new to the language, even if they have other programming experience. We will first introduce the language and developer environment, and explain how to adjust the appearance to suit your needs. We will then go through examples of *R* syntax for fundamental programming concepts such as scripts, data objects, functions, packages, conditional statements, and loops. Finally, we will discuss how to import data and solve problems, finishing with a table of useful *R* functions and some practice questions.

Note that this chapter is intended to cover the basics of *R* in enough detail for you to understand the rest of the book. However, if it seems overwhelming you might also find it helpful to look at some online resources that teach the basics of *R* at a slower pace. There are many free resources available (including instructional videos), as well as some paid-for courses, and both online and print textbooks. This is a rapidly evolving area, so it is worth seeing what is available at the time of reading.

What is *R*?

R is a statistical programming language. It has been around since the early 1990s, but is heavily based on an earlier language called *S*. It is primarily an interpreted language, meaning that we can run programs directly, rather than having to translate them into machine code first (though a compiler which does this is also available). Its particular strengths lie in the manipulation, analysis, and visualization of data of various kinds. Because of this focus, there are other tasks that it is less well suited to—you probably wouldn't use it to write a computer game, for example. However, it is rapidly increasing in popularity and is currently the language of choice for statisticians and data analysts. We will use it exclusively throughout this book for all practical examples.

The core *R* language is maintained by a group of around 20 developers known as the *R Development Core Team*. The language is freely available for all major operating systems (Windows, Mac, and various flavours of Linux), and the *R Foundation* is a not-for-profit organization. This means that *R* is an inherently free software project—nobody has to pay to use it. This is quite different from many other well-known programming languages and statistical software packages that you may have come across. *R* is available to anyone with a computer and internet connection, anywhere in the world, regardless of institutional affiliation or financial circumstances. This seems to me an inherently good thing.

You can download *R* from the *R project* website (**https://www.r-project.org/**). It is hosted by the Comprehensive R Archive Network (CRAN), a collection of around 100 servers, mostly based at universities across the world. The CRAN servers all mirror the same content, ensuring that if one goes down the software is still available from the rest. The CRAN mirrors also contain repositories of *R* packages, which we will discuss further in 'Packages' later in the chapter. If you do not have *R* installed on your computer and plan to start using it, now would be a good time to download it. Exactly how the installation works depends on your computer and its operating system, but instructions are available for all systems at the *R* project website.

RStudio

In parallel with the development of the core *R* language, a substantial amount of work has been done by a company called *RStudio*. This is a public benefit corporation that makes some money from selling things like 'pro' and enterprise versions of its software, web hosting, and technical support. However its primary product, the *RStudio* program, is free and open source. *RStudio* (the program) is an integrated development environment (IDE) for *R*. It has a number of user-friendly features that are absent from the core *R* distribution, and is now the most widely used *R* environment (again it is available for all major operating systems). I strongly recommend downloading and installing it from the *RStudio* website (**https://www. rstudio.com/**). Note that *RStudio* requires that you already have a working *R* installation on your computer, as it is a separate program that sits 'on top of' *R* itself. So you need to first install *R*, and then install *RStudio* in order for it to work. If you have insurmountable problems installing *R* on your computer, there are now web-based versions that run entirely through a browser and do not require installation, such as RStudio Cloud (**https://rstudio.cloud/**) and rdrr.io (**https://rdrr.io/snippets/**), though these services may not be free.

The *RStudio* company has also produced a number of very well-written and useful *R* packages designed to do various things. For example, the *Shiny* package can produce dynamic interfaces to *R* code that can run in web browsers. The *RMarkdown* package can produce documents that combine text, computer code, and figures (discussed further in 'Using R Markdown to combine writing and analysis' in Chapter 19). A version of it (*bookdown*) was used to create the first draft of this book. Finally, a suite of tools collectively referred to as the *Tidyverse* offer a uniform approach to organizing and manipulating data (essentially storing even very complex data structures in a spreadsheet-like format). Although many introductions to *R* now focus on these tools, they are advanced-level features that are not required to implement the examples in this book, and so we will not discuss them further.

Finding your way around *RStudio*

On my (Apple Macintosh) computer, the default window for *RStudio* looks like that shown in Figure 2.1. In the lower left corner is the Console. You can use this section to type in commands, which *R* will execute immediately. One way to think of it is a bit like a calculator. In fact, you can type in basic sums and it will give you the answer. Try it now—type in 1 + 1 and press return. Hopefully it will give you the answer, 2. All code that is executed during an *R* session is echoed to the Console, along with any output such as the results of statistical tests.

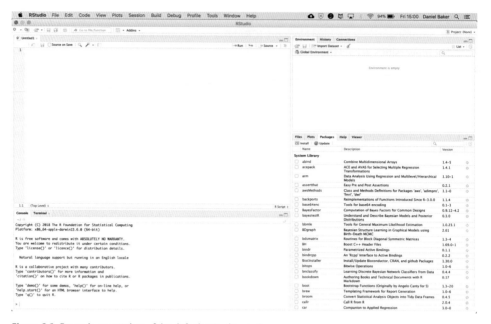

Figure 2.1 Example screenshot of the default *RStudio* window.

An alternative to entering commands directly in the Console is to write a *script* containing instructions. We will go into more detail about scripts in the section 'Scripts' later in the chapter, but for now note that they appear in the upper left corner of the *RStudio* window.

In the upper right corner is the *Environment* tab, which contains a summary of all of the information currently stored in *R*'s memory. This will usually be empty when you first launch *R*, but will fill up with things like data sets and the results of various analyses. There are other tabs in the upper right corner, including the *History* tab which keeps track of all commands executed in the current *R* session. The *Environment* and *History* tabs can be saved to files (using the disk icon), and also loaded back in from a previous session. The broom icon, which appears in several panels, empties the contents of the *Environment* (or other section).

The lower right corner of the *RStudio* window contains several more tabs. The first is the *Files* tab, which contains a file browser, allowing you to open scripts and data files from within *RStudio*. The second tab is the *Plots* tab, which will display any graphs you create. You can also export graphs from this window, and navigate backwards and forwards through multiple graphs using the left and right arrow buttons.

The next tab in the lower right pane is the *Packages* list, which contains all of the packages of *R* code currently installed on your computer. We discuss packages in more detail in the 'Packages' section later in the chapter. The *Help* tab is also in this section of the window, and is used to display help files for functions and packages when you request them.

As with most other desktop applications, there are a number of drop-down menus that allow you to do various things such as load and save files, copy and paste text, and so on. I use these much less frequently than in other programs, largely because many of the functions are duplicated in the various toolbars in the main *RStudio* window. Overall, the *RStudio* package provides a user-friendly environment for developing and running *R* code.

Customizing *RStudio* for your own needs

In common with other contemporary software, *RStudio* has several customizable options to improve accessibility. For example, in the *Preferences* menu (accessed through the *File* menu) there is a section called *Appearance*. This allows you to change the font and text size, and also to zoom the entire window in and out. You can choose between a wide range of different colour themes for the application. These are helpful because they change the text colour for different components of an *R* script, which can make code easier to navigate. It is also possible to change the background colour, for users who find reading easier against a particular background. Finally, there are options to change the layout of the different panels in the *RStudio* window, and features for spellchecking and autocompleting text. Since different users will have different requirements, it is best to have a look at these settings and see what works well for you.

Scripts

A script is a text file that contains computer code. Instead of entering commands one at a time into the *R* Console, we can type them into a script. This allows us to save our work and return to it later, and also to share scripts with others. Although this sounds like a simple concept, it has enormous benefits. First of all, we have an exact record of exactly what we did in our analysis that we can refer back to later. Second, someone else could check the code for errors, reproduce your analysis themselves, and so on. This is very different from the way most people use statistical packages with graphical interfaces like SPSS (although technically it is possible to record the analysis steps in these packages, researchers rarely do this in practice). Sharing scripts online through websites like the *Open Science Framework* is becoming commonplace, as part of a drive for openness and reproducibility in research, as we will discuss in greater detail in Chapter 19.

In *RStudio* we can create a new script through the *File* menu. These traditionally have the file extension .R—however they are just plain text files, so you can open and edit them in any normal text editor if you want to. It is a good idea to create scripts for everything you do in *R* and save them somewhere for later reference. *R* works slightly differently from some other programming languages, in that you can easily run chunks of code from a script without having to run the whole thing. You do this by highlighting the section of code you wish to execute (using the mouse) and then clicking the *Run* icon (the green arrow at the top of the script panel). You will see the lines of code reproduced in the Console, along with any output.

Another way to run a script is to use the *source* command, and provide it with the location and file name of the script you wish to run. This will execute the entire script with a single command, as follows:

```
source('~/MyScripts/script.R')
```

You can use the *source* command from the Console, and also include it in a script so that you execute the code from multiple scripts in sequence.

Data objects

A key concept in all programming languages is the *data object* (sometimes called a *variable*, though we will not this term here because of the potential for confusion with variables in the context of statistics). You can think of a data object as a container that you can keep information in. The information will usually be numbers or text. In some programming languages it is important to specify in advance what the data object is going to contain, but *R* will usually sort this out for you without you having to worry about it. Every data object has a name, which can be anything you like as long as it doesn't start with a number, or contain any spaces or reserved characters (some symbols such as @). We will discuss choosing good names in the section 'Writing code for others to read' in Chapter 19.

In general it is advisable to give data objects a meaningful name. A data object containing your birthday could be called *MyBirthday*, rather than something arbitrary such as *a*. To create a data object and store some information in it, we traditionally use the <- operator, though current versions of *R* also permit the = operator. Here is a simple example:

```
a <- 10
```

In this line of code, we have created a data object called *a* and stored the number 10 in it. We read the <- operator as 'is given the value'. So this example means '*a* is given the value 10'. This data object will then appear in the *Environment* panel of the *RStudio* window. If we want the data object to contain a string of text, we enclose the text in inverted commas (quotes) so that *R* knows it is text data, rather than a reference to another data object:

```
b <- 'Hello'
```

Once a data object has been created, we can use it in calculations, in much the same way that mathematicians use letters to represent numbers in algebraic expressions. For example, having stored the number 10 in the object *a*, we can multiply it by other numbers as follows:

```
a*2    # multiply the contents of 'a' by 2
## [1] 20
```

Data objects can contain more than one piece of information. A list of numbers is called a vector, and can be generated in several ways. A sequence of integers can be defined using a colon:

```
numvect <- 11:20
numvect
##  [1] 11 12 13 14 15 16 17 18 19 20
```

Or we can combine several values using the *c* (concatenate) operation, which here has the same result:

```
numvect <- c(11,12,13,14,15,16,17,18,19,20)
numvect
##  [1] 11 12 13 14 15 16 17 18 19 20
```

Just as for a single value, we can perform operations on the whole vector of numbers. For example:

```
numvect^2  # raise the contents of 'numvect' to the power 2
##  [1] 121 144 169 196 225 256 289 324 361 400
```

The above code calculates the square of each value in the data object *numvect* (the ^ symbol indicates raising to a power). We can also request (or *index*) particular values within a vector, using square brackets after the name of the data object. So, if we want to know just the fourth value in the *numvect* object, we can ask for it like this:

```
numvect[4]   # return the 4th value in 'numvect'
## [1] 14
```

If we want a range of values we can index them using the colon operator:

```
numvect[3:8] # just entries 3 to 8 of the vector
## [1] 13 14 15 16 17 18
```

And if we want some specific entries that are not contiguous, we can again use the *c* (concatenate) function:

```
numvect[c(1,5,7,9)] # some specific entries from the vector
## [1] 11 15 17 19
```

Finally, we can use other data objects as our indices. For example:

```
n <- 6
numvect[n]
## [1] 16
```

Data objects can also have more than one dimension. A two-dimensional data object is like a spreadsheet with rows and columns, and is referred to as a *matrix*. We need to tell R how big a matrix is going to be so that it can reserve the right amount of memory to store it in. The following line of code generates a matrix with ten rows and ten columns, storing the values from 1 to 100:

```
d <- matrix(1:100, nrow=10, ncol=10)
d
##        [,1] [,2] [,3] [,4] [,5] [,6] [,7] [,8] [,9] [,10]
##  [1,]    1   11   21   31   41   51   61   71   81    91
##  [2,]    2   12   22   32   42   52   62   72   82    92
##  [3,]    3   13   23   33   43   53   63   73   83    93
##  [4,]    4   14   24   34   44   54   64   74   84    94
##  [5,]    5   15   25   35   45   55   65   75   85    95
##  [6,]    6   16   26   36   46   56   66   76   86    96
##  [7,]    7   17   27   37   47   57   67   77   87    97
##  [8,]    8   18   28   38   48   58   68   78   88    98
##  [9,]    9   19   29   39   49   59   69   79   89    99
## [10,]   10   20   30   40   50   60   70   80   90   100
```

Again, we can index a particular value. For example, if we want the number from the 8th row and the 4th column, we can ask for it by adding the indices 8,4 in square brackets after the data object name:

```
d[8,4]
## [1] 38
```

This is very similar to the way you can refer to rows and columns in spreadsheet software such as Microsoft Excel, except that in *R* rows and columns are both indexed using numbers (whereas Excel uses letters for columns). If we want all of the rows or all of the columns, omitting the number (i.e. leaving a blank before or after the comma) will request this. For example:

```
d[8,] # row 8 with all columns
##   [1]  8 18 28 38 48 58 68 78 88 98
d[,4] # column 4 with all rows
##   [1] 31 32 33 34 35 36 37 38 39 40
```

We can also request a range of values for rows and/or columns using the colon operator:

```
d[1:3,5:7] # rows 1:3 of columns 5:7
##       [,1] [,2] [,3]
## [1,]   41   51   61
## [2,]   42   52   62
## [3,]   43   53   63
```

Data objects are not limited to having only two dimensions. In *R*, objects with three or more dimensions are called *arrays*, and will be introduced when required throughout the book.

A particularly useful class of data object in *R* is the *data frame*. This is very similar to a matrix, but each column can contain a different type of data (whereas in a matrix we cannot combine numbers and text in different columns). Columns and rows can have headings to help identify what they contain. This is very similar indeed to spreadsheets in software packages like Excel, and this is a helpful way to think about them. The following code produces a simple data frame containing the top five greatest songwriters according to *Rolling Stone* magazine:

```
Position <- 1:5
Songwriter <- c('Bob Dylan','Paul McCartney',
      'John Lennon','Chuck Berry','Smokey Robinson')
chart <- data.frame(Position, Songwriter)
chart
##   Position        Songwriter
## 1        1         Bob Dylan
## 2        2    Paul McCartney
## 3        3       John Lennon
## 4        4       Chuck Berry
## 5        5   Smokey Robinson
```

In the above code, we first created two vectors called *Position* and *Songwriter*. Then we used the *data.frame* function to combine these together into a single data object that I've called *chart*. When we look at the *chart* object (by typing its name), we can see that the data are organized into two columns and five rows. The first column contains the chart position, and the second column contains the name of the songwriter. The *Environment* panel will also contain the two vectors and the data frame, as shown in Figure 2.2.

Data frames can have as many rows and columns as you like, so they are a very general and flexible way to store and manipulate data of different kinds. There are numerous other classes of data object in *R*, and new classes can be defined when required, so we will discuss any other data types as they come up throughout the book.

It is important to have a good conceptual understanding of data objects, because this is how *R* stores information. If you load in a data set from an external file, this will be stored in

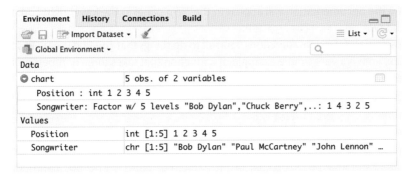

Figure 2.2 Example screenshot of the *Environment* panel, showing two vectors, and also a data frame that contains both vectors.

a data object. You might, for example, have a spreadsheet file containing questionnaire responses, which you could load into *R* and store in a data frame called *questionnaire*. Similarly, when you run a statistical test, the results will typically be stored in a data object as well. If you run a t-test (see Chapter 4 for details of how to do this), this will produce a data object that you might call *ttestresults*, and will contain the t-statistic, the p-value, the degrees of freedom, and lots of other useful information.

Functions

The real power of high-level programming languages is the use of *functions*. A function is a section of code that is wrapped up neatly. It accepts one or more *inputs*, and produces one or more *outputs*. Just like data objects, every function also has a name (subject to similar restrictions, in that they cannot start with a number, or contain any reserved characters or spaces). Indeed, any functions that you create will also appear in the *Environment*, and can be considered a type of object. There are many hundreds of functions built into *R*, which do different useful things. A very simple example is the *mean* function, which calculates the mean (average) of its inputs. The following code calculates the mean of the numbers from 1 to 10:

```
mean(1:10)
## [1] 5.5
```

Notice that the inputs to the function are inside the brackets after the function name. Functions can also take data objects as their inputs, and store their outputs in other data objects. Earlier on, we stored the numbers from 11 to 20 in a data object called *numvect*. So we can *pass* these values into the *mean* function, and store the output in a new object as follows:

```
averageval <- mean(numvect)
averageval
## [1] 15.5
```

You can find out more about a function using the *help* function, and passing it the name of the function you are interested in:

```
help(mean)
```

In *RStudio* the documentation will appear in the *Help* panel in the lower right corner of the main window.

It is possible to define new functions yourself to do things that you find useful. To show you the syntax, here is a very simple example of a function that takes three numbers as inputs. It adds the first two together, and divides by the third:

```
addanddivide <- function(num1,num2,num3){
  output <- (num1 + num2)/num3
  return(output)
}
```

The name of the function (*addanddivide*) is defined as being a function with three inputs (*num1*, *num2*, and *num3*). These inputs become data objects inside the function, and can be used in the same way as any other data object, though they do not appear in the *Environment*, and are available only from within the function.

All of the operations involved in the function then appear inside the curly brackets. In this example, we have a line of code that adds two numbers together and divides by the third number, and stores this in a new data object called *output*. The *return* argument tells *R* to pass the *output* object back out of the function so it is available in the *Environment*. We create the function by executing (running) the code that defines it. Once it has been created, it appears in the *Environment*, and we can call our new function just like we'd call any other built-in function:

```
addanddivide(7,5,3)
## [1] 4
```

With these inputs (7, 5, and 3) we get the output 4, because (7+5)/3 = 4. But now that we have created the function, we can call it as many times as we like, with any inputs. You can also call functions from within other functions, meaning that operations can be nested within each other, producing sequences of arbitrary complexity.

Packages

Sets of useful functions with a common theme are collected into *packages*. There are many of these built into the basic *R* distribution, which you can see by clicking on the *Packages* tab in the lower right panel of the main *RStudio* window (an example is shown in Figure 2.3(a)). You can also install new packages to perform specific functions. Those meeting some basic quality standards are available through CRAN (at the time of writing over 10000 packages), but it is also possible to download other packages and install them from a package archive file. In *RStudio*, the package manager has a graphical interface for installing and updating packages (click the Install button in the *Packages* tab). The window that pops up (see Figure 2.3(b)) allows you to install any packages by typing the package name into the dialogue box. However you can also install packages from the Console or within a script using the *install.packages* command. For example, the following code will download and install the *zip* package used for compressing files:

```
install.packages('zip')
```

(a)

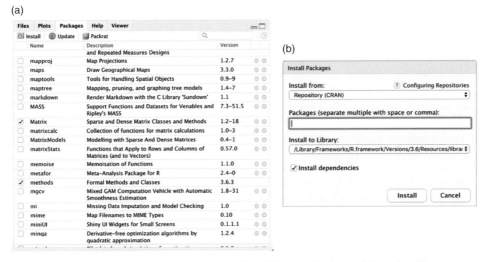

(b)

Figure 2.3 Example screenshot of the *Packages* panel (a) and the Install Packages dialogue box (b).

Once installed, packages need to be activated before they are visible to *R* and their functions become available for use. This can be done manually by clicking the checkbox in the packages list (see examples in Figure 2.3(a)—the *Matrix* and *methods* packages are active), or using the *library* function in the Console or in a script. For example, we can activate the *zip* package that we just downloaded like this:

```
library(zip)
```

One consequence of the large number of available packages is that sometimes there are name conflicts, where two packages contain different functions with the same name. To specify which package a function comes from, we can include the package name before the function call, separated by two colons. For example, the following code specifies the *median* function from the *stats* package (though it is unlikely that this particular function would have a name conflict):

```
stats::median(1:10)
##  [1] 5.5
```

We will use several packages throughout the book, and some of the more important ones are summarized in the alphabetical list of key *R* packages at the end of the book.

Conditional statements

Sometimes in a computer program, we want to run some lines of code only if particular conditions are met, and skip them otherwise. We achieve this using something called a *conditional statement*. The most common is the *if* statement, which has the form 'if X then run this code'. 'X' in this context will be a logical statement, like $(a > 5)$, meaning '*a* is greater than 5', or (a == 1), meaning 'a is equal to 1'. If the statement is evaluated as TRUE, the subsequent code will be run. If the statement is FALSE, the code will be ignored. Here is a short example:

```
a <- 1

if (a==1) {
  print('a is equal to 1')
}
## [1] "a is equal to 1"
```

If a has a different value, the code inside the *if* statement (inside the curly brackets) will not execute (note no output is printed):

```
a <- 0

if (a==1) {
  print('a is equal to 1')
}
```

We can augment an *if* statement by telling *R* what to do if the condition is *not* met—the *else* argument:

```
a <- 0

if (a==1) {
  print('a is equal to 1')
} else {
  print('a is not equal to 1')
}
## [1] "a is not equal to 1"
```

This form is known as an *if...else* statement, and can even be extended with many other conditional statements as follows:

```
a <- -1

if (a>0) {
  print('a is a positive number')
} else if (a<0) {
  print('a is a negative number')
} else {
  print('a is zero')
}
## [1] "a is a negative number"
```

This sequence first checks if the value of a is positive (a > 0). If it is, it produces a message to tell us. Next it checks if it is negative (a < 0), and tells us if it is. If neither of these conditions is met, it concludes that a must equal zero, and tells us that. Of course, in a real computer program, we would do something more meaningful inside our *if* statement. I often use them to select whether to export figures as files rather than draw them to the plot window, or whether to load in raw data or data that have already been processed.

Logical statements can involve calls to various functions. Two useful ones are the *is.na* and *is.infinite* functions. These check if data are classed as *not a number* (for example if values are missing, or are irrational numbers such as the square root of −1), or infinite values.

The functions return TRUE or FALSE values, which are interpreted appropriately by the *if* statement. These functions are useful for preventing operations that will cause a script to crash, for example if a missing or infinite number is used in a calculation.

Loops

Another powerful feature of programming languages is the *loop*. A loop instructs the computer to repeat the same section of code many times, which it will typically do extremely fast. The simplest type of loop is the *for* loop. This repeats the operations inside the loop a fixed number of times. For example, the following code will print out the word 'Hello' 10 times:

```
for (n in 1:10) {
  print('Hello')
}
```

```
## [1] "Hello"
## [1] "Hello"
## [1] "Hello"
## [1] "Hello"
## [1] "Hello"
## [1] "Hello"
## [1] "Hello"
## [1] "Hello"
## [1] "Hello"
## [1] "Hello"
```

The terms in the brackets (n in 1:10) define the behaviour of the loop. They initialize a *counter*, which can have any name, but here is called *n*. The value of *n* increases by one each time around the loop, between the values of 1 and 10. We could change these numbers to span any range we like. The instructions within the curly brackets {print('Hello')} tell *R* what to do each time around the loop.

We can also incorporate the value of the counter into our loop instructions, because it is just a data object containing a single value. In the following example, we print out the square of the counter:

```
for (n in 1:10) {
  print(n^2)
}
```

```
## [1] 1
## [1] 4
## [1] 9
## [1] 16
## [1] 25
## [1] 36
## [1] 49
## [1] 64
## [1] 81
## [1] 100
```

A really useful trick is to use the counter to index another data object. In the following example, we store the result of an equation in the nth position of the data object called *output*:

```
output <- NULL # create an empty data object to store the results
for (n in 1:10) {
  output[n] <- n*10 + 5
}
output
##   [1]  15 25 35 45 55 65 75 85 95 105
```

Of course we can have as many lines of code inside a loop as we like, and these can call functions, load in data files, and so on. It is also possible to embed loops inside other loops to produce more complex code structures.

There are two other types of loop available in R, called *while* and *repeat* loops. These do not always repeat a fixed number of times. Their *termination criteria* are defined so that the loop exits when certain conditions are met. For example, a *while* loop will continue repeating as long as a particular logical statement is satisfied. The following code is a *while* loop that adds a random number to a counter object on every iteration. The loop continues while the counter value is less than 5, and exits when the value of the counter is greater than (or equal to) 5. If you run this code several times, the loop will repeat a different number of times on each execution.

```
counter <- 0
while (counter<5) {
  counter <- counter + runif(1)
  print(counter)
}

## [1] 0.4270285
## [1] 0.5372249
## [1] 0.9300624
## [1] 1.41297
## [1] 2.092478
## [1] 3.04567
## [1] 3.17457
## [1] 3.936555
## [1] 4.25802
## [1] 4.520581
## [1] 4.762705
## [1] 5.5585
```

A *repeat until* loop is very similar, except that the conditions for terminating the loop are evaluated (checked) at the end of the loop rather than at the start (as with a *while* loop). These are used less frequently but are appropriate in some circumstances.

Importing data

Most of the time, you will want to use R to process and analyse your own data. There are many functions for importing data into R, including packages to deal with specific file formats. For example the *readxl* package allows you to load Microsoft Excel files, and the *R.matlab* package reads the Matlab .mat file format. A really useful built-in function is *read.csv*, which reads in comma-separated-values text files. This kind of file is a plain text spreadsheet format, which can be exported from various software packages, where the distinct values (cells in the spreadsheet) are separated by commas. By default, the data you read in are stored in a data frame, with the option of treating the first row as the column headings. The following line of code will do this:

```
data <- read.csv('filename.csv',header=TRUE)
```

Whereas if you want to treat the first row as data values (and specify your column headings separately), you would enter:

```
data <- read.csv('filename.csv',header=FALSE)
```

The spreadsheet contents will be stored as a data frame in the *Environment* (the memory) of *R*. As noted earlier, there are so many *R* packages now that you will most likely be able to find a function to read in virtually any file format you need, even including specialist data formats like MRI images.

A helpful feature of *RStudio* is the *Import Dataset* option from the *File* menu. This provides a graphical interface for importing data that is stored in widely used formats (including Excel and SPSS). What is especially clever is that the *R* code for loading in the data is automatically generated and sent to the Console. This means you can copy the code into a script so that loading data is automated in the future. In Chapter 3, we will discuss how to inspect and clean up data that you have imported.

How to find out how to do something

As mentioned, the *help* function is a great way of finding out more about a specific *R* function. However this is not much use if you don't already know the name of the function, or the package it is part of. Back in the early years of programming, people used printed coding manuals for a given language. These were pretty hard to navigate, and could only give you information about the core functions of a language. But that was before the internet. All of the documentation for every *R* package is available online, and there are many support forums (websites like *Stack Overflow*: **https://stackoverflow.com/**) where you can read the answers to coding questions posted by others. All of this information means it is now possible to just type a question into a search engine and get a useful answer almost immediately.

People who are new to programming often feel like this is cheating somehow, and that they should magically know the answer to their question already. This is not true. Everyone uses search engines to find out how to do something, or to remind themselves of the name of a function, or the specific syntax they need. I do this all the time—almost everything in this book I have worked out how to do by reading about it online. Expert professional programmers do it all the time too. Often if you ask someone who is a more experienced programmer for help, they will actually just search for the answer on the internet. This is such a truism that there is a whole subgenre of internet memes about how programmers all have to Google things all the time. There is no shame in this—it's the best way to learn.

Table of useful core *R* functions

Table 2.1 lists some core *R* functions that I find particularly useful. The table omits plotting functions, which are discussed in Chapter 18. At the end of the book there is also a list of the main packages we use throughout the book, with some key functions highlighted.

Table 2.1 Table of useful core R functions with brief descriptions.

Function	Description
mean	Calculates the average of a vector of numbers
sd	Calculates the standard deviation of a vector of numbers
sqrt	Calculates the square root of its inputs
nrow	Returns the number of rows in a matrix
ncol	Returns the number of columns in a matrix
dim	Returns the dimensions of a data object
rowMeans	Calculates the mean of each row in a matrix
colMeans	Calculates the mean of each column in a matrix
seq	Generates a sequence of numbers with specified spacing
rep	Generates a repeating sequence of numbers
abs	Calculates the absolute value (removes the sign)
sign	Returns the sign of the input (−1, 0, or 1)
pmatch	Partial matching of strings
which	Returns the indices of items satisfying a logical condition
is.nan	Returns TRUE for any values that are not a number
is.infinite	Returns TRUE for any values that are infinite
apply	Applies a function over specified dimensions of a matrix
%%	Returns the remainder (modulus) for integer division
%/%	Returns the quotient (the bit that's not the remainder)
unique	Removes duplicate values from a vector
paste	Combines two or more strings into a single string

Practice questions

1. What is the name of the language that R is based on?
   A) SPSS
   B) S
   C) Matlab
   D) BASIC

2. In RStudio, you can see the data objects currently held in memory using the:
   A) Environment tab
   B) History tab
   C) Viewer tab
   D) Console

3. Which of the following is **not** a benefit of using scripts?
   A) You can share your code with others
   B) Your code will run in other programming languages such as Python
   C) You can reproduce your own analysis in the future
   D) You can save your work and return to it later

4. Which of the following is **not** a legal object name in R?
   A) Jonathan
   B) a
   C) var125
   D) 9thNumber

5. In *R*, a data object with rows and columns can be called a:
   A) Vector or scalar
   B) Array or list
   C) Matrix or data frame
   D) Spreadsheet variable

6. To refer to a particular value within a data object in *R*, we index it using:
   A) Normal brackets ( )
   B) Square brackets [ ]
   C) Curly brackets { }
   D) Pointy brackets < >

7. Collections of functions on a common theme are called:
   A) Repositories
   B) Functions
   C) Toolboxes
   D) Packages

8. How many times will the following loop repeat? *for (n in 31:35){}*
   A) 1
   B) 5
   C) 31
   D) 35

9. What will be the output of the following lines of code?
   ```
   a <- 0
   for (n in 1:3) {
      a <- a + 4
   }
   if (a>10) {print('Bananas')}
   if (a<10) {print('Apples')}
   if (a==10) {print('Pears')}
   if (a==12) {print('Oranges')}
   ```
   A) Bananas
   B) Apples
   C) Oranges
   D) Bananas and Oranges

10. What is the name of the *R* function used for transposing a matrix? (Hint: the answer is not given anywhere in this chapter, but the section on how to find out how to do something might help!)
    A) *tp*
    B) *trans*
    C) *tr*
    D) *t*

Answers to all questions are provided in the answers to practice questions at the end of the book.

3 Cleaning and preparing data for analysis

Why do data need to be 'cleaned'?

Real data are messy. In any study a number of factors, such as equipment malfunction, participant variability, experimenter error, environmental noise, data entry mistakes, or file corruption, can have an effect on your results. Hopefully such problems are not too severe, and will affect relatively few of your measurements. But there is no way to know this without somehow checking your data. In this chapter we recommend initial visual inspection, and illustrate two common methods for checking data visually (histograms and scatterplots). We will then discuss how to identify and deal with outliers more formally, and discuss several useful methods of checking distributions and rescaling data. However, we begin with the practical issue of how to format data appropriately for use in *R*.

Organizing data in *wide* and *long* formats

There are two main conventions for storing data in *R*. They are known as *wide* and *long* formats. The idea of wide format is that each cell in a matrix stores a single observation (i.e. a measurement of the dependent variable), and the location in the matrix gives us information about where that observation has come from. So, consider an experiment where eight participants each complete five repetitions of some task. Their data could be stored in an 8 (rows) by 5 (columns) matrix, much as we might arrange them in a spreadsheet. Each row will represent a different participant, and each column will represent a different repetition. The data might look something like this:

```
##         [,1] [,2] [,3] [,4] [,5]
## [1,]    122   90  174  149  125
## [2,]     87  117  168  152  157
## [3,]    102   77  163  139  159
## [4,]    187  181  116  143  108
## [5,]     31   93  127  133  161
## [6,]     88   45  126   41  148
## [7,]    155  122  142    1   88
## [8,]     44  111  147   68  103
```

An alternative way to lay out the data is long format, in which each observation has its own row. In each row, one column will store the value of the observation, and any other columns will store associated condition information about that observation. When stored in a data frame, the column headings should indicate the information stored in them (much like variable names in SPSS or Excel). The data set from the above example laid out in long format would look (for the first 10 entries) like this:

```
##  Participant Repetition Measurement
## 1            1          1         122
## 2            2          1          87
## 3            3          1         102
## 4            4          1         187
## 5            5          1          31
## 6            6          1          88
## 7            7          1         155
## 8            8          1          44
## 9            1          2          90
## 10           2          2         117
```

Notice that the measurement values are the same as those in the wide table. The numbers in the participant and repetition columns correspond to the row and column numbers from the wide version of the data set.

Readers familiar with running ANOVA in the widely used SPSS package may recall that, in that software, repeated measures data are stored in wide format, whereas between subjects data are stored in long format. In R, the decision about how to lay out your data will vary depending on your experimental design, and also how you are planning to analyse the data. Many R functions assume long format, as we will see in Chapter 4, and so for running tests like regression and ANOVA this may be preferred. However, some of the more sophisticated methods we will look at in later chapters, which require the user to store data in an efficient way, might tend to work better in wide format. A good example might be an image—representing each pixel intensity in a two-dimensional matrix is much more natural than representing the whole image as a long list of intensities.

To facilitate switching between long and wide formats, there are several helper functions. The *reshape2* package contains functions called *melt* (for converting from wide to long) and *acast* (for going the other way). Alternatively, the newer *tidyverse* framework offers functions called *gather* and *spread* which fulfil analogous roles as part of the *tidyr* package. Since requirements will differ depending on the idiosyncrasies of a data set, the help documentation for these functions will likely prove useful if your data need restructuring. A final useful function for restructuring data is the transpose function, *t*. This function turns rows into columns, and columns into rows, much more straightforwardly than can be done using spreadsheet software, where copying and pasting with special options is required.

Inspecting data: histograms and scatterplots

The best way to understand what is happening in a data set is usually to look at it. There are many methods of visualization that are useful, but two of the most commonly used are histograms (for each individual dependent variable) and scatterplots (for pairs of variables). A histogram counts the number of observations within *bins* of a particular size. Figure 3.1(a)

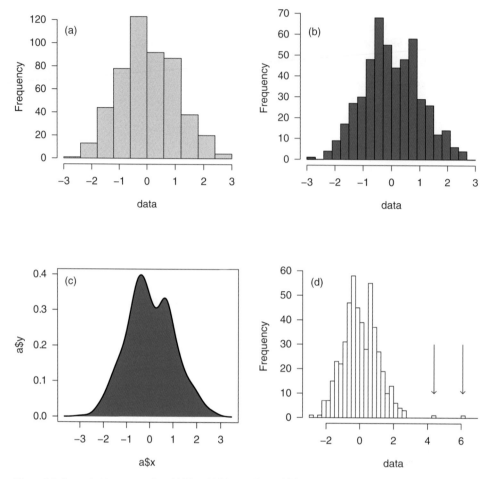

Figure 3.1 Example histograms. Panel (a) has 10 bins, and panel (b) has 20 bins. Panel (c) shows a kernel density function, and panel (d) includes two extreme outliers.

and (b) show two histograms with different bin sizes, generated using the *hist* function with the following code:

```
hist(data,breaks=seq(-3,3,length=11),col='grey')
hist(data,breaks=seq(-3,3,length=21),col='cornflowerblue')
```

The entry to the *breaks* argument defines the start and end points of each bin. The *hist* function will add up the total number of values in the data set between the lower and upper boundaries of each bin. To define these boundaries, the *seq* function generates a sequence of numbers between two points, here with a specified length. For example:

```
seq(-3,3,length=11)
##   [1] -3.0 -2.4 -1.8 -1.2 -0.6 0.0 0.6 1.2 1.8 2.4 3.0
```

So this line of code generates a sequence of numbers between −3 and 3, with 11 evenly spaced values. The *hist* function will then use this sequence to bin the data. Note that there will always be one bin fewer than the length of the sequence, and that the midpoints of the bins are the means of successive pairs of break points.

A popular alternative to showing histograms with discrete bins is to plot a smoothed *kernel density function*. The smoothing can sometimes obscure discontinuities in the data, but smoothed functions are often more visually appealing. An example is shown in Figure 3.1(c), created using the *density* function as follows:

```
a <- density(data)
plot(a$x,a$y,type='l',lwd=2)
```

The histograms in Figure 3.1(a)–(c) show data that are approximately normally distributed, and have no obvious outliers. But real data are often much less clean. For example, in Figure 3.1(d) there are two outlier points, at x-values of 4.5 and 6.1, as indicated by the arrows. These data points are far outside of the range of values in the rest of the sample, and are likely to have been caused by some kind of error. In the section 'Identifying outliers objectively' we will discuss some methods for identifying such outliers more objectively, and also consider how to replace them in an unbiased way.

Histograms are informative regarding the shape of a distribution, and will usually make any outliers clear. However it is often also helpful to plot each individual data point using a scatterplot. Usually the term scatterplot makes us think of bivariate plots such as that shown in Figure 3.2(b), which show two variables plotted against each other. But if we have only a single dependent variable, it is still important to inspect the individual data points. We can use a *univariate scatterplot* to do this, in which the x-position is arbitrary, as shown in Figure 3.2(a). This plot shows every individual observation on some measure (shown on the y-axis), but within a given condition the x-position of a data point is not informative.

The scatterplot in Figure 3.2(b) contains three outliers, which are highlighted in blue. These values are not particularly remarkable in either their x- or y-values, so the grey kernel density histograms along the margins do not reveal them. However they clearly differ from the rest of the population (grey points), and might therefore be considered to be outliers. Both of the graphs in Figure 3.2 were produced using the generic *plot* functions (see Chapter 18 for more detailed discussion of creating plots in R), though alternative methods are also available, for example using the *geom_point* function in the *ggplot2* library. Remember that the code to produce all figures is also available in the book's GitHub repository at: **https://github.com/bakerdh/ARMbookOUP**.

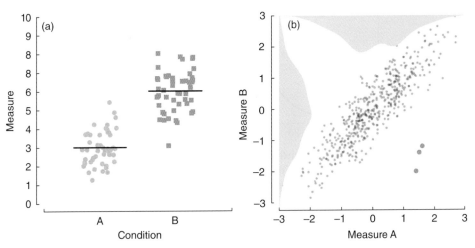

Figure 3.2 Examples of univariate (a) and bivariate (b) scatterplots. Kernel density functions are shown for each measure along the margins of panel (b) (grey curves).

Identifying outliers objectively

Plotting data will often reveal obvious outlying values. But in many situations, we would ideally like a more quantitative method for identifying outliers that are more ambiguous. There have been many approaches proposed for this, some of which we will describe in the following sections. The goal of outlier detection is to identify values that were generated by a different process from the rest of the data set. This is most often due to some form of equipment malfunction or other unforeseen event. For example, imagine that you set up a camera to count the number of times a bird leaves its nest each day. Most days the bird leaves between 20 and 30 times. But on one day a leaf fell on the camera early in the morning and blocked the lens—on this day only the five exits before the leaf fell were recorded. The value 5 is a clear outlier, and is far outside the range for the other days. But how might we identify such an outlier objectively?

Tukey's 'fence' method

A widely used method for outlier identification was proposed by the American mathematician John Tukey (1977). This involves calculating two *fences* using the interquartile range of a data set. The interquartile range (sometimes abbreviated to IQR) is the distance between the two points that encompass 50% of the values in a data set. In other words, it is the distance between the 25% and 75% *quantiles* of the data set (see Figure 3.3). The *inner fence* is placed 1.5 times the IQR below the first quantile, and 1.5 times the IQR above the third quantile. Data points falling outside of these values are considered outliers. The *outer fence* is defined similarly using 3 times the IQR, and values exceeding these limits are considered extreme outliers.

   A major practical advantage of Tukey's method is that the fences are determined based on the interquartile range. This is calculated using only the central 50% of data points, so it is

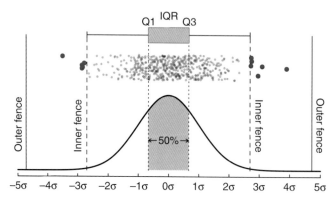

Figure 3.3 Illustration of the inner and outer fences, and the interquartile range. The x-axis is scaled in standard deviation (sigma) units, relative to the mean. The cloud of grey data points shows representative data sampled from the black population distribution, but also features several outliers (blue points). The vertical dotted lines, horizontal bar, and blue shaded region of the distribution illustrate the interquartile range, between which 50% of data points lie. The dashed lines and error whiskers indicate the inner fence, which extends 1.5 times the IQR above and below Q1 and Q3 respectively. The outer fence extends 3 times the IQR above and below Q1 and Q3, and is shown by the solid vertical lines.

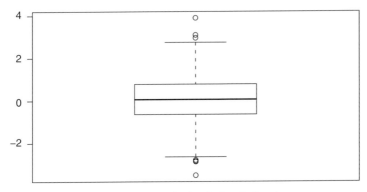

Figure 3.4 Example rudimentary boxplot, generated using the *boxplot* function.

not influenced by extreme outliers. For example, imagine we had a sample of 50 data points with values in the range 0 to 150. We calculate the interquartile range as lying between 63 and 85, with the inner fence extending from 28 to 119. These estimates would be unaffected if an equipment malfunction meant that two or three data points took on extreme values of >2000. And of course these extreme values would certainly be identified as outliers by Tukey's method!

Outliers identified in this way are often illustrated visually using box and whisker plots (see Figure 3.4). The box encompasses the interquartile range, and the whiskers (error bars) show the limits of the inner fence. Individual outliers are traditionally plotted using different symbols from the rest of the data set, and in some conventions non-outlying points are not plotted at all. Sometimes extreme outliers (beyond the outer fence) use a different symbol from near outliers (between the inner and outer fences). Rudimentary boxplots can be created automatically using the *boxplot* function from the *graphics* package in R. For example, the following line of code produces the boxplot in Figure 3.4:

```
boxplot(ndata)
```

Standard deviations and related criteria

An alternative to calculating fences is to exclude data points that lie beyond some multiple of the sample standard deviation (actually, for normally distributed data, the inner fence is at around 2.7 times the standard deviation, so this method is similar to just shifting the fence!). Conventions vary, and should ideally be decided (and preregistered) in advance before seeing the data. Classifying as outliers any data points that lie beyond 3 standard deviations from the mean would, for normally distributed data, count fewer than 0.3% of data points as being outliers. So for a sample size below 100, any data points that exceed this threshold are very likely to be outliers.

In the following R example, we generate 100 random samples from a normal distribution (how computers generate random numbers is explained in Chapter 8), and add a clear outlier at a known location (entry 57). We then calculate our threshold criterion as 3 times the standard deviation of the sample. To work out which values exceed this distance from the mean, we subtract the mean from each data point, and for convenience take the absolute values (i.e. any negative numbers become positive). Any values in this normalized data set that exceed the criterion value will be picked up as outliers. We can find the indices of these values using the *which* function.

```
data <- rnorm(100) # generate some random data
data[57] <- 100       # replace the value in row 57 with an outlier
criterion <- 3*sd(data)      # calculate 3 times the standard deviation
normdata <- abs(data-mean(data))     # subtract the mean and take the
absolute value
which(normdata>criterion)    # find the indices of any outlier values
## [1] 57
```

Two more formal variants of this approach include Chauvenet's criterion (Chauvenet 1863) and Thompson's Tau (Thompson 1985). These methods use the total sample size to determine the threshold for deciding that a given data point is an outlier, based on properties of either the normal (Gaussian) distribution or the T-distribution.

The procedure for Chauvenet's criterion is to convert all values to absolute z-scores (subtracting the mean and scaling by the standard deviation) and again identify data points that exceed a criterion. This time the criterion is calculated by taking the quantile of the normal distribution at $1/(4N)$, where N is the sample size. This means that the criterion becomes more stringent (i.e. larger) as the sample size increases, because with larger samples we expect a greater number of extreme values. A rudimentary implementation is given by the following function:

```
d_chauv <- function(data){
    i <- NULL      # initialize a data object to store outlier indices
    m <- mean(data) # calculate the mean of the data
    s <- sd(data) # calculate the standard deviation
    Zdata <- abs(data-m)/s # convert data to absolute z-scores
    dmax <- abs(qnorm(1/(4*length(data)))) # determine the criterion
    i <- which(Zdata>dmax) # find indices of outliers
return(i)}
```

This function uses the *qnorm* function to estimate the appropriate quantile, and the *which* function to return the indices of any outlier values. Using this function with the example data from above (where we added an outlier at entry 57) again correctly identifies the outlier:

```
d_chauv(data)
## [1] 57
```

The modified Thompson's Tau method is conceptually similar, except that the critical value is obtained from the T-distribution using a given α value, and then converted to the tau statistic with the equation:

$$\tau = \frac{t_{\alpha/2}(N-1)}{\sqrt{N}\sqrt{\left(N-2+t_{\alpha/2}^2\right)}} \tag{3.1}$$

where $t_{\alpha/2}$ is the critical t-value with $N-2$ degrees of freedom, using the significance criterion α (typically $\alpha = 0.05$), and N is the sample size. Another difference is that the tau method is iterative, with only the value that deviates most from the mean being compared with the

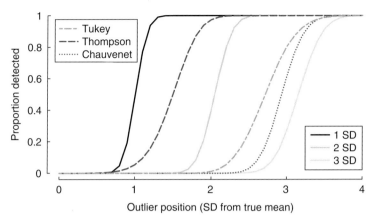

Figure 3.5 The proportion of outliers detected as a function of outlier position, expressed in standard deviation units. The grey curves were calculated by rejecting all data points that exceed 1, 2, or 3 standard deviations from the sample mean. Tukey's fence method produces a curve equivalent to approximately 2.7 standard deviations (light blue dash-dotted curve). The blue dashed and dotted curves were calculated using Thompson's Tau and the Chauvenet criterion respectively. All simulations involved 99 values drawn from a normal distribution, and an added 'outlier', with values given along the x-axis. Note that changes in sample size affect the relative positions of the curves.

critical value on each iteration. The τ value is recalculated on each iteration, using only the data remaining after removing outliers on previous iterations.

Figure 3.5 illustrates the results of simulations that show how sensitive each test is for detecting a single outlier of known value. The Thompson's Tau test identifies the known outlier with a similar sensitivity to methods that reject values exceeding 1.5 times the sample standard deviation. The Chauvenet criterion is more conservative, and similar to rejecting values exceeding about 2.8 times the standard deviation. Note that these values are dependent on the sample size; for example the Chauvenet criterion becomes more conservative as sample size increases.

One distinct danger posed by the availability of several different outlier-detection algorithms is that they provide an experimenter with hidden degrees of freedom in their analysis. An unscrupulous researcher could easily engineer a desired result by choosing an outlier-rejection algorithm after their data have been collected. This is highly unethical and strongly discouraged. To avoid such issues, analysis plans should ideally be preregistered before the data have been collected, and must detail how outliers will be handled. In general, a criterion of around 2.5 or 3 standard deviations is usually appropriate for univariate data, though it is worth checking standard practice in a given research area.

The Mahalanobis distance for multivariate data

Methods based on the properties of the normal distribution are usually sufficient for identifying outliers in univariate data sets, where we have only a single dependent variable. But in multivariate data sets, outliers can sometimes occur that are within the mid-range of each individual variable, yet occupy a distinct part of the possible space of values compared with the rest of the data set (see the example in Figure 3.2(b)).

In such cases, a useful statistic is the *Mahalanobis distance, D,* proposed by the prolific Indian mathematical physicist and statistician Prasanta Mahalanobis (1936). Intuitively, this is similar to the Euclidean distance (shortest straight line) between each point and the centre of mass (centroid) of the whole data set. However, it is more sophisticated than the Euclidean distance because the Mahalanobis distance for a given data point is scaled by the variance of the data set in the direction of that data point. This allows us to take into account any correlations between the two (or more) dependent variables.

To understand how this works, consider first the data set shown in Figure 3.6(a). Here, the two variables are uncorrelated, and the variable plotted on the x-axis has a smaller variance than the one plotted on the y-axis. The two white outlier points are both the same Euclidean distance from the centroid of the data (black point), because the blue vector lines are the same length. But if we express their distance in standard deviation units along the appropriate axis, the square outlier is clearly a more extreme outlier than the triangle outlier (because the standard deviation is larger in the y direction). This is what the Mahalanobis distance calculates, and for this example the square point would have a Mahalanobis distance of around $D = 6$, and the triangle point a distance of around $D = 3$. So the square is a more extreme outlier than the triangle when expressed with this metric, which fits with our intuitions.

Next let's consider the data set shown in Figure 3.6(b). This time, there is a clear correlation between the two variables. Again, the two outliers are the same Euclidean distance from the centroid. As before, the Mahalanobis distance is greater for the square point, because the variance is estimated in the direction of the blue vector line. This is somewhat harder computationally than calculating the standard deviation for one or other dependent measure. Happily, we do not need to perform these calculations by hand. The *mahalanobis* function is

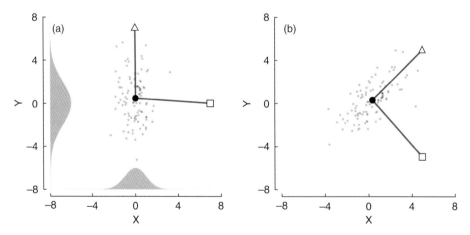

Figure 3.6 Illustration of outliers in two-dimensional (bivariate) data sets. The black point is the centre of mass, and the white triangle and square symbols are outliers. The blue lines depict the Euclidean distances between the centroid and each outlier, and are of identical length.

part of the core *stats* package in *R*. Imagine we have a data frame containing around 500 rows of bivariate data (from Figure 3.2(b)) structured as follows:

```
head(bdata)
##               x          y
## 1   0.685092067   0.6039170
## 2  -0.005550195   0.2395705
## 3  -0.777641329  -1.0976698
## 4   1.875702830   1.5417293
## 5  -0.377129105   0.3195294
## 6  -0.454686991   0.3052273
```

We can pass this data frame into the *mahalanobis* function, along with the means of each variable (calculated inline using the *colMeans* function), and the covariance matrix (calculated inline using the *cov* function). The function returns a squared distance estimate for each x–y pair of points:

```
D2 <- mahalanobis(bdata, colMeans(bdata), cov(bdata))

D2[1:6]
## [1] 0.4343403 0.2757397 0.9850249 3.5338519 1.8662220 2.2249707
```

Plotting these distance estimates (in Figure 3.7) reveals the three clear outliers that we highlighted in blue in Figure 3.2(b). Because the Mahalanobis distance is calculated in units of the standard deviation, we can apply similar threshold criteria as described in the earlier section 'Standard deviations and related criteria'. However, the output of the *mahalanobis* function is a squared distance, so we must either first take the square root (with the *sqrt* function), or square our standard deviation threshold criterion. A threshold of 3 times the standard deviation is a squared distance value of $3^2 = 9$, shown by the vertical dotted line in Figure 3.7. Our three outliers (indicated by the arrows) clearly exceed this criterion. Note also that the squaring avoids any negative distance values.

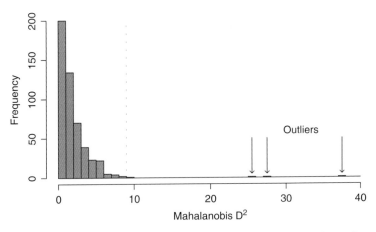

Figure 3.7 Histogram of squared Mahalanobis distance measures. The dotted vertical line indicates a threshold equivalent to 3 standard deviations. The positions of three clear outliers are indicated by arrows.

One final benefit of the Mahalanobis distance is that it generalizes easily to any number of dimensions. If you have a large multivariate data set with more than two dependent variables, it is still a useful method for identifying outliers. Importantly, regardless of the number of dimensions, a single distance statistic will be computed for each case (e.g. participant). If you plan on using any of the multivariate methods discussed in Chapters 11 to 14, a working knowledge of the Mahalanobis distance will prove useful.

Identifying missing and out-of-range values

In addition to the techniques we have discussed, it is also worth being aware of some useful R functions for identifying completely missing values, or those that are outside of some specified range. The *is.na* function returns TRUE for any values stored in a data object that are classed as 'not a number'. This is especially useful as missing values loaded in from spreadsheets are assigned NA by default. We can combine the *is.na* function with the *which* function to return the indices of the missing values.

To give an example, suppose we have a data set comprising 10 values, three of which are missing:

```
## [1]  4  6 NA  1  7  8 NA  3 NA  5
```

The *is.na* function will return TRUE for the missing values, and FALSE for the others:

```
is.na(nandata)
## [1] FALSE FALSE TRUE FALSE FALSE FALSE TRUE FALSE TRUE FALSE
```

Wrapping the *which* function around this will return the indices of the NA values, at positions 3, 7 and 9:

```
which(is.na(nandata))
## [1] 3 7 9
```

One way to remove the NA values is to invert the output of the *is.na* function using the ! operator (making TRUE become FALSE, and vice versa), to return only the rows that contain real numbers:

```
trimmeddata <- nandata[which(!is.na(nandata))]
trimmeddata
## [1] 4 6 1 7 8 3 5
```

A related function is the *is.infinite* function, which returns TRUE for positive and negative infinity—these often appear if values are inadvertently divided by zero. However the *which* function can take any logical argument, so it can also be used to restrict values to within a certain range. For example:

```
trimmeddata[which(trimmeddata<5)] # return only values < 5
## [1] 4 1 3
```

It might be useful to do this if your data have a natural range. For example, many real-world measurements such as height, weight, heart rate, and so on, must take on positive values. If some sort of error has resulted in negative values for some observations of these types of variables, it would be reasonable to consider these observations as out of range.

Finally, we can combine these logical statements with other functions, such as the *mean*, *sum*, or *sd* functions. To calculate the mean of the *nandata* object, excluding the NA values, we can nest the *which* statement inside the indexing of the variable as follows:

```
mean(nandata[which(!is.na(nandata))])
## [1] 4.857143
```

Although this is a good general solution, it can be rather cumbersome. In some functions, there is an alternative implementation specifically for NA values known as the *na.rm* flag. By setting this to TRUE, we instruct the function to remove the NA values automatically. The *mean*, *sum*, and *sd* functions (along with other core functions) have this facility, for example:

```
mean(nandata, na.rm=TRUE)
## [1] 4.857143
```

Notice that this returns the same result as the more involved solution. Not all functions have an *na.rm* flag, but you can check the help files to work out which ones do.

The above tools are extremely useful for cleaning up data and dealing with missing or problematic values. They can be used in combination with several of the other outlier-detection criteria that were described in the first three subsections of 'Identifying outliers objectively'. However, this leads us to a more philosophical question of what we should be doing with outliers and missing values in the first place.

What should we do with outliers?

Once outliers and missing values have been identified, we must decide what to do with them. There are three main possibilities: we can delete the outliers, we can replace them with estimated values, or we can leave them in the data set. Opinions differ regarding the best course of action, and this will typically depend on the nature of your data set and what you are trying to do with it. In general, it is not thought to be a good idea to exclude outliers simply because they look different from the other values. One should instead have some independent rationale for removing them, for example if there is evidence that the outliers were caused by a substantively different data-generating process from the rest of the data.

Deleting a small number of outliers from a large data set is unlikely to make much difference to our ability to detect an effect. However, when sample sizes are smaller, removing several outliers will substantially reduce our statistical power (see Chapter 5). Outlier removal is a particular issue with multivariate data sets, because it can often be the case that many individual participants will be missing observations from at least one of the dependent variables. As we discuss further in Chapter 12, this can drastically reduce effective sample sizes for methods such as structural equation modelling. On the other hand, approaches such as mixed-effects models (see Chapter 7) are able to cope well with missing values, and might be a better alternative to more traditional methods such as analysis of variance (ANOVA), for which a single missing value usually requires complete removal of a participant.

Alternatively, techniques exist to estimate (impute) plausible values to replace those that are missing, which can make some statistical models more stable. These include mean imputation (replacing a missing value with the sample mean), Windsorization (replacing an outlying value with a neighbouring value), and clipping (setting extreme values to a prespecified maximum or minimum threshold). If you have a solid rationale for replacing missing or outlying values, and have ideally specified this in advance through preregistration, then these methods avoid many of the shortcomings of outlier deletion.

Finally we can decide to leave outliers in a data set. This might affect our choice of statistical test, as data sets with substantial outliers are unlikely to meet the assumptions of parametric statistics. We could instead use methods that make fewer assumptions, such as non-parametric alternatives, or the 'bootstrapping' approach we will introduce in Chapter 8. There is also a class of statistics called *robust statistics* that are designed to be used with noisy data. A simple example of a robust statistic is the *trimmed mean*, in which some percentage of extreme values is removed from the data set, and the mean calculated using the remainder. For example, a 10% trimmed mean would involve rejecting the lowest and highest 10% of values from a data set, and using the remaining 80% of values to estimate the mean. Other variants forgo the assumption that data are normally distributed, and instead use other distributions such as t-distributions, which have longer tails (this approach is also common in Bayesian statistics: see Kruschke 2014). A full discussion of robust statistics is beyond the scope of the current text, but the interested reader is referred to the book *Robust Statistics* by Peter Huber (2004).

Normalizing and rescaling data

For some analysis methods, it is appropriate to normalize data, either by subtracting the mean, and/or by scaling by the variance (also sometimes called standardizing). This is especially important for multivariate techniques such as *k*-means clustering (see Chapter 13) and multivariate pattern analysis (see Chapter 14), where different dependent variables might have very different ranges. By rescaling, we level the playing field, so that each variable is weighted equally in the analysis. Of course for other types of analysis this would be very inappropriate, and might even remove any legitimate differences between groups!

When appropriate, we can rescale data using the *scale* function:

```
rescaleddata <- scale(data)
```

By default this function performs the following operations on each column of a matrix:

1. subtracts the column mean from each data point
2. divides each data point by the standard deviation of the column

It is possible to perform only one or the other of these operations by setting either the *center* or *scale* arguments to FALSE as follows:

```
rescaleddata <- scale(data,scale=FALSE) # only subtract the mean
rescaleddata <- scale(data,center=FALSE) # only divide by the SD
```

It is important to remember that the units of rescaled data will not be the same as the units of the original measurements. One way to think of normalized data is as being akin to z-scores, where each data point is expressed in standard deviation units. The univariate scatterplots in Figure 3.8 illustrate the effect of rescaling a data set.

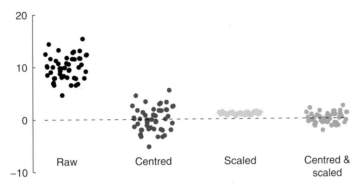

Figure 3.8 Illustration of rescaling using an example data set (black points). The dark blue points show centring, with no change in variance. The grey points illustrate scaling by the standard deviation—this also affects the mean, but does not centre the mean on 0. Finally, the light blue points show the effects of centring and then scaling.

Transforming data and testing assumptions

A key assumption underlying parametric statistical tests is that data should be normally distributed—in other words a histogram of the data should follow the bell-shaped curve of a Gaussian distribution. But what if, even after dealing with outliers, this is still not the case? If this happens, it is sometimes possible to *transform* the data by applying mathematical operations to the full data set. The most common of these are logarithmic transforms, which pull in the long tail of a positively skewed distribution, and squaring, which has a similar effect on negatively skewed distributions (exponential transforms can also be used for dealing with negative skew). Examples of how each of these transforms can make skewed data conform more closely to a normal distribution are shown in Figure 3.9. A more in-depth discussion on the advantages of data transforms is given by Bland (2000).

As many introductory statistics texts will explain in detail, there are two main tests to assess whether data are consistent with a normal distribution. These are the Kolmogorov–Smirnov test (Smirnov 1948; Kolmogorov 1992) and the Shapiro–Wilk test (Shapiro and Wilk 1965). The Kolmogorov–Smirnov test involves comparing the cumulative distribution functions of the data and a reference distribution (i.e. a normal distribution), and is implemented in *R*'s built-in *stats* package by the *ks.test* function. For example:

```
ks.test(data1,'pnorm',mean(data1),sd(data1))
##
## One-sample Kolmogorov-Smirnov test
##
## data: data1
## D = 0.17671, p-value = 0.003879
## alternative hypothesis: two-sided
```

Note that the *ks.test* function requires as inputs the sample of data and the name of the cumulative distribution you wish to compare it to (*pnorm* is a cumulative normal distribution).

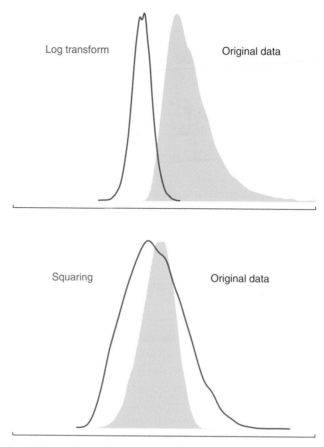

Figure 3.9 Examples of data transforms. The upper panel shows some positively skewed data (shaded grey), and a more normal distribution following a log transform (blue). The lower panel shows some negatively skewed data (shaded grey), and a more normal distribution following a squaring transform (blue).

By default the data will be compared to a distribution with a mean of 0 and a standard deviation of 1. So it is necessary either to provide the actual mean and standard deviation of your data as additional arguments (as above), or to first normalize your data using the *scale* function (see 'Transforming data and testing assumptions'):

```
ks.test(scale(data1),'pnorm')
##
## One-sample Kolmogorov-Smirnov test
##
## data: scale(data1)
## D = 0.17671, p-value = 0.003879
## alternative hypothesis: two-sided
```

The data we have tested here were generated using a lognormal distribution (the grey distribution in the top panel of Figure 3.9), so the test is significant, indicating a deviation from normality. As with most assumption tests, a non-significant Kolmogorov–Smirnov would mean that there was no significant difference between the data and reference distribution, so we could assume the data are approximately normal.

In common with other statistical tests, increasing sample size will increase statistical power (see Chapter 5), so small data sets are less likely to be significant (and thus more likely to pass the test) than larger ones, all else being equal. Simulation studies (e.g. Yap and Sim 2011) have demonstrated that the Kolmogorov–Smirnov test is lower in power than the Shapiro–Wilk test, which works in a similar way but using a regression framework. In R, the *shapiro.test* function is conducted as follows:

```
shapiro.test(data1)
##
## Shapiro-Wilk normality test
##
## data: data1
## W = 0.886, p-value = 3.285e-07
```

Again, a significant result ($p < 0.05$) implies a deviation from normality. Notice that the p-value for the Shapiro–Wilk test is generally smaller than that for the Kolmogorov–Smirnov test using the same data, illustrating its greater power. However, most implementations of Shapiro–Wilk are limited to samples of less than 5000 data points, whereas the Kolmogorov–Smirnov test has no such restrictions.

Finally, Q–Q plots can be very informative in visually assessing and understanding deviations from normality. These are constructed by plotting the quantiles of the data against the quantiles of a reference distribution (e.g. a normal distribution). The *qqnorm* function generates a plot, and the *qqline* function adds a reference line. Examples are shown in Figure 3.10 for normally distributed data (left) and positively skewed data (right). Note the substantial deviation from the diagonal reference line at the upper end of the skewed distribution.

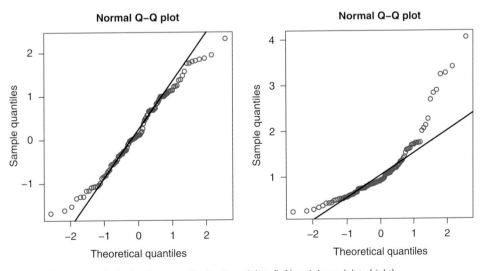

Figure 3.10 Example Q–Q plots for normally distributed data (left) and skewed data (right).

Typically, one would use tests of normality and Q–Q plots on a raw data set first. If this deviates from normality, it is worth trying an appropriate transform, and then running the normality test again. Unfortunately, even after applying a transform, some data sets simply cannot be manipulated sufficiently to meet the assumptions of parametric statistics. This might rule out the use of some tests—however there are also many non-parametric alternatives that can be used instead, just like for data sets with outliers. In general, non-parametric methods involve rank-ordering a data set, and performing calculations on the rankings rather than the raw data. This avoids any problems with outliers and deviations from normality, but at the expense of statistical power (the ability of a test to detect a true effect). Common examples are the Spearman correlation coefficient, the Mann–Whitney U test, and Friedman's ANOVA, all of which are implemented in R. Non-parametric methods are not the main focus of the current text, but in Chapter 8 we will discuss bootstrap resampling techniques, which can be used as a flexible non-parametric approach to statistical testing.

Recoding categorical data and assigning factor labels

So far we have mostly talked about data that are numerical and continuous. But often we also need to deal with categorical data, for example reflecting group membership, sex, condition information, or responses on a categorical scale (such as A, B, C or low, medium, high). In R, such data are usually stored as *factors*—data objects with a set of possible text labels, which are also (behind the scenes) given numerical values. Factor variables are particularly useful when conducting statistical tests to compare different groups. For example in analysis of variance (ANOVA) the independent variables are categorical, and these should be stored as factors (see the section 'ANOVA' in Chapter 4 for more details on how to run ANOVA in R).

To demonstrate how factors work, let's create a data object containing sex information for seven rats:

```
ratsex <- c('M','M','F','M','F','F','F')
ratsex
## [1] "M" "M" "F" "M" "F" "F" "F"
```

By default, this data object contains a list of character strings. We can convert it to a factor using the *factor* function:

```
ratfactor <- factor(ratsex)
ratfactor
## [1] M M F M F F F
## Levels: F M
```

Now we see that R has defined two *levels* for the factor, F and M. An integer code is also assigned to each level, by default in alphabetical order. We can see these numerical values using the *as.numeric* function:

```
as.numeric(ratfactor)
## [1] 2 2 1 2 1 1 1
```

Notice that all values of M are coded as 2, and all values of F are coded as 1. If we want a particular ordering of the numerical values associated with each level of the factor, we can specify this when we create the data object, using the factor command:

```
ratfactor <- factor(ratsex, levels = c('M','F'))
as.numeric(ratfactor)
## [1] 1 1 2 1 2 2 2
```

Now we have coded M as 1, and F as 2. What if we wanted to change the labels from 'M' and 'F' to 'male' and 'female'? We can do this using the *levels* function as follows:

```
levels(ratfactor)[levels(ratfactor)=='M'] <- 'male'
levels(ratfactor)[levels(ratfactor)=='F'] <- 'female'
ratfactor
## [1] male   male   female male   female female female
## Levels: male female
```

Notice that in the above code we used a logical statement inside the square brackets to find the levels of the *ratfactor* object that were either 'M' or 'F'. For example, the code *levels(ratfactor)=='M'* is interpreted as 'find the levels of the ratfactor object that have the value M'. This is the most straightforward way to recode a factor using base *R* commands, though there are also functions in other packages that achieve the same effect, including the *revalue* and *mapvalues* functions in the *plyr* package, and the *recode_factor* function in the *dplyr* package.

Putting it all together—importing and cleaning some real data

In this section we will load in and analyse some data based on a real experiment I conducted using the online *Qualtrics* platform. I have actually replaced the original data with some simulated values, to better illustrate the points we have made in this chapter. However it is close enough to the sort of thing one might download from an online platform for collecting data (many of which now exist). The experiment was quite light-hearted—it was a general knowledge quiz with 10 questions. Before seeing the questions, participants rated their own general knowledge ability on a scale from 0 to 100. The idea was to see if these ratings predicted actual performance in the quiz.

We can load in the data using the *read.csv* function (the data can be downloaded from the book's GitHub repository), and take a look at the first few rows as follows:

```
quizdata <- read.csv('data/qualtricsexample.csv')
head(quizdata)
##    StartDate           EndDate         Progress Duration..in.seconds. Finished
## 1 17/03/2021 03:08 17/03/2021 03:08      100           34             TRUE
## 2 17/03/2021 03:08 17/03/2021 03:09      100           20             TRUE
## 3 23/03/2021 11:05 23/03/2021 11:05      100           54             TRUE
## 4 23/03/2021 11:05 23/03/2021 11:14      100          564             TRUE
## 5 23/03/2021 11:05 23/03/2021 11:14      100          555             TRUE
## 6 23/03/2021 11:05 23/03/2021 11:14      100          568             TRUE
```

```
##    RecordedDate ResponseId DistributionChannel UserLanguage Q1 Q2 Q3
## 1 17/03/2021 03:08 R_1jx28z4EYzdLmNd      anonymous      EN 84  D  A
## 2 17/03/2021 03:09 R_20U6cF4f7pFfbxb      anonymous      EN 77  C  D
## 3 23/03/2021 11:05 R_oXHNzSI4aAISjxn      anonymous      EN 51  D  C
## 4 23/03/2021 11:14 R_29iR3XxZbzDWizy      anonymous      EN 83  B  A
## 5 23/03/2021 11:14 R_1dBvXJRckeXpMrZ      anonymous      EN 88  C  A
## 6 23/03/2021 11:14 R_1qURia8GbrDPfAE      anonymous      EN 68  A  B
##   Q4 Q5 Q6 Q7 Q8 Q9 Q10 Q11
## 1 C  A  A  A  B  A   C   C
## 2 A     A  B  C  A   A   C
## 3 D  B  A  D  D  A   A   A
## 4 D  A  C  A  C  A   A   A
## 5 A  D  A  A  B  C   A   A
## 6 A  B  A  C  D  A   C   A
```

The first 10 columns are things we aren't really that interested in—they contain metadata about the running of the experiment that we can ignore. We can strip these columns out by restructuring the data frame to only include the final 11 columns as follows (sometimes it is also necessary to remove the first few rows in a similar way):

```
quizdata <- quizdata[,10:20]
head(quizdata)
##    Q1 Q2 Q3 Q4 Q5 Q6 Q7 Q8 Q9 Q10 Q11
## 1 84  D  A  C  A  A  A  B  A   C   C
## 2 77  C  D  A     A  B  C  A   A   C
## 3 51  D  C  D  B  A  D  D  A   A   A
## 4 83  B  A  D  A  C  A  C  A   A   A
## 5 88  C  A  A  D  A  A  B  C   A   A
## 6 68  A  B  A  B  A  C  D  A   C   A
```

After restructuring, the first column contains the ratings from 0 to 100. Let's take a closer look at these values using a histogram—see Figure 3.11(a):

```
hist(quizdata$Q1)
```

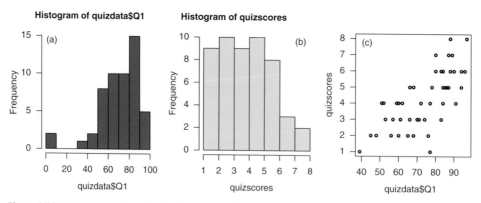

Figure 3.11 Histograms and scatterplot for the example Qualtrics data. Panel (a) shows the histogram of self-ratings of general knowledge, panel (b) shows the histogram of actual quiz performance, and panel (c) shows the correlation between the two measures.

Two features are clear from the histogram: there are two outlier points with a value of 0, and overall the distribution looks negatively skewed. Perhaps the ratings of 0 were genuinely participants who thought they had very poor quiz ability. But it could also be that 0 was the default rating on the scale used, and these participants did not change it for whatever reason. We can use the code from earlier in the chapter to identify data points that are more than 3 standard deviations from the mean, as follows:

```
criterion <- 3*sd(quizdata$Q1)     # calculate 3 times the standard
deviation
normdata <- abs(quizdata$Q1-mean(quizdata$Q1))     # subtract the
mean and take the absolute value
which(normdata>criterion)     # find the indices of any outlier values
## [1] 21 30
```

This code identifies the participants in rows 21 and 30, and indeed these are the ones who produced ratings of 0:

```
quizdata$Q1[c(21,30)]
## [1] 0 0
```

Given our concerns about the possibility that these participants did not use the rating scale correctly, and their distance from the rest of the scores, we might be justified in removing them from the data set. We can do this using another *which* statement, but this time to include only the participants whose ratings are less than 3 standard deviations from the mean, as follows:

```
quizdata <- quizdata[which(normdata<criterion),]
```

To see whether the data are normally distributed, we can run the Shapiro–Wilk test:

```
shapiro.test(quizdata$Q1)
##
## Shapiro-Wilk normality test
##
## data: quizdata$Q1
## W = 0.96021, p-value = 0.08514
```

Although the data show some evidence of negative skew, the test is (just) non-significant ($p = 0.085$), so we can proceed assuming a normal distribution.

Next we can look at the answers to the quiz questions themselves. These are all four-option multiple choice questions, with the answers stored as A, B, C, and D. When we loaded in the data, R converted the responses to factors. However, some of the questions have not been answered by certain participants, and we have some missing values. Just as we might in an exam, we will mark the questions with missing data as incorrect.

To score the quiz, we will use two loops (see 'Loops' in Chapter 2), one inside the other. The outer loop will run through each participant in turn, and the inner loop will run through each question for a given participant. For this example, we will assume that the correct answer for each question was 'A', and we will count up the number of questions each participant got right and store this in a new data object called *quizscores*:

```
# make a list of zeros to store the scores
quizscores <- rep(0,nrow(quizdata))

for (participant in 1:nrow(quizdata)){ # loop through participants
    for (question in 1:10){ # loop through questions
        # check if the answer to this question was correct (A)
        if (as.character(quizdata[participant,question+1])=='A'){
            # if so, increase the score for this participant by 1
            quizscores[participant] <- quizscores[participant] + 1}
    }
}
```

We can inspect a histogram of these scores, again using the *hist* function, as shown in Figure 3.11(b), and also run the Shapiro–Wilk test:

```
hist(quizscores)
shapiro.test(quizscores)
##
## Shapiro-Wilk normality test
##
## data: quizscores
## W = 0.95898, p-value = 0.07544
```

Despite the somewhat unusual form of the data, with evidence of a floor effect at the lower end, the Shapiro–Wilk test is again non-significant. Finally, we can inspect a scatterplot of the two variables plotted against each other, to see if there is evidence that participants were able to predict their own general knowledge ability:

```
plot(quizdata$Q1,quizscores,type='p')
```

The graph produced by this code is shown in Figure 3.11(c). It does appear to be the case that individuals who rated their ability more highly also obtained generally higher test scores, which we might go on to test using correlation, or other statistical tests described in Chapter 4. We can also check for multivariate outliers using the Mahalanobis distance as follows:

```
bothscores <- data.frame(quizdata$Q1,quizscores)
D <- mahalanobis(bothscores,colMeans(bothscores),cov(bothscores))
sort(round(D,digits=2))
## [1] 0.02 0.18 0.18 0.23 0.49 0.51 0.51 0.51 0.61 0.64 0.72 0.74
0.74 0.79 0.90
## [16] 0.99 1.01 1.04 1.07 1.12 1.17 1.19 1.30 1.30 1.32 1.35 1.47
1.63 1.66 1.72
## [31] 1.84 1.86 1.89 2.21 2.21 2.30 2.47 2.52 2.55 3.11 3.15 3.34
3.46 3.91 3.96
## [46] 3.98 4.36 4.57 5.21 5.71 8.31
```

By sorting the distances using the *sort* function, we can see that the largest value is 8.31, which does not exceed our criterion of $D^2 = 9$ (or $D = 3$).

The Qualtrics quiz data are a deliberately simple example—most data sets would involve multiple conditions, experimental manipulations, or independent variables. However this has hopefully given a good indication of how we can import and clean a data set in preparation for further analysis.

Practice questions

1. Two conventions for arranging data in *R* are known as:
   A) Wide and tall
   B) Wide and long
   C) Melted and spread
   D) Spread and long

2. The whiskers on a boxplot conventionally show:
   A) The standard error
   B) The interquartile range
   C) The inner fence
   D) 95% confidence intervals

3. Which of the following is *not* a method for replacing outliers?
   A) Windsorization
   B) Mean imputation
   C) Clipping
   D) The interquartile range

4. Which method would be most suitable for identifying outliers in a correlation analysis?
   A) The modified Thompson's Tau statistic
   B) Tukey's outer fence method
   C) The Chauvenet criteria
   D) The Mahalanobis distance

5. For normally distributed data, Tukey's inner fence is approximately equivalent to which multiple of the standard deviation?
   A) 1.5
   B) 1.96
   C) 2.7
   D) 3

6. A 20% trimmed mean would be calculated using:
   A) The central 60% of values
   B) The central 20% of values
   C) The central 80% of values
   D) The central 90% of values

7. Which of the following is *not* a method that can be used when the assumptions of parametric statistics are violated?
   A) Bootstrap tests
   B) The Kolmogorov–Smirnov test
   C) Non-parametric statistics
   D) Robust statistics

8. Which command would normalize a data set by its standard deviation only?
   A) *scale(data)*
   B) *scale(data,scale=FALSE)*
   C) *scale(data,center=FALSE)*
   D) *scale(data)/sd(data)*

9. To assess deviations from a normal distribution, you could use:
   A) The Kolmogorov–Smirnov test
   B) A Q–Q plot
   C) The Shapiro–Wilk test
   D) All of the above

10. To make positively skewed data more normal, you could use a:
    A) Logarithmic transform
    B) Exponential transform
    C) Squaring transform
    D) Additive transform

Answers to all questions are provided in the answers to practice questions at the end of the book.

Statistical tests as linear models

This chapter will demonstrate *R* implementations of statistical tests that are commonly taught on introductory statistics courses. I would anticipate that most readers will be familiar with such tests, including t-tests, regression, ANOVA, and correlation. If you have not come across these before, you might find it helpful to read an introductory statistics text (e.g. Navarro 2019; Field, Miles, and Field 2012), which will explain them in much more detail. I will introduce the tests using a slightly different approach from that traditionally taken—we will see in this chapter that many such tests can be considered in terms of explaining data using a *linear model* (i.e. a straight line).

Many statistical tests involve comparing different models

Statistical tests were developed by different people at different times. For example, the t-test (which compares the means of two groups) was developed by William Sealy Gosset (writing as *Student*) who was employed by the Guinness corporation during the early twentieth century to analyse data relating to brewing beer. Student needed to establish whether different brewing methods affected the end product, despite having only small samples of noisy data to work with. Analysis of Variance (ANOVA) was developed by Ronald Fisher for analysing data on crop yields (Fisher unfortunately held some abhorrent views on issues such as race and eugenics). The way these methods are usually taught makes them seem like totally distinct techniques, with different underlying assumptions and operations. But really, they are doing fundamentally the same thing—fitting straight lines to data and checking how much variance (or error) is left over that the straight lines can't explain.

Considered from this perspective, we will see that many statistical tests involve fitting a model to try and explain our data. Usually the model assumes that our measurements (known as the dependent variable) can be predicted to some extent by one or more other factors (known as independent variables). We can then compare the fit of this statistical model to a 'null model', in which those other factors do not predict our measurements. If our model explains the data better than the null model (according to some criterion), we consider it to be statistically significant (we will discuss these assumptions in more detail in Chapter 17). The clearest way to demonstrate the model comparison approach is by starting with linear regression (where it is most explicit), and then applying the same logic to other situations.

Regression (and correlation)

In linear regression, we look at the influence of a predictor (independent variable) on some sort of outcome measure (dependent variable). Both variables are continuous, rather than being split into discrete categories. We will demonstrate regression using an example data set from the literature, on growth rates of moso bamboo. Bamboo is a fast-growing plant, and so has the potential for rapid carbon sequestration, which has important environmental implications. A study by Yen (2016) measured the relationship between the diameter at breast height (DBH—the width of the plant at around 140 cm from the ground) and the overall culm (stem) height. There is a very straightforward linear relationship across a sample of 30 plants, shown by the data points in Figure 4.1.

The data (which I extracted from a figure in the original paper) are stored in a data frame called *bamboodata*. It has two columns, containing the DBH and culm height data for 30 plants. The first few rows look like this:

```
head(bamboodata)
```

```
##     DBH      culmheight
## 1 6.39      7.92
## 2 6.19      8.39
## 3 6.88      8.03
## 4 6.69      8.74
## 5 6.48      9.06
## 6 7.18      8.33
```

It is clear from Figure 4.1 that the data are highly correlated, which we can confirm by calculating a correlation coefficient by passing the data to the *cor* function:

```
cor(bamboodata)
```

```
##                 DBH        culmheight
## DBH       1.0000000    0.9314495
## culmheight 0.9314495   1.0000000
```

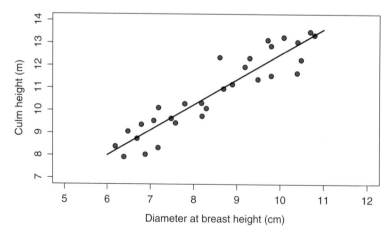

Figure 4.1 Linear regression between diameter at breast height and culm height for moso bamboo, based on figure 1 of Yen (2016).

If we were interested in statistical significance, there is also a test version (*cor.test*), which requires us to pass the two variables (columns) in as separate arguments:

```
cor.test(bamboodata$DBH,bamboodata$culmheight)
```

```
##
##   Pearson's product-moment correlation
##
## data:  bamboodata$DBH and bamboodata$culmheight
## t = 13.545, df = 28, p-value = 8.122e-14
## alternative hypothesis: true correlation is not equal to 0
## 95 percent confidence interval:
##   0.8596580 0.9671648
## sample estimates:
##      cor
## 0.9314495
```

The output gives us the same correlation coefficient on the final line (along with 95% confidence intervals just above), and also calculates a t-statistic and *p*-value to assess statistical significance (on the fifth line). Notice that the *p*-value is expressed in scientific notation as 8.122e-14. This is how *R* represents very small numbers: in this case it means 8.122×10^{-14}, or 0.00000000000008122 (the easiest way to think of this is that you shift the main number by the value given after the *e*—here 14 places to the right).

In regression, we want to fit a linear model (i.e. a straight line) that allows us to predict values of the dependent variable (culm height) using the values of the independent variable (DBH). (Note that in this example these are both things that have been measured, and so DBH might not meet strict criteria for being an independent variable because it will involve measurement error, but this is just an example.) To do this, we will use the *lm* (linear model) function in *R*. The *lm* function (in common with other related functions, including those for running ANOVA) uses a syntax to specify models that has the general form DV ~ IV. The tilde symbol (~) means *is predicted by*. In other words, we're saying that the dependent variable is predicted by the independent variable. For our example, we want to run the model *culmheight ~ DBH*, and we will also tell the *lm* function the name of the data frame containing our data (*bamboodata*). Finally, we will store the output of the model in a new data object called *bamboolm*. We do this with a single line of code, and then have a look at the output using the generic *summary* function:

```
bamboolm <- lm(culmheight ~ DBH, data=bamboodata)
summary(bamboolm)
##
## Call:
## lm(formula = culmheight ~ DBH, data = bamboodata)
##
## Residuals:
##     Min      1Q  Median      3Q     Max
## -1.26249 -0.50913 -0.02087 0.45275 1.43376
##
## Coefficients:
##             Estimate Std. Error t value Pr(>|t|)
## (Intercept) 1.27726   0.71223   1.793   0.0837 .
## DBH         1.12081   0.08274 13.545 8.12e-14 ***
```

```
##   ---
##   Signif. codes:  0 '***' 0.001 '**' 0.01 '*' 0.05 '.' 0.1 ' ' 1
##
##   Residual standard error: 0.6488 on 28 degrees of freedom
##   Multiple R-squared: 0.8676,  Adjusted R-squared: 0.8629
##   F-statistic: 183.5 on 1 and 28 DF,  p-value: 8.122e-14
```

The output tells us what we have done (repeating the function call), and then gives us a table of (unstandardized) coefficients. These are the intercept (1.28, given in the *Estimate* column for the (*Intercept*) row) and the slope of the fitted line (1.12, given in the *Estimate* column for the *DBH* row). We can use these to plot a straight line with the equation $y = \beta_0 + \beta_1 x$ (where β_0 and β_1 are the model coefficients that correspond to the y-intercept and gradient), which in this case is *culmheight* = 1.28 + 1.12*DBH. That is the line that is shown in Figure 4.1, and which gives an excellent fit to the data. If we need to obtain standardized regression coefficients we can either standardize the data first (see 'Transforming data and testing assumptions' in Chapter 3), or run the *lm.beta* function from the *QuantPsyc* package on the model output object (e.g. *lm.beta(bamboolm)* for the above example).

The summary table also provides some other useful statistics. There is an R-squared value (R^2 = 0.87), telling us the proportion of the variance explained, and an overall F-statistic (F = 183.5) and *p*-value for the regression model. The *p*-value is the same as the one for the correlation, and for the DBH coefficient (in the Coefficients table), because we only have one predictor. For multiple regression models the table will indicate the significance of each predictor, and the F-statistic will tell you about the full model. We are also given some information about the residuals (the left over error that the model can't explain), which can be used to check the assumptions of the test (more on this in 'Assumptions of linear models').

So that's how to run a straightforward linear regression in R. But to set things up for the rest of this chapter, we should dig a little bit deeper into what linear regression is actually doing. We have fitted a line to describe our data, but what does the *p*-value indicate? In regression, we are effectively comparing two different models. In the null model, the line has a slope of 0, which means there is no effect of the value of *DBH* on *culmheight*. The best we can do to predict *culmheight* is to use the overall mean (sometimes known as an *intercept-only* model). In the alternative model, the slope of the line is allowed to vary to try and fit the data better. Usually in regression we don't bother to show both lines, but it's worth making them explicit so that you can see the difference—they are plotted in Figure 4.2.

In both panels of Figure 4.2 I have also added some thin vertical lines that join the (thick black) fitted line to each individual data point. These are called the *residuals*—they are the error between the data and the model prediction. One way of thinking about residuals is that they represent how well the model (thick line) is able to describe the data (points). If the fit is poor, the residual lines will be long (as in the null model). If the fit is good, the residual lines will be short (as in the regression model). The proportion of the variance explained (R^2) and the statistical comparison between the null and alternative models are both based on the lengths of these lines (though we will not go into further details here about precisely how this works). I think of it as the left over variance (i.e. differences between points) that the models cannot explain. The *p*-value in regression is really telling us whether the alternative model can explain significantly more of the total variance than the null model.

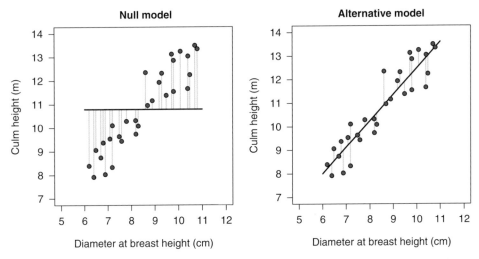

Figure 4.2 Null and alternative model fits for the bamboo data. The null model has a slope of 0, while the alternative model can have any slope value. Thin vertical lines show the residuals between model and data.

T-tests

It is rarely made explicit in introductory texts that this basic idea—of assessing the fits of two models by comparing the left over variances—is also what is happening in other tests such as t-tests. The t-test is used to compare the means of two groups to see if they differ from each other. Usually, this is explained as taking the mean difference and dividing by the pooled standard error (which is the equation of the t-statistic). However, we can also think of a t-test as comparing two models: a null model where the two means are the same, and an alternative model where they differ. Moreover, we can consider group membership to be a predictor variable just like the independent variable in regression—the only difference being that it takes on discrete values (1 and 2), rather than continuous values.

To demonstrate this, we will split our example data set in half. All of the bamboo plants with a DBH value greater than 8.5 cm will be in the *wide* group, and all of the plants with a DBH value below 8.5 cm will be in the *narrow* group. In general it is a bad idea to throw away information in this way, but it might be justified in some contexts, and besides this is just an example. We can perform this split by creating a grouping variable in *R*, and adding it to the *bamboodata* data frame:

```
sizegroup <- NULL   # make an empty data object to store the group
labels
sizegroup[which(bamboodata$DBH<8.5)] <- 1 # set the narrow plants to
be in group 1
sizegroup[which(bamboodata$DBH>8.5)] <- 2 # set the wide plants to be
in group 2
bamboodata$sizegroup <- as.factor(sizegroup) # add to data frame as a
factor
sizegroup
##   [1] 1 1 1 1 1 1 1 1 1 1 1 1 1 1 1 2 2 2 2 2 2 2 2 2 2 2 2 2 2 2
```

The *which* function in this code returns the indices of the DBH vector that satisfy the conditional statement (e.g. it tells us which entries in the DBH vector are less than 8.5, or greater than 8.5). The *as.factor* command tells *R* that the data should be treated as categorical (factor) labels for the purposes of conducting statistical tests (see 'Recoding categorical data and assigning factor labels' in Chapter 3). The data with the group split are plotted in Figure 4.3.

In *R*, we can then run a t-test using the *t.test* function. The standard way to do this is to split the *culmheight* data into two separate data objects for the narrow and wide groups, and then plug them into the t-test function:

```
group1 <- bamboodata$culmheight [bamboodata$sizegroup==1]
group2 <- bamboodata$culmheight [bamboodata$sizegroup==2]
t.test(group1, group2, var.equal=TRUE)
##
##  Two Sample t-test
##
## data:  group1 and group2
## t = -10.016, df = 28, p-value = 9.298e-11
## alternative hypothesis: true difference in means is not equal to 0
## 95 percent confidence interval:
## -3.66897 -2.42303
## sample estimates:
## mean of x mean of y
##  9.267333 12.313333
```

The fifth line of the output gives us a large t-value (–10), and a very small *p*-value, indicating a highly significant group difference. An alternative syntax would be to use the same formula structure that we used for regression, which the *t.test* function also accepts. This time we are predicting the height values using group membership, so the appropriate formula is *culmheight ~ sizegroup*:

```
t.test(culmheight ~ sizegroup, data=bamboodata, var.equal=TRUE)
##
##  Two Sample t-test
##
```

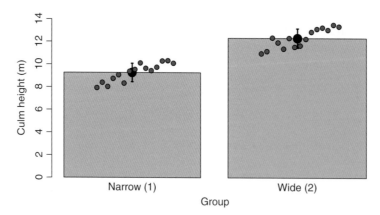

Figure 4.3 Culm height data split into narrow and wide groups by DBH value.

```
## data:  culmheight by sizegroup
## t = -10.016, df = 28, p-value = 9.298e-11
## alternative hypothesis: true difference in means is not equal to 0
## 95 percent confidence interval:
## -3.66897 -2.42303
## sample estimates:
## mean in group 1 mean in group 2
##      9.267333     12.313333
```

This is a different way of achieving exactly the same result, and you can see that the outcomes are identical. But we could also run the test explicitly as a linear model (using the *lm* function), again with the *sizegroup* variable as the predictor and the *culmheight* variable as the outcome.

```
lmttest <- lm(culmheight ~ sizegroup, data=bamboodata)
summary(lmttest)
##
## Call:
## lm(formula = culmheight ~ sizegroup, data = bamboodata)
##
## Residuals:
##     Min      1Q  Median      3Q     Max
## -1.35333 -0.74333 0.06967 0.79667 1.18667
##
## Coefficients:
##      Estimate Std. Error t value Pr(>|t|)
## (Intercept)  9.2673   0.2150  43.09 < 2e-16 ***
## sizegroup2   3.0460   0.3041  10.02 9.3e-11 ***
## ---
## Signif. codes: 0 '***' 0.001 '**' 0.01 '*' 0.05 '.' 0.1 ' ' 1
##
## Residual standard error: 0.8329 on 28 degrees of freedom
## Multiple R-squared: 0.7818, Adjusted R-squared: 0.774
## F-statistic: 100.3 on 1 and 28 DF, p-value: 9.298e-11
```

The output looks different from the output of the t-test function, as it has the same layout as the regression output we saw in the previous section. But you can see that the values of the t-statistic and p-value in the table of coefficients are exactly the same as the ones we got from the t-test function (the minus sign is missing from the t-statistic, but this is arbitrary anyway because it depends on the order in which the groups are entered).

Now, this consistency across methods prompts us to think about the t-test in the context of regression. Just like with regression, we can conceptualize the t-test as comparing two models. The null model is one in which the means do not vary with group, given by the horizontal black line in the left panel of Figure 4.4. The alternative model is one where the means can vary with group, given by the diagonal black line in the right panel of Figure 4.4.

Just as with regression, we can calculate the residual error between each data point and the accompanying model prediction for its group. The model prediction for the null model is the grand mean (horizontal black line). The model prediction for the alternative model is the group mean for each condition. Then we compare the two model fits statistically to see if the alternative model describes significantly more of the variance than the null model. This is another way of thinking about what a significant t-test means: conceptually it is exactly the same as linear regression.

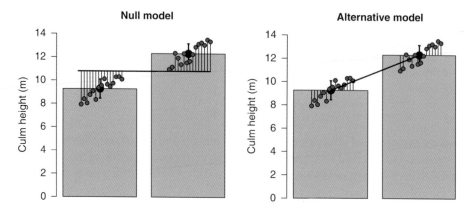

Figure 4.4 T-tests conceptualized as a comparison between a null model where the group means do not differ (left) and an alternative model where they do (right).

ANOVA

Finally, we can extend the same regression logic to analysis of variance (ANOVA), where the independent variable has more than two levels. In the bamboo paper by Yen (2016), the data set is split into five groups by DBH value, in 1 cm increments as shown in the two graphs in Figure 4.5.

Again, we can think of ANOVA as comparing a null model where the predicted values are not affected by group (left panel), with an alternative model where the predicted values change across groups (right panel). Notice that the alternative model involves specifying four separate lines (thick lines joining the means), which can have different slopes. This is why the number of degrees of freedom for the independent variable in a one-way ANOVA

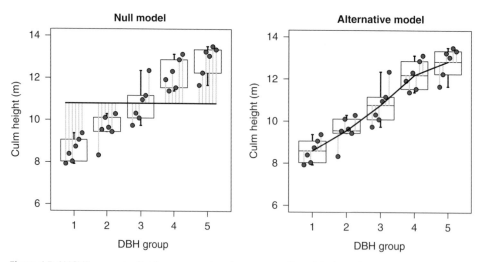

Figure 4.5 ANOVA conceptualized as a comparison between a null model where the group means do not differ (left) and an alternative model where they do (right).

is always one less than the number of levels. To conduct the ANOVA in *R*, we can use the *aov* function (the *DBHgroup* column contains the groupings):

```
anovamodel <- aov(culmheight ~ DBHgroup, data=bamboodata)
summary(anovamodel)
##              Df Sum Sq Mean Sq F value  Pr(>F)
## DBHgroup      4  75.42 18.854   34.68 7.28e-10 ***
## Residuals    25 13.59  0.544
## ---
## Signif. codes:  0 '***' 0.001 '**' 0.01 '*' 0.05 '.' 0.1 ' ' 1
```

Or we can achieve the same result using the linear model (*lm*) function:

```
anovalm <- lm(culmheight ~ DBHgroup, data=bamboodata)
summary(anovalm)
##
## Call:
## lm(formula = culmheight ~ DBHgroup, data = bamboodata)
##
## Residuals:
##     Min      1Q  Median      3Q     Max
## -1.2283 -0.5837  0.1033  0.4988  1.5800
##
## Coefficients:
##             Estimate Std. Error t value Pr(>|t|)
## (Intercept)   8.5850     0.3010  28.520  < 2e-16 ***
## DBHgroup2     0.9733     0.4257   2.286   0.031 *
## DBHgroup3     2.1850     0.4257   5.133 2.64e-05 ***
## DBHgroup4     3.6100     0.4257   8.480 7.99e-09 ***
## DBHgroup5     4.2583     0.4257  10.003 3.19e-10 ***
## ---
## Signif. codes:  0 '***' 0.001 '**' 0.01 '*' 0.05 '.' 0.1 ' ' 1
##
## Residual standard error: 0.7373 on 25 degrees of freedom
## Multiple R-squared: 0.8473, Adjusted R-squared: 0.8229
## F-statistic: 34.68 on 4 and 25 DF, p-value: 7.277e-10
```

Notice that the F-statistic in the ANOVA summary table, and the final line of the regression output, are identical (F = 34.68). This is because the underlying calculations for the tests we call ANOVAs, t-tests, and regressions are fundamentally the same thing (a linear model).

If required, it is also possible to conduct factorial and repeated measures ANOVAs in *R*. The formula syntax for linear models can be easily extended to include multiple independent variables, using the format: *DV ~ IV1 * IV2*. The asterisk denotes factorial combination, so that interaction terms will also be calculated. Repeated measures components can be included as *random factors*, which we will discuss in more detail in Chapter 7. This can be achieved using either the *ezANOVA* function from the *ez* package, or the *lme* function from the *nlme* package, by providing a column of participant ID numbers and specifying which independent variables are repeated measures.

Table 4.1 provides example *R* functions for popular tests, using both the generic function and the linear model form. Traditionally, we would use regression when our independent variable is continuous, a t-test when it is discrete with two levels, and ANOVA when it has more levels. But as Table 4.1 illustrates, these separate tests are really all part of the wider family of the general linear model, and can all be implemented within the same framework.

Assumptions of linear models

It is only appropriate to fit linear models when their assumptions are met. A key assumption of parametric statistics is that the data are normally distributed (conforming to a Gaussian, or bell-shaped, curve), which we can test using the Kolmogorov–Smirnov and Shapiro–Wilk tests we encountered in 'Transforming data and testing assumptions' in Chapter 3. Actually, for linear models it is more accurate to say that we test whether the *residuals* are normally distributed, as this allows us to make the comparison on a full data set, taking the effect of the independent variable(s) into account. For tests comparing different groups, it is also necessary to check whether the variances are equivalent for each group (the homogeneity of

Table 4.1 Table summarizing how common statistical tests can be implemented in R, using both a dedicated function and a linear model.

Test	Generic function call
One-sample t-test	t.test(DV)
Independent t-test	t.test(DV[which(IV==1)], DV[which(IV==2)], var.equal=TRUE)
Dependent (paired) t-test	t.test(DV[which(IV==1)], DV[which(IV==2)], paired=TRUE)
Linear regression	summary(lm(DV ~ IV, data=dataset))
Multiple regression	summary(lm(DV ~ IV1 + IV2, data=dataset))
One-way independent ANOVA	summary(aov(DV ~ IV, data=dataset))
One-way repeated measures ANOVA	ezANOVA(dataset, dv=DV, wid=ID, within=IV)
Factorial independent ANOVA	summary(aov(DV ~ IV1 * IV2, data=dataset))
Factorial repeated measures ANOVA	ezANOVA(dataset, dv=DV, wid=ID, within=c(IV1,IV2))
Mixed design ANOVA	ezANOVA(dataset, dv=DV, wid=ID, within=IV1, between=IV2)

Test	Linear model call
One-sample t-test	summary(lm(DV ~ 1, data=dataset))
Independent t-test	summary(lm(DV ~ IV, data=dataset))
Dependent (paired) t-test	summary(lm(DV[which(IV==1)] - DV[which(IV==2)] ~ 1, data=dataset))
Linear regression	summary(lm(DV ~ IV, data=dataset))
Multiple regression	summary(lm(DV ~ IV1 + IV2, data=dataset))
One-way independent ANOVA	summary(lm(DV ~ IV, data=dataset))
One-way repeated measures ANOVA	anova(lme(DV ~ IV, random = ~1|ID/IV, data=dataset))
Factorial independent ANOVA	anova(lm(DV ~ IV1 * IV2, data=dataset))
Factorial repeated measures ANOVA	anova(lme(DV ~ IV1 * IV2, random = ~1|ID/IV1/IV2, data=dataset))
Mixed design ANOVA	anova(lme(DV ~ IV1 * IV2, random = ~1|ID/IV1, data=dataset))

The terms DV and IV (IV1, IV2) are assumed to be column names of the dependent (DV) and independent (IV) variables in a data object called 'dataset', and also to exist as independent data objects.

The term ID indicates a participant identification variable for repeated measures tests.

The *lme* function is part of the *nlme* package, and the *ezANOVA* function is part of the *ez* package.

Note that alternative implementations for more complex designs can produce different results, and do not necessarily test appropriate assumptions or make the same corrections.

variances assumption), which is typically achieved using Levene's test (the *leveneTest* function in the *car* package). Amazingly, even Levene's test can be conceptualized as a linear model, but one using the absolute residuals rather than the original data points. Finally, for repeated measures designs we should test whether the pairwise differences between conditions also have equal variance (the *sphericity* assumption) in a similar way, using Mauchly's test (the *mauchly.test* function). If the assumptions are not met, it is possible to run alternative versions of many statistical tests instead, including non-parametric tests. As it turns out, many non-parametric statistics can even be thought of as linear models, but using the ranks of the data instead of the data themselves!

Is everything just a linear model then?

Despite the ubiquity of linear models throughout this chapter, not everything can be considered in this framework. For example, in Chapter 9 we will discuss fitting non-linear curves to data, and in Chapter 10 we will describe data as the sum of multiple sine waves. However, the general concept of linear models is relevant to much of the other material in the book, particularly Chapter 7 in which we will introduce (linear) mixed-effects models, and Chapter 11 where we extend the t-test to cope with multiple dependent variables. More generally, linear regression can be extended to the case of multiple independent (predictor) variables, which is called multiple regression. We will not consider multiple regression further in this book, but there are many good texts available that do so in *R*, including Fox and Weisberg (2018), Lilja (2016), and Field, Miles, and Field (2012).

Practice questions

1. T-tests were originally developed to analyse data relating to:
   A) Brewing beer
   B) Bamboo growth
   C) Crop yields
   D) Counting students

2. To test the effect of age on brain volume, the appropriate linear model formula would be:
   A) age ~ brainvolume
   B) brainvolume ~ age
   C) age - brainvolume
   D) brainvolume - age

3. In *R*, the function to run a t-test is called:
   A) ttest
   B) t-test
   C) t.test
   D) Ttest

4. In regression, the residuals indicate:
   A) The total variance in the data set
   B) The differences between each pair of data points
   C) The amount of the variance explained by a model
   D) The error between the data and the model fit

5. The null hypothesis produces a model line with a slope of:
   A) 1
   B) -1
   C) 0
   D) It depends on the data

6. The alternative hypothesis produces a model line with a slope of:
   A) 1
   B) -1
   C) 0
   D) It depends on the data

7. For a one-way ANOVA with three levels, how many regression coefficients would we expect (not including the intercept)?
   A) 1
   B) 2
   C) 3
   D) 4

8. The *as.factor* function is used to:
   A) Define a dependent variable
   B) Turn numeric data into text
   C) Tell *R* that a data object is categorical
   D) Round a number so that it is an integer

9. For a categorical independent variable with four levels, which *R* functions could you use to analyse the data?
   A) *lm* or *t.test*
   B) *lm* or *aov*
   C) *aov* or *t.test*
   D) *lm* only

10. For factorial ANOVA designs, we indicate an interaction between two independent variables using:
    A) An asterisk symbol ($*$)
    B) A plus symbol ($+$)
    C) A tilde symbol ($\sim$)
    D) A slash symbol ($/$)

Answers to all questions are provided in the answers to practice questions at the end of the book.

5 Power analysis

What is statistical power?

The *power* of a statistical test is its ability to detect an effect when it is actually there. Another way of putting this is to say it is the ability of a test to correctly reject the null hypothesis. Power is a probability and so it always has a value between 0 and 1, though it is also sometimes expressed as a percentage. One way to think about it is to imagine that we ran the same experiment lots of times, and then counted how often the experiments produced significant results.

Consider an experimental design with a very strong effect and a very large sample size. The Stroop effect (Stroop 1935) is a neat example. Participants view coloured letters that are arranged to spell out the names of colours in the participant's native language. They are asked to press a button to indicate the colour of the letters. On average, people are slower to report the colour of the letters when it is incongruent with the word the letters spell (i.e. the word 'green' written in red letters) than when it is congruent (i.e. the word 'green' written in green letters).

If we were to repeat this experiment 100 times on 100 different people each time (fortunately this is just hypothetical!), and 98 of the repetitions produced a significant difference in reaction time between congruent and incongruent words, we would say that this experimental design had a power of 0.98 (or 98%). Now let's imagine that we reduced the sample size to 10 participants, and ran the experiment 100 more times. Because the sample size is smaller, we are much less likely to find statistically significant effects. Perhaps now only 40 of the repetitions are significant: our power would be 0.4 (or 40%).

Rather than laboriously running hundreds of replications of an experiment, we can estimate the power of a study design if we know three things: the sample size, the effect size, and our criteria for statistical significance. Sample size is straightforward—it is the number of separate observations we are planning to make. These observations might be human participants, animal subjects, cells, plants, or higher-order entities like schools or countries. Traditionally in the life sciences, $p < 0.05$ is our criterion for statistical significance (called the α level). So the last thing we need to know about is the effect size. What's that?

Effect size

An effect size is a statistic that quantifies, in standardized units, some effect of interest. There are two main types of effect size: those that summarize the amount of variance explained by a statistical model, and those that quantify the difference between group means. Examples in the former category include the correlation coefficient, r, the coefficient of determination R^2 calculated during linear regression, and the η^2 and partial-η^2 effect sizes often used when reporting ANOVA. All of these measures are based on the proportion of the total variance in the data set that we can explain using a linear statistical model (see Chapter 4). For example, an R^2 value of 0.9 tells us that 90% of the variance in our dependent variable is explained by the independent variable. These measures are standardized, so they can be easily compared across different studies, giving us an idea of how important an effect is in both relative and absolute terms. In Chapter 6 we will introduce meta-analysis—a method for pooling effect sizes across studies.

The other type of effect size is based on the standardized difference in means. The general idea is that we want to be able to quantify the difference of group means in a way that also considers the variability of the data. This results in values that are independent of the units of measurement, and so can again be easily compared across studies. For t-tests and related statistics, Cohen's d is a widely used measure of effect size, which we will focus on here. It is defined as the mean difference scaled by the standard deviation (Cohen 1988):

$$d = \frac{\bar{x}_1 - \bar{x}_2}{\sigma}$$
(5.1)

where $\bar{x}_1$ and $\bar{x}_2$ are the group means, and σ is the pooled standard deviation. This is a standardized score, conceptually similar to the z-score, but for means rather than individual observations. Because the denominator is the standard deviation (and not the standard error), the effect size is independent of sample size, although effect size estimates do become more accurate as sample size increases. Other related effect sizes include Hedge's g and Glass's δ, which slightly vary the denominator term. For multivariate statistics, the Mahalanobis distance (Mahalanobis 1936) extends Cohen's d to the multivariate case (see 'The Mahalanobis distance for multivariate data' in Chapter 3).

As a heuristic, Cohen (1988) suggested that effect sizes (d) of 0.2, 0.5, and 0.8 correspond to small, medium, and large effects respectively. Let's think about how this applies to some hypothetical data sets. Imagine we want to know if tortoises can run faster than hares. We time a group of hares and a group of tortoises running along a racetrack. In the first race, the mean times are 57 seconds for the hares, and 108 seconds for the tortoises. The difference (51 seconds) seems large. But when we look at the raw data, we see that the individual animals are quite variable in how long they take—maybe some of them get distracted eating grass, whereas others are more on task. We calculate the standard deviation as being 200 seconds, meaning that our effect size is $d = (108 - 57)/200 = 0.26$—quite a small effect according to Cohen's heuristics.

Next suppose we rerun the race but we remove all of the distractions so that the animals stay focused. The mean times are rather shorter overall: 32 seconds for the hares, and 80 seconds for the tortoises. Notice that the mean difference is about the same as it was before—48 seconds this time. But this time the standard deviation is much smaller, at just 54 seconds.

The smaller standard deviation means that our effect size ends up being much larger: $d = (80 - 32)/54 = 0.89$. So even though the raw difference in means has stayed the same, the precision of the measurement has improved.

One way to think about Cohen's d is to consider the underlying population distributions implied by different values of the statistic. Figure 5.1 shows four pairs of distributions, with various mean differences and standard deviations. The figure illustrates that one can increase d either by increasing the difference between the group means, or by reducing the variance (spread) of the distributions. Although in many situations these properties are fixed, and we must simply measure them as best we can, in some experimental contexts it may be possible to influence them to increase an effect size and boost statistical power. For example, to increase the effect of a drug or intervention, one could apply a higher drug dose, or a longer intervention duration. In 'Measurement precision impacts power' below, we will discuss how collecting more data for each individual in a study can often reduce the overall standard deviation.

How can we estimate effect size?

In order to estimate the power of a study we plan to run, we need to know the likely effect size. But how can we estimate this before we've run the study? If we are running a replication study, we can use the effect size from the original study. However we should note that this

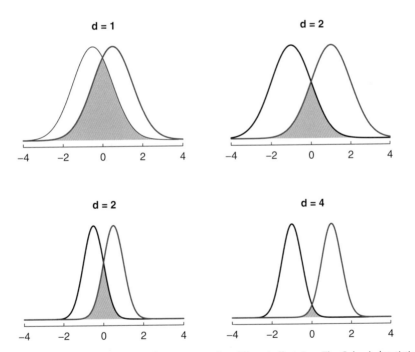

Figure 5.1 Example pairs of distributions that correspond to different effect sizes. The Cohen's d statistic for each pair is given above the panel. The shaded blue region highlights the overlap between the curves; more shading implies a smaller effect size.

is likely to be inflated if the original study was underpowered (Ioannidis 2008). We can understand why this should be so by thinking about the distribution of observed effect sizes across many hypothetical repetitions of a study.

When power is low, effect size estimates are very variable. Only the very largest effect sizes are significant, and these are much bigger than the true effect size, so must be an overestimate. This is illustrated in the left panel of Figure 5.2, which shows the results of a simulation (see Chapter 8 for discussion of how to run such simulations) in which t-tests were run on 1000 synthetic data sets, each comprising N = 10 participants. The true (generative) effect size is given by the white diamond, and the individual points are the estimated effect sizes for each data set. Grey points below the horizontal line are non-significant, and blue points above it are significant. If we imagine that only the significant 'studies' were published (a phenomenon known as *publication bias*), we might estimate a mean effect size around d = 1 (blue diamond), much higher than the true effect size of d = 0.5 (white diamond).

If the study design has a larger sample size (N = 50), the estimates of effect size become more accurate and regress to the mean. The spread of effect sizes becomes tighter about the true value, and most repetitions return an effect size close to the actual effect size (see right panel of Figure 5.2). Note that because of the larger sample size, the power is higher, and a smaller observed effect size is required for statistical significance (i.e. the horizontal line moves down). Effect size estimates from underpowered studies should therefore be treated with caution because they are more likely to overestimate the true effect size.

Another approach is to find or do a meta-analysis on the topic in question. We will cover meta-analysis in more detail in Chapter 6, but in brief it is a technique for calculating the average effect size across a number of studies. Because this increases the overall power, the effect size estimate is likely to be more accurate. Note that the effects of publication bias can still influence meta-analyses, as non-significant results are less likely to be available for inclusion in the analysis.

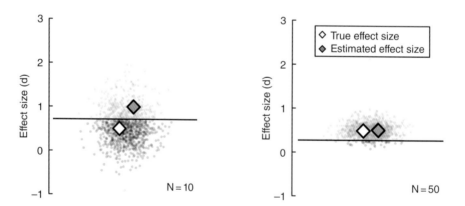

Figure 5.2 Simulations to demonstrate effect size inflation resulting from underpowered studies. 1000 data sets were generated using a mean of 1 and a standard deviation of 2, for 10 participants per data set (left) or 50 participants per data set (right). Points above the horizontal lines indicate significant effects, and points below the lines indicate non-significant effects. White diamonds are the true effect sizes, and blue diamonds are effect size estimates calculated only from significant studies. The position of each point along the x-axis is arbitrary.

When conducting novel or exploratory research, there may not be any suitable studies on which to base our effect size estimates. One common solution is to run a pilot study with a smaller sample. This is commonplace in clinical trials, where the eventual sample size is very large indeed (hundreds or thousands of participants), and large cost savings can potentially be made by running a smaller-scale pilot study first, perhaps on a few dozen participants. Although for many lab-based studies this might not be practical, piloting a new experimental paradigm is always worthwhile if possible, as it provides much additional information besides a possible effect size.

If there really are no existing data to estimate the likely effect size, one can use Cohen's heuristics for small, medium, or large effect sizes to perform power calculations. An important concept is the *smallest effect size of interest*. The idea here is that effects smaller than some value would be of no theoretical or practical importance. For example, if a drug treatment had an effect size of $d = 0.01$, it would provide no meaningful benefit to patients and would not be worth the expense of developing. So it might be practical to power a study to detect a larger effect size that we think would be clinically meaningful. Of course, as expected effect sizes get smaller, the sample size required to achieve adequate power will increase, and there is a balance to be struck with practical considerations around resource allocation for a given study. A useful way to think about these issues is to plot power curves, as we describe in the next section.

Power curves

For a given effect size and α level, we can calculate power as a function of sample size (N). The resulting power curves are shown in the left panel of Figure 5.3, and are instructive and perhaps surprising to many the first time they see them. Although the general trend fits with our intuitions (i.e. that we are more likely to find a significant effect if we test more individuals, at least in cases where there is a real effect), the actual numbers involved depart from the

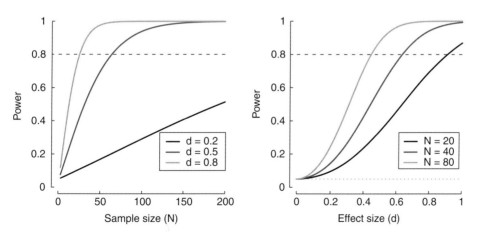

Figure 5.3 Power curves for different combinations of sample size and effect size. The criterion for statistical significance was fixed at 0.05 in all cases.

sample sizes of typical studies in many areas of research. It is generally considered desirable to aim for a power of 0.8 (shown by the horizontal dashed line). For a large effect (of $d = 0.8$) this will require testing at least 26 individuals per group (e.g. 52 in total in a between subjects design). For a medium effect (of $d = 0.5$) we would need to test 64 individuals per group. For a small effect (of $d = 0.2$), our sample size would be almost 400 per group. Relatively few experimental studies have sample sizes this large, yet most effects in the life sciences tend to be in the small to medium range. We discuss the consequences of underpowered studies in the following section, 'Problems with low power'.

The right panel of Figure 5.3 shows analogous functions for fixed sample sizes as a function of effect size. These echo the results of the left panel, showing for example that a sample of 20 individuals (per group) can only detect effect sizes of $d > 0.9$ with satisfactory power. Such large effect sizes are unusual in most areas of life science research (indeed, many consider large effects to be trivial and not worth investigating at all), yet many published studies across diverse disciplines tend to have samples around this size. Similarly, effect sizes in the small to medium range will always have low power with double-digit sample sizes. A consequence of all of this is that many studies are *underpowered*.

Problems with low power

Why are low-powered studies bad? First of all, an underpowered study is unlikely to detect an effect even when one exists. This means returning a null result that, within the framework of frequentist inferential statistics, is inconclusive (see Chapter 17 for further discussion of this point). So, we spend time and resources collecting data, but get back a result that doesn't tell us much. This is disappointing, but it often interacts with a second issue, *publication bias*, to produce even more serious problems.

Publication bias is the general tendency for significant results to be more likely to be published than non-significant ones. This is a systemic issue with the current scientific publishing model, whereby journals are more likely to accept papers reporting significant results (which are seen as more 'interesting'). Researchers are more likely to write up significant findings, with non-significant results languishing in a notional 'file drawer' (hence the moniker, the *file drawer effect*).

Now, any study with $\alpha = 0.05$ has a fixed false positive rate of 0.05, meaning that 1 in 20 studies in which there is no real effect will incorrectly identify significant effects (known as a Type I error). If our power is low and we observe a positive result, the likelihood that it is a false positive increases dramatically. In the extreme case where there is no true effect (when $d = 0$), any 'significant' effect we find is by definition a false positive (see right panel of Figure 5.3—at the far left, the curves converge at a power of 0.05). If these results are more likely to get published than true null findings, the literature will quickly fill up with spurious effects.

There is substantial evidence that this has been happening for many years. For example, an attempt to replicate many preclinical cancer biology results found that only 11% of findings were still significant on replication (Begley and Ellis 2012). In neuroscience, most studies have surprisingly low power, averaging around 0.3 (Button et al. 2013). Recent attempts to replicate large numbers of studies in experimental psychology (Open Science

Collaboration 2015) have found that around 65% of reported effects cannot be replicated, even when using much larger sample sizes than the original studies. This situation is termed the *replication crisis*; because spurious effects tend not to replicate, and many reported effects do not replicate, it is very likely that many reported effects are spurious. One solution to these issues is to increase statistical power, so power analysis has become increasingly important, and is now often required when seeking ethical approval and grant funding.

Problems with high power

The above discussion might lead us to conclude that the more power we have, the better. But there can be issues with high statistical power as well. One potential problem of *overpowered* studies is that many effects will become significant, even if they have trivially small effect sizes. A very small effect size might have no practical or theoretical implications, and yet with a huge sample size it will still be statistically significant. This point is related to the idea of the smallest effect size of interest—any smaller effect sizes may simply not be of interest to anyone. Additionally, for complex factorial designs it can be rather tedious to write up an overpowered study, as the expectation is that every significant effect will need to be commented on and discussed (for a humorous take on this issue, with a more serious appendix, see Friston 2012).

Historically, overpowered studies were not something that many researchers needed to worry about. But in the era of *big data*, it is quite possible to amass enormous data sets, perhaps over the internet, or through national testing programmes such as the UK Biobank (see **https://www.ukbiobank.ac.uk/**). When data sets involve tens of thousands of individuals, standard frequentist inferential tests can become essentially meaningless, as almost everything is significant. In such cases, standardized effect sizes become a key statistic for understanding the structure of the data, and modelling approaches that are not primarily concerned with null hypothesis significance testing may be more appropriate (see Chapters 9 and 12 for examples).

Measurement precision impacts power

Most work on power analysis implicitly assumes that effect sizes are fixed. However, experimenters often have some control over effect size via the precision of their measurement of the dependent variable. When the measurements we make are noisy (as is typically the case in most experiments), it is intuitively understood that running more repetitions (trials) for each individual delivers 'better' data (i.e. data that are less noisy, and more likely to produce statistically significant results). Formally, the effect of running more trials is to reduce the contribution of within-participant variance to the overall sample standard deviation. The sample standard deviation is defined as:

$$\sigma_s = \sqrt{\sigma_b^2 + \frac{\sigma_w^2}{k}}$$

$$(5.2)$$

where σ_w and σ_b are the within- and between-participant standard deviations, and k is the number of trials (Baker et al. 2021). Because the sample standard deviation appears on the denominator of the effect size equation for Cohen's d (equation 5.1), running more trials on each individual can increase effect size, and therefore drive up statistical power. Of course this is only meaningful when the within-participant variance is high compared to the between-participants variance, but this appears to be the case for many experimental paradigms in psychology and neuroscience, and by extension other areas of the life sciences.

A useful way to assess the joint impact of trial number (k) and sample size (N) on statistical power is to produce a two-dimensional contour plot, as shown in Figure 5.4. Each curve illustrates the combinations of N and k that lead to a given level of statistical power. Researchers can therefore optimize their study design by trading off these factors—if individual participants are hard to recruit, each one could be tested for longer; if individual participants are plentiful, each one could be tested for less time. A *Shiny* application to generate power contours is available at: **https://shiny.york.ac.uk/powercontours**.

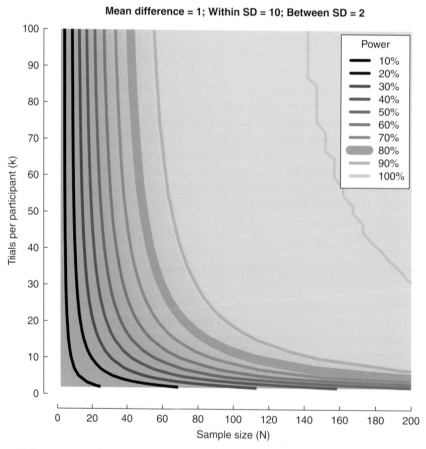

Figure 5.4 Power contour plot. Curves show combinations of N and k that give a constant level of statistical power. This example assumes a true group mean of 1, within-participant standard deviation of 10, and between-participants standard deviation of 2.

Reporting the results of a power analysis

When reporting the results of a power analysis, it is important to include the values you have used to perform the calculations, and also to give some justification or rationale for why you have chosen those specific numbers. For example, imagine we are applying for ethical approval to compare a new drug treatment with a placebo in a group of mice, and need to conduct a power analysis to decide how many subjects to test. The literature on similar drugs reports a range of effect sizes from $d = 0.2$ to $d = 0.52$. We might decide that we should power our study to detect the largest effect that is likely to be of interest, because we only care if our drug is better than those already available. So we'll go with the upper bound here ($d = 0.52$). The typical α level in this field is likely to be 0.05, so we would probably just stick with this (though it is a good idea to think more deeply about our choice of criterion—see Lakens et al. 2018). Finally, we choose the standard power of 0.8 as our target. The power calculations tell us we need 60 subjects in each of the control and experimental groups. We might report these results as follows:

> We conducted a power analysis based on data from previous studies. Our main objective is to determine whether this new treatment is superior to the best-in-class treatment, which has an effect size of $d = 0.52$. Power calculations with $\alpha = 0.05$ indicate that a sample size of 60 mice per group is required to achieve 80% power for this effect size. Our total sample size is therefore 120.

For a second example, let's suppose we wish to replicate a reaction time experiment from the literature (for examples of replication studies on a large scale, see Open Science Collaboration 2015). In the original study, participants made responses in two conditions, and there was a significant difference between the conditions using a repeated measures design: participants were 30 ms faster to respond in condition A than in condition B. The original study reported a significant effect size of $d = 0.44$, with 95% confidence intervals ranging from 0.38 to 0.51. We wish to power our replication study to be able to detect effects at the lower end of this range, so we use $d = 0.38$ for power calculations, along with a power of 0.9 and $\alpha = 0.05$. Because we have an a priori expectation about the direction of the effect, it is reasonable to use a one-sided statistical test. The power calculations tell us that 61 participants would be required for this replication. Note that the design and assumptions are important here—using a two-sided test and a between participants design would require almost five times as many participants! We might report the power analysis like this:

> To design a well-powered replication of the experiment reported by Roy and Kim (2018), we conducted a power analysis. The lower bound effect size from the original study was $d = 0.38$. We planned to detect this using a repeated measures design, assuming a one-sided test in the direction of the original effect. For a criterion of $\alpha = 0.05$, this will require N = 61 participants to reach 90% power.

Post-hoc power analysis

So far, we have talked about trying to determine how many individuals we should test in order to achieve a target level of power. Another use of power analysis is to calculate the power of a study that has already been conducted. This is known as post-hoc power analysis, or *observed power*, and is available as an option in commercial statistics packages such as SPSS. Indeed, this is the method that has been used to estimate the level of power in particular areas of the literature. It is also sometimes used to interpret null results—determining whether an effect was likely to be non-significant because a study was underpowered. However, there are some theoretical concerns with interpreting observed power (Hoenig and Heisey 2001). Most of these centre around the fact that observed power is inversely proportional to the p-value (with low p-values equating to high power). In other words, a non-significant result is likely to have low power *because* it is non-significant. This means that calculating observed power provides no additional information beyond a properly reported p-value and is therefore misleading. A more fruitful approach to interpreting null results is offered by Bayesian statistics, as discussed in Chapter 17.

Doing power analysis in *R*

Power analysis can be conducted using a number of online tools, as well as specialist software such as the free G*Power application (Faul et al. 2007, 2009). There is also an *R* package called *pwr* that we will use for the examples here. It contains dedicated functions for different types of statistical test. They all have a similar structure: we specify all the values apart from the one we want the function to calculate, and it works out the missing value. If we want to know the power, we specify the effect size, the sample size, and the α level. If we want to know the sample size, we specify the power, the effect size, and the α level, and so on. You can install the *pwr* package either using the Packages tab in *RStudio* (see 'Packages' in Chapter 2), or by typing *install.packages('pwr')* in the console.

We will begin with the function for a t-test. Let's conduct a power analysis to estimate the required sample size to detect an effect of $d = 0.5$ with power of 0.8 and $\alpha = 0.05$. Note that we do not specify the sample size, as this is what we want the function to calculate for us.

```
library(pwr)
pwr.t.test(d=0.5, power=0.8, sig.level=0.05)
##
##      Two-sample t test power calculation
##
##              n = 63.76561
##              d = 0.5
##      sig.level = 0.05
##          power = 0.8
##    alternative = two.sided
##
## NOTE: n is number in *each* group
```

The output returns all of the values we have just entered, and also tells us that we will require a sample size of N = 63.77. Of course it is not practical to test 0.77 of a participant, so we always round up to the nearest whole number. Therefore a sample size of N = 64 *per group* is required (so N = 128 in total). Variants for one-sample and paired t-tests are also available, as detailed using the help function.

A similar function can conduct power analysis for correlations. Let's say we want to know the smallest correlation coefficient that can be detected with a power of 0.8 and a sample size of 30 participants.

```
pwr.r.test(n=30, power=0.8, sig.level=0.05)
##
##      approximate correlation power calculation (arctangh
transformation)
##
##             n = 30
##             r = 0.4866474
##     sig.level = 0.05
##         power = 0.8
##     alternative = two.sided
```

This output tells us that a correlation coefficient of $r = 0.49$ or larger can be detected with 80% power.

Next let's look at power calculations for a one-way ANOVA. Here we use a different measure of effect size, called f (also known as Cohen's f). Note that, importantly, this is very different from the F-ratio usually reported in an ANOVA summary table. The effect size f is closely related to d, such that $f = \frac{d}{2}$. It is calculated by taking the standard deviation across the population (i.e. group) means and dividing it by their pooled standard deviation (i.e. across participants). We also need to know how many groups there are in the study design (the input k). Let's calculate the power of a study design with an effect size of $f = 0.1$, and N = 30 participants in each of five groups:

```
pwr.anova.test(k=5, n=30, f=0.1, sig.level=0.05)
##
##      Balanced one-way analysis of variance power calculation
##
##             k = 5
##             n = 30
##             f = 0.1
##     sig.level = 0.05
##         power = 0.1342476
##
## NOTE: n is number in each group
```

This design has a very low power indeed, at only 0.13.

We can use these functions to produce power curves, such as those shown in Figure 5.3, by entering a range of effect sizes or sample sizes in a loop (see 'Loops' in Chapter 2). For example, we can produce a very instructive power curve for correlations as follows:

```
N <- 4:100
r <- NULL
for (n in 1:length(N)){output <- pwr.r.test(n=N[n], power=0.8, sig.
level=0.05)
r[n] <- output$r
}

plot(N,r,type='l',lwd=3,xlim=c(0,100),ylim=c(0,1))
```

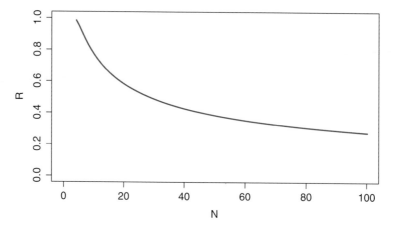

Figure 5.5 Curve showing the minimum correlation coefficient that can be detected at 80% power, as a function of sample size.

This curve (plotted in Figure 5.5) shows the minimum r value that can be detected with 80% power at a range of sample sizes. Even studies with N = 100 cannot reliably detect small correlations where $r < 0.25$.

Power calculations for more complex and sophisticated designs, or those using statistical techniques not covered by the *pwr* package, are best done by simulation. An excellent introduction to power analysis by simulation is given by Colegrave and Ruxton (2020), and we will discuss some of the stochastic methods required for this in Chapter 8.

Practice questions

1. If we ran 10000 simulations of an experiment, and 2000 were significant, what would the statistical power be?
   A) 0.8
   B) 0.2
   C) 0.02
   D) 0.08

2. Calculate Cohen's d for group means of 15 and 12, with a standard deviation of 20.
   A) 0.75
   B) 0.6
   C) 0.15
   D) 0.42

3. As the sample size increases, what happens to effect size estimates?
   A) They increase
   B) They decrease
   C) They become more accurate
   D) It depends on the magnitude of the effect

4. A low-powered study:

   A) is very likely to produce a significant effect

   B) has a higher false positive rate than a high-powered study

   C) typically has a large sample size

   D) is very unlikely to produce a significant effect

5. Effect sizes are often inflated in published studies with low power because:

   A) for a result to be significant, it must have a large effect size

   B) low power reduces the standard deviation

   C) low power increases the standard deviation

   D) replications are usually more accurate

6. How many participants would be required to reach 80% power for a one-sample t-test with an effect size of 0.8, assuming an α level of 0.05?

   A) 20

   B) 25

   C) 26

   D) 15

7. What is the power of a study testing 24 participants to detect a correlation of $r = 0.3$?

   A) 0.30

   B) 0.29

   C) 0.23

   D) 0.26

8. What is the observed power of a one-way ANOVA with an effect size of $f = 0.33$, 8 groups, and 30 participants per group, assuming an α level of 0.01?

   A) 0.99

   B) 0.91

   C) 0.98

   D) 1.00

9. What is the smallest effect size (w) that can be detected using a Chi-squared test with 12 participants, 10 degrees of freedom, and a power of 0.5 (assume an α level of 0.05)?

   A) 1.16

   B) 0.91

   C) 0.88

   D) 0.99

10. The function *pwr.f2.test* calculates power for factorial ANOVAs using the general linear model. Assuming numerator and denominator degrees of freedom of 2 and 12, what is the smallest effect size that can be detected with 80% power and α level of 0.05?

    A) 0.83

    B) 24.4

    C) 0.69

    D) 0.60

Answers to all questions are provided in the answers to practice questions at the end of the book.

Meta-analysis

Meta-analysis is a method for combining the results of several studies computationally. Usually, it is some measure of effect size (see 'Effect size' in Chapter 5) that we choose to combine, such as Cohen's *d*, or the *r* value from a correlation. Of course, the simplest way to do this is just to average the effect sizes from a bunch of studies that all measure the same thing. But often there are differences in study design, sample size, and other features, that make a straightforward average inappropriate. Imagine combining three studies. Two of them are of high quality, testing hundreds of participants using state-of-the-art methods. The other is a rather shoddy affair that should probably never have been published in the first place. It would hardly seem fair to give them all equal weight in our calculations. The tools of meta-analysis allow us to take factors such as this into account. The main outcome of a meta-analysis is an aggregate estimate of effect size, which is used to determine whether, on the balance of evidence, a real effect exists.

Why is meta-analysis important?

The scientific literature is huge, with over a million new articles published every year (e.g. for 2020, *PubMed* lists 1.6 million publications, just in the journals they index). Studies often aim to replicate the key findings from previous work, and many researchers tackle the same problem using similar methods. But these individual studies do not always come to the same conclusions, and are individually limited by constraints on sample size and other aspects of the methodology. By combining results across multiple studies, we can increase the overall statistical power of our observations, and hopefully come up with a more reliable answer to important questions.

The importance of meta-analysis first became apparent in the early 1990s. The classic example is the use of corticosteroids to treat complications arising from premature birth. During the 1970s and 1980s, several studies were published on the topic, but the evidence from reading them individually appeared mixed. For this reason, corticosteroids were not routinely prescribed in cases of premature birth. In 1990 a meta-analysis was published (Crowley, Chalmers, and Keirse 1990) that showed a clear benefit of the drugs (a reduction in mortality of 30–50%), and their use became mainstream clinical practice.

There are two ways of interpreting this story. On the one hand, many thousands of babies suffered and died unnecessarily during the years when the evidence supporting corticosteroid use was available but had not been synthesized together. On the other hand, over the past three decades, many thousands of babies have been treated using this method, and many lives have been saved. Either way, the importance of meta-analysis is clear—unambiguous answers to medical questions can save lives.

The corticosteroid example led to the creation in 1993 of the Cochrane Collaboration, an international charity organization dedicated to coordinating meta-analyses on a range of topics. These are freely available in the Cochrane Library (**https://www.cochranelibrary. com/**). Most of the Cochrane meta-analyses are on medical topics, including a substantial number on mental health and psychiatric conditions. They do not focus only on medications though—many analyses are concerned with dietary and lifestyle factors, and other therapeutic techniques. The logo of the Cochrane Collaboration is a stylized version of the corticosteroid data.

In addition to medical reviews, the tools of meta-analysis can be applied to other topics, including basic experimental laboratory science. These might be less obviously life-saving, but they have become increasingly important in recent years for establishing whether or not reported effects are robust. This has led to some interesting conclusions about entire subfields of research, and is an important aspect of the *replication crisis* being widely discussed in many fields. Overall, a meta-analysis should represent the strongest form of evidence on a particular topic, as it synthesizes all of the available data in a systematic and quantitative way.

Designing a meta-analysis

The stages involved in conducting a meta-analysis (particularly on clinical topics) can be quite prescribed. For example the *Cochrane Handbook* (Higgins and Green 2011) is a substantial official guide that describes the process in detail. I will not attempt to replicate too much of this content here. However, the key stages are:

1. Decide the scope of the analysis—what are you interested in?
2. Decide on search terms and inclusion/exclusion criteria
3. Search online databases using the key terms and build a database of articles
4. Sift through the articles and apply the inclusion/exclusion criteria
5. Extract an effect size from each included study
6. Combine the effect sizes and produce a summary plot

The literature search (stages 1–4) can be succinctly summarized using a PRISMA diagram, which we will introduce in the next section. The remainder of the chapter will mostly focus on the final two stages, as these comprise the numerical and computational parts of the process.

Conducting and summarizing a literature search

A meta-analysis is a quantitative analysis that results from a systematic review of the literature. It is therefore important to be very clear about the scope of the underlying review. We might be interested in a particular disease, or a particular experimental paradigm, for example. We implement these decisions by deciding on a set of *search terms* that we will use when searching online databases, and also some *inclusion* and *exclusion* criteria. Let's imagine we want to conduct a meta-analysis looking at the effectiveness of chlorophyllin (chlorophyll dissolved in water) for treating fish parasites. Our search terms might be *chlorophyllin* and *fish parasites*, and we might decide only to include experimental studies that look at the effect on parasitic algae—studies looking at crustaceans and protozoa would be excluded. It is advisable to *preregister* the protocol for your meta-analysis online, for example using the OSF website (**https://osf.io/**) or the AsPredicted website (**https://aspredicted.org/**). This gives your readers confidence that you are sticking to your original intentions, and not changing your inclusion criteria or other protocols based on the results.

When conducting the literature search, it is important to be systematic and record all decisions about including or excluding a study. A useful tool to guide this process is the PRISMA diagram. This is a flow chart detailing how many studies were involved at each stage of the literature search. An example PRISMA diagram is shown in Figure 6.1, and it is good practice

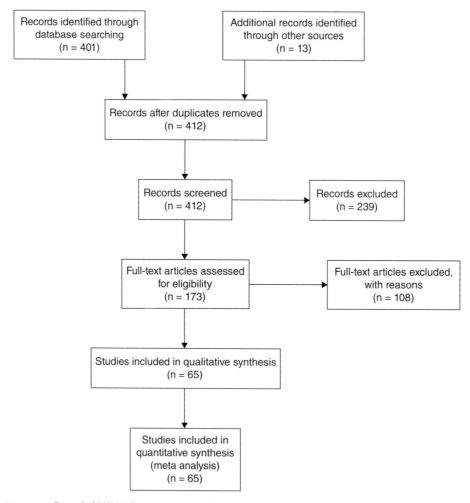

Figure 6.1 Example PRISMA diagram, reporting the number of studies included and excluded at each stage of a literature review.

to include one in the write-up of any meta-analysis. PRISMA stands for the *Preferred Reporting Items for Systematic Reviews and Meta-Analyses*, which are a set of standard protocols that have been widely adopted. They include the diagram, as well as a 27-item statement described in an article by Moher et al. (2009). Templates are available from the PRISMA website (**http://www.prisma-statement.org/**), and the diagram can also be created using several *R* packages including *PRISMAstatement* and *prismadiagramR*.

The PRISMA diagram is helpful because it proposes a structure for conducting a meta-analysis. We would start by searching online databases, such as *PubMed*, *Google Scholar*, *Web of Science*, and other subject-specific databases (e.g. *PsychInfo*, *ArXiv*, *CINAHL*). Deciding key terms for literature searches is quite idiosyncratic, and so will differ widely depending on the topic (see Field and Gillett (2010) for some suggestions). Occasionally, there will also be relevant studies that are not returned by these searches, but which you already know about—these can be included as 'other sources'. One example might be very recently published work, including preprints, that is yet to be indexed. Another example might be if you contact researchers in the field to ask about unpublished data that you subsequently include—this is a good way to help reduce the 'file-drawer' effect that can cause publication bias (see 'Problems with low power' in Chapter 5).

The various sources of information will usually have overlapping content, so it is important to develop a principled method for removing duplicate records. Ideally this can be automated using unique identifiers such as DOI (digital object identifier) numbers or PubMed IDs. Where this is not possible, identifying duplicates using titles, author names, and journal page or issue numbers might be necessary. For very large literatures there will be some benefit to automating this process, and in general, scripting analyses is preferred as this is reproducible, and less prone to data entry errors. However for a smaller corpus of studies it might be better done manually, perhaps using a spreadsheet or some reference manager software.

The next stage is to screen all non-duplicate records, usually by reading the brief description of each study given in the abstract. There will be many records that are clearly not within the remit of the analysis, and these should be excluded at this stage without needing to inspect the full text. Good reasons for exclusion might be studies that use a completely different method, species, or paradigm, or that are review articles rather than primary research. All of the remaining records that pass this screening process should then be obtained and inspected. At this stage it might again become clear that some studies are not within the remit of the meta-analysis. However this time a note should be made of the reasons for exclusion. These reasons need not be lengthy or particularly detailed, but they should reflect the inclusion criteria you originally specified before starting to look through the literature.

The results should then be extracted from each of the remaining studies. Normally effect sizes are used in meta-analysis to summarize the findings of each study, and we will discuss these in the next section. It is always worth keeping detailed notes of the whole process. This will help anyone trying to reproduce the decisions that were taken during the analysis, including your future self if you have forgotten what you did! In particular, you should keep a record of the numbers of studies involved at each stage, which you can then use to generate the PRISMA diagram.

Different measures of effect size

As mentioned in Chapter 5, there are several types of effect size. Many studies use ordinal scales or continuous data. Examples of ordinal data include rating scales and questionnaire data—these can take on fixed values, but the order is meaningful (e.g. a pain rating of 7 means more pain than a rating of 2). Examples of continuous data include physical properties of the world or of an organism: things like heart rate, temperature, brain activity, reaction time, and so on. These can take on any value within some reasonable range. These types of data are well described by effect size measures based on differences in means (such as Cohen's *d*), or those indicating the proportion of the overall variance explained by some predictor (e.g. the correlation coefficient *r*, and the ANOVA effect size measure η^2).

In much of the clinical literature, other types of effect size are common, which you will come across if you read materials on meta-analysis. These are based on the concepts of *risk* and *odds*, which are important ideas to know about. They are used most often for dichotomous (binary) data, which have obvious relevance in medicine—is the patient dead or alive; are they infected or cured? In fact any type of data can be arbitrarily made dichotomous, for example by deciding on a criterion or cut-off. For example, continuous measurements of blood pressure can be categorized into high and low blood pressure groups by choosing some threshold (currently 120/80 for stage 1 hypertension). So the risk and the odds can in principle be calculated for any type of data, though this should only be done when it is a theoretically meaningful thing to do.

The risk is defined as the number of events divided by the sample size. This is a familiar concept: if one out of every thousand people get a particular disease, the risk is 1/1000 or 0.001. The odds is very closely related, but subtly different. It is the number of events divided by the number of non-events. So, in the example of one in a thousand people, the odds would be 1/999, which will be very similar indeed to the risk. However the numbers start to

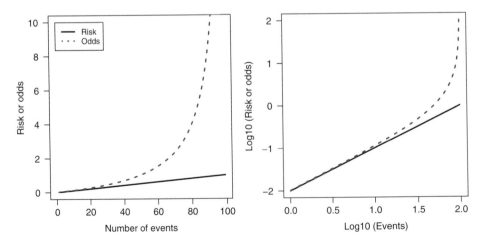

Figure 6.2 Comparison of odds and risk scores for events of different probabilities, assuming a total sample of 100. The right-hand panel shows the log transform of the same values.

diverge as events become more common. Consider a condition that affects half of a sample of 100 people: the risk will be 50/100 = 0.5, but the odds will be 50/50 = 1. A risk score can never exceed 1, but an odds score can take on any positive number. Figure 6.2 shows how the risk and odds diverge as events become more common.

Although raw risk and odds scores are sometimes clinically meaningful, in clinical trials it is more common to report the *risk ratio* or the *odds ratio*. These are the ratios of risk or odds values comparing a treatment group and a control group (e.g. the risk for the treatment group, divided by the risk for the control group). They will tell you, for example, how much a treatment or drug *changes* your risk of some outcome, such as recovering from a disease. Because these are ratios of event counts, they will always be positive numbers, and a value of 1 will always mean there is no difference between the treatment and control groups. However, whether values above or below 1 indicate a positive outcome will depend on exactly what is being measured. An odds ratio of 3 might be good news if it means a drug makes you more likely to recover from an illness, but very bad news if it makes you more likely to have a heart attack!

Lastly, you will also see that some studies report the *log odds ratio*. This is just a log transform of the odds ratio, which is a sensible thing to do given the range of possible odds ratios. After the log transform, odds ratios >1 will have positive values, and odds ratios <1 will have negative values. Ratios in general are often more appropriately represented in log units, and thinking 'logarithmically' is something that gets easier with practice.

Converting between effect sizes

To conduct our meta-analysis, we need the same type of effect size from each study we include. But not all studies report the same measures of effect size, and some report no effect size measures at all. For this reason, we need tools to calculate effect sizes, and to convert between them. We have already encountered (in Chapter 5: 'Effect size') the equation for Cohen's *d* (Cohen 1988), which is the difference in means scaled by the standard deviation:

$$d = \frac{\bar{x}_1 - \bar{x}_2}{\sigma} \tag{6.1}$$

where $\bar{x}_1$ and $\bar{x}_2$ are the group means, and σ is the pooled standard deviation. If these values are reported in a source paper, we can use them to calculate *d*. If the means and standard deviation are not available, we can also convert from a correlation coefficient (*r*):

$$d = \frac{2r}{\sqrt{1 - r^2}} \tag{6.2}$$

or from a t-statistic (where *df* is the degrees of freedom):

$$d = \frac{2t}{\sqrt{df}} \tag{6.3}$$

In the section 'Calculating and converting effect sizes in R' later in the chapter, we will discuss how to convert between effect sizes and common statistics using *R*. However tools also exist online to perform these calculations—for example a number of tools are available at the website: **http://www.psychometrica.de/effect_size.html** (Lenhard and Lenhard 2016).

Fixed and random effects

There are two main types of meta-analysis, which involve slightly different assumptions and calculations. In a *fixed effects* meta-analysis, the underlying effect size is assumed to be constant across all studies included. This is particularly appropriate for things like clinical trials, where the dosage of a drug, and the outcome measure, might be the same across all studies. The alternative is a *random effects* meta-analysis. This has provision for the underlying effect size to vary across studies, and so is more flexible, and often more applicable to lab-based research topics that are less standardized. It is usually clear which type of meta-analysis is appropriate for a given topic, though this can become more apparent as the data are collected. If in doubt, the random effects design is a safer bet as it has fewer assumptions.

Forest plots

Once we have collated the effect sizes from all of the studies we are including in a meta-analysis, it is typical to represent them using a *forest plot*. This graph plots the effect size along the x-axis, and the authors of each study (usually in chronological or alphabetical order) along the y-axis. Figure 6.3 shows a simple example, using a built-in data set from the *rmeta* package in R (these are the corticosteroid data that form the Cochrane Collaboration logo). Forest plots can become much more elaborate than this; for example, I recently published a meta-analysis (Baker et al. 2018) involving 65 studies, for which the forest plot was much larger.

    Forest plots have several additional features. The vertical dashed line at an odds ratio of 1 is known as the *line of no effect* (recall that this means equal odds in the treatment and control groups). If we were using other types of effect size measure, such as *d* or *r*, this would be at *x* = 0. The individual studies can then be compared to this value. Each individual study is represented by a rectangle. The size of the rectangle is typically proportional to the sample size in the study, such that larger sample sizes produce larger rectangles. We can therefore

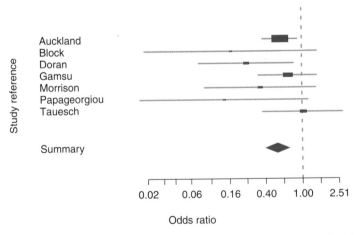

Figure 6.3 Example forest plot of corticosteroid data from Crowley, Chalmers, and Keirse (1990), available as part of the *rmeta* package.

place more confidence in studies represented by larger rectangles, as they should give us a more reliable estimate of the true effect size.

Each rectangle has an accompanying error bar. These are traditionally 95% confidence intervals. Confidence intervals are useful because they work a bit like a t-test. If the confidence intervals overlap the line of no effect for a particular study, the chances are it is not significant. Also, if the confidence intervals from one study do not overlap the mean effect size for another study, the two studies can be considered to have different effect sizes. If our data meet parametric assumptions, the confidence intervals are usually calculated using the approximation 1.96*SE, though they can also be derived by bootstrap resampling (see Chapter 8) if the original data are available.

At the foot of the plot is the summary effect—this is the grand average effect size across all studies. It is traditionally represented by a diamond. The middle of the diamond corresponds to the mean, and the left and right corners are the 95% confidence intervals. The effect is deemed to be significant if the error bars do not overlap the line of no effect, as is clearly the case for the example in Figure 6.3. As mentioned before, the grand average effect size is not simply the arithmetic mean of the individual studies. To understand how it is calculated, we need to discuss the concept of *weighted averaging*.

Weighted averaging

To average three numbers, we add them up and divide by how many numbers we have. For example:

$$mean = \frac{x_1 + x_2 + x_3}{3} = \frac{2 + 7 + 3}{3} = 4 \tag{6.4}$$

Another way of thinking about this is to assume that each number has a *weight* of 1, which it is multiplied by, with the denominator being the sum of the weights:

$$wmean = \frac{\omega_1 \times x_1 + \omega_2 \times x_2 + \omega_3 \times x_3}{\omega_1 + \omega_2 + \omega_3} = \frac{1 \times 2 + 1 \times 7 + 1 \times 3}{1 + 1 + 1} = 4 \tag{6.5}$$

In this example, because the weights are all set to 1, the end result is the same. But if we weight some values differently from others, it will change the outcome. For example, if we assign the second number a higher weight it will bring the average up (because it is a bigger number than the others):

$$wmean = \frac{1 \times 2 + 3 \times 7 + 1 \times 3}{1 + 3 + 1} = 5.2 \tag{6.6}$$

Note that each weight appears twice: to multiply the value on the numerator, and also as part of the sum of the weights on the denominator. In general terms, a weighted average is defined as:

$$wmean = \frac{\Sigma(\omega_{1:i} \times x_{1:i})}{\Sigma \omega_{1:i}} \tag{6.7}$$

where ω_i is the ith weight, and x_i is the ith value in the list of numbers we wish to average. This is the procedure typically used to calculate the grand mean effect size. But what are the weights?

In meta-analysis, we want to use weights that give an indication of the quality of each study. One very simple way to do this is to use the sample size as the weights—a study testing 10 participants would have a weight of 10, and a study testing 100 participants would have a weight of 100. This treats each study as though it were part of a single monolithic study (i.e. it is a fixed effects approach). Other alternatives are to use uniform weights (e.g. to give each study a weight of 1), or to choose some predetermined criteria based on the methodology used. For example, one might decide to weight studies using a state-of-the-art recording device more highly than those using older technology (for example in neuroscience, MEG has lower recording noise than EEG; in genetics, PCR is better than older methods like RFLP).

These options are reasonable and defensible in some situations, but they are not what is typically done in meta-analysis. Instead, the weights are derived from the *variance* for each study. Specifically, we use the inverse variance, $1/\sigma^2$. This will be a large value for studies with small variance (i.e. very reliable studies), and a small value for studies with large variances (i.e. unreliable studies). Note that the σ term represents the standard deviation of the sampling distribution, which is the sample standard error. As such, the sample size contributes to the inverse variance weights (because the standard error calculation includes the sample size).

As well as calculating the weighted average of the effect sizes, we also need to derive confidence intervals (to tell us where the corners of the diamond should go). One way to do this is to calculate the variance of the weighted average using the squares of the weights to combine the variances. Another option is to use stochastic simulations (see Chapter 8) to estimate the variance (Sánchez-Meca and Marin-Martinez 2008). Fortunately, these rather complex calculations are done automatically in meta-analysis software, so we will not consider them further here.

Publication bias and funnel plots

Once the forest plot has been created, we can conclude whether or not there is an overall effect. But there is a problem here: the results of the meta-analysis depend entirely on the studies we include, and we know (from biases in the scientific publication process) that statistically significant studies are more likely to be published (see 'Problems with low power' in Chapter 5). So does this mean that meta-analysis is pointless, because due to publication bias we are almost guaranteed to get a positive result, given that the majority of our source studies will report significant effects?

Fortunately, meta-analysis provides a useful tool to assess the likely impact of publication bias— the *funnel plot*. Like the forest plot, the funnel plot represents effect size on the x-axis. But this time, the studies are ordered along the y-axis, usually according to either sample size or inverse variance. Studies with large samples appear towards the top, and studies with small samples appear near the bottom. An ideal funnel plot looks like the example in the left panel of Figure 6.4.

The symmetrical funnel plot gets its triangular shape because studies (points) with large sample sizes (at the top) are more likely to produce estimates of effect size close to the true

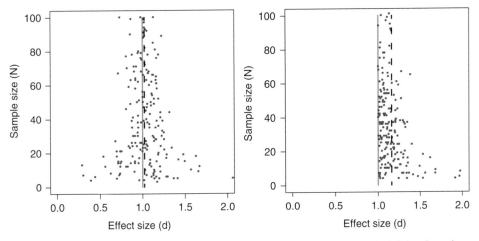

Figure 6.4 Example funnel plots, in which each point represents a simulated study. In the left-hand panel, a symmetrical funnel (triangle) shape is apparent, with large-sample studies producing estimates close to the true effect size (top) and small-sample studies producing more variable estimates (bottom). In the right-hand panel, studies with estimates below the true mean are suppressed (i.e. remain unpublished). The plot becomes asymmetric, and the mean effect size (dashed line) is overestimated.

mean (solid line), whereas studies with small sample sizes (at the bottom) will produce more variable effect sizes. If a funnel plot looks like this, it is unlikely that publication bias is a big problem for the area under study.

Now let's think about what would happen if studies that were non-significant did not get published. This might happen for nefarious reasons (such as a pharmaceutical company deliberately suppressing a study that shows its drug is ineffective), but it is also likely just to occur as a consequence of human nature, and the current incentive system in scientific publication. Non-significant results are much harder to get published, as many journals will simply reject them out of hand as being 'uninteresting'. Furthermore, most researchers have limited time, and will often prioritize publishing studies with significant results, which might be more likely to get published in prestigious journals and so be better for their career.

The right-hand panel of Figure 6.4 shows an asymmetrical funnel plot, in which all studies with effects below the true mean of $d = 1$ are omitted. The effects of this are clearest in the small-sample studies, which now skew out to the right. Studies with a larger effect size were close to the true mean anyway, so these look much the same as before. One consequence is that the mean effect size across all the studies (shown by the dashed line) now overestimates the true mean (solid line).

Funnel plots can be used to test for publication bias, and this is routinely done as part of a meta-analysis. In situations where publication bias is detected, techniques exist to estimate what the true underlying effect size is likely to be. One striking example is a meta-analysis by Shanks et al. (2015) that looked at priming studies of consumer choice. The funnel plot they produced (see their figure 2) was highly asymmetrical, whereas a funnel plot of replication studies was symmetrical about an effect of $d = 0$. This is strong evidence for publication bias, or other types of questionable research practice (such as p-hacking) in this particular paradigm.

Some example meta-analyses

To illustrate the diverse range of topics that are amenable to meta-analysis, this section will discuss three examples from the literature. The first is a study that investigated whether sexual orientation is a risk factor for suicidal behaviour in young people (Miranda-Mendizábal et al. 2017). The PRISMA diagram from this study is quite remarkable, as it shows how an initial set of over 30000 database records was reduced to only seven that were suitable for inclusion in the quantitative analysis. The odds ratios from these seven studies all exceeded 1, indicating an increased risk of suicidal behaviour in individuals identifying as homosexual, though three of the original studies were non-significant. The random effects meta-analysis resulted in an overall odds ratio of 2.26—this means that lesbian, gay, and bisexual adolescents have more than twice the risk of suicidal behaviours compared with their heterosexual counterparts. A funnel plot showed no evidence of publication bias in the seven included studies, which was supported by a non-significant Harbord test (a quantitative test of asymmetry (Harbord, Egger, and Sterne 2006)). The study concludes that public health strategies should be developed to explicitly support the LGB community, given that the analysis indicates they are a high-risk group.[1]

Glasziou and Mackerras (1993) investigated the effect of vitamin A supplementation on the likelihood of dying from infectious diseases. They included only controlled trials in their meta-analysis, involving measles, respiratory diseases, or diarrhoea in children in developing countries. They pooled odds ratios across studies, with aggregate effects being calculated separately for different diseases, and in community studies. The largest result was for measles: across three studies, the average odds ratio was 0.34 (i.e. a reduced risk of death of 66% following supplementation). A subset of five community studies also found a mortality reduction of 30%. To assess the potential impact of publication bias, the authors calculated a statistic called the 'failsafe N' (Rosenthal 1979) for the community study result (with five studies). This statistic tells us the number of non-significant studies that would need to exist (yet remain unpublished) for there to be no effect overall. For this example it was 53, which is an implausibly large number of studies to remain unpublished. Overall, these results suggest a strong benefit of either vitamin A supplementation, or a well-balanced diet, in reducing the risk of death from infectious disease.

An amazing online resource is the website *Neurosynth*. This uses an automated meta-analysis approach to function magnetic resonance imaging (fMRI) data, combining brain activation across thousands of studies (Yarkoni et al. 2011). One can choose a search term, such as *acoustic* or *emotion*, and the system will show the brain regions most strongly associated with that term in the fMRI literature. It also works the other way around—if you choose a brain region, it will list studies reporting significant activation there. Finally, there are many thousands of whole-brain maps showing gene expression, that link back to the term-based meta-analyses. So if you are interested in a particular gene, you can generate hypotheses about the sorts of cognitive functions it might affect. The *Neurosynth* website showcases the enormous potential of expanding the basic concept of meta-analysis to large modern data sets, as well as automating much of the computation.

---

[1] Note that the use of the acronym LGB here reflects the identities of the participants included in the study. This is not intended to exclude other groups (for example, from the extended form of the acronym LGBTQIA+). However, the extent to which the results might generalize beyond the population studied is unclear.

Calculating and converting effect sizes in *R*

The *compute.es* package in *R* provides a set of 16 tools for calculating and converting between effect sizes. All of these have a common output, so the only thing that differs is the information you feed in. The simplest is the *mes* function, which converts means to effect sizes:

```
mean1 <- 14
mean2 <- 12
sd1 <- 5
sd2 <- 6
n1 <- 100
n2 <- 100

mes(mean1, mean2, sd1, sd2, n1, n2)
## Mean Differences ES:
##
##   d [ 95 %CI] = 0.36 [ 0.08 , 0.64 ]
##    var(d) = 0.02
##    p-value(d) = 0.01
##    U3(d) = 64.14 %
##    CLES(d) = 60.11 %
##    Cliff's Delta = 0.2
##
##   g [ 95 %CI] = 0.36 [ 0.08 , 0.64 ]
##    var(g) = 0.02
##    p-value(g) = 0.01
##    U3(g) = 64.09 %
##    CLES(g) = 60.07 %
##
##   Correlation ES:
##
##   r [ 95 %CI] = 0.18 [ 0.04 , 0.31 ]
##    var(r) = 0
##    p-value(r) = 0.01
##
##   z [ 95 %CI] = 0.18 [ 0.04 , 0.32 ]
##    var(z) = 0.01
##    p-value(z) = 0.01
##
##   Odds Ratio ES:
##
##   OR [ 95 %CI] = 1.93 [ 1.16 , 3.2 ]
##    p-value(OR) = 0.01
##
##   Log OR [ 95 %CI] = 0.66 [ 0.15 , 1.16 ]
##    var(lOR) = 0.07
##    p-value(Log OR) = 0.01
##
##   Other:
##
##   NNT = 8.64
##   Total N = 200
```

For a single line of code, the output is very extensive. The idea is to provide all the various measures of effect size you might need to use. These include Cohen's *d* in the first section (which has a value of 0.36), Hedge's *g* in the second section, the equivalent correlation coefficient (*r*) and z-score, odds ratios, and number needed to treat (NNT). Similar functions exist if you have a *p*-value, t-value, *r* value, and so on—for example:

```
pvalue <- 0.03
pes(pvalue, n1, n2)

tvalue <- 2.5
tes(tvalue, n1, n2)

rvalue <- 0.24
res(rvalue, n = n1)
```

I have hidden the output from the above functions, but it is in exactly the same format as for the previous example. What we usually want, though, is just to extract the single effect size we are interested in. We can do this by assigning the output of the function call to a data object as follows:

```
output <- mes(mean1, mean2, sd1, sd2, n1, n2)
```

The object called *output* then contains all of the numbers you might need, in fields with sensible names. For example, you can request Cohen's *d* and its variance as follows:

```
output$d
## [1] 0.36
output$var.d
## [1] 0.02
```

And we can convert the variance to a standard deviation by taking the square root:

```
sqrt(output$var.d)
## [1] 0.1414214
```

The effect size, its standard deviation, and the sample size are the values you will need to enter into a meta-analysis. It will often help to store them in another data object so they can be easily accessed and entered into the meta-analysis functions.

Conducting a meta-analysis in *R*

Imagine that we have compiled effect sizes (and their standard errors) from a list of studies as follows:

```
effectsizes <- c(0.7, 0.4, 2.1, 0.9, 1.6)
standarderrors <- c(0.2, 0.3, 0.9, 0.3, 0.5)
```

The effect sizes might be values of Cohen's *d*, and the standard errors will be the square root of the variance estimates that are returned when the effect size is calculated (see previous section). The *rmeta* package contains functions that will use these values to conduct a meta-analysis. There are several varieties of meta-analysis available, but we will use the *meta.summaries* function to conduct a random effects meta-analysis using the effect size measures.

```
meta.summaries(effectsizes,standarderrors,method='random')
## Random-effects meta-analysis
## Call: meta.summaries(d = effectsizes, se = standarderrors, method
= "random")
## Summary effect=0.852    95% CI (0.456, 1.25)
## Estimated heterogeneity variance: 0.078  p= 0.149
```

The output from this function tells us the summary effect size (0.852) and its 95% confidence intervals. This is useful, but it's more helpful if we save the output of the function into a data object, which we can then pass into the *metaplot* and *funnelplot* functions to produce graphical summaries of the results:

```
metaoutput <- meta.summaries(effectsizes,standarderrors,method='ran
dom')

# this line of code tells R to put the next two plots side by side
par(mfrow=c(1,2), las=1)

metaplot(effectsizes,standarderrors,summn= metaoutput$summary,
        sumse=metaoutput$se.summary,sumnn= metaoutput$se.summary^-2,
        xlab='Effect size (d)',ylab="Study",summlabel='')

funnelplot(metaoutput, plot.conf=TRUE)
```

These plots (shown in Figure 6.5) are quite rudimentary, but can be improved by specifying additional input arguments. For example, the author names can be specified using the *labels* argument, and different colours chosen with the *colors* argument (see further details in the help files). The funnel plot can be automatically mirrored about its midpoint by adding the argument *mirror=TRUE* to the function call.

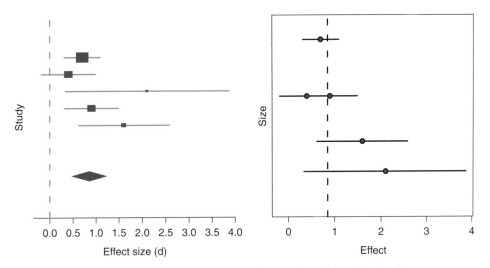

Figure 6.5 Auto-generated forest and funnel plots, using the *metaplot* and *funnelplot* functions.

It is also helpful to use the generic *summary* function to get a more detailed summary of the meta-analysis, which includes everything you would need to generate your own forest plot manually (e.g. in another plotting package):

```
summary(metaoutput)
## Random-effects meta-analysis
## Call: meta.summaries(d = effectsizes, se = standarderrors, method
= "random")
##  -----------------------------------------------------
##    Effect (lower  95% upper) weights
## 1    0.7     0.31       1.09    1.7
## 2    0.4    -0.19       0.99    1.2
## 3    2.1     0.34       3.86    0.2
## 4    0.9     0.31       1.49    1.2
## 5    1.6     0.62       2.58    0.6
##  -----------------------------------------------------
## Summary effect:  0.85 95% CI ( 0.46,1.25 )
## Estimated heterogeneity variance: 0.078   p= 0.149
```

Those are the basics of doing a meta-analysis in R. There is much more functionality in the *rmeta* package, and other packages are available for specific types of meta-analysis and other variations on the analysis.

Practice questions

1. Calculate Cohen's *d* for a study with means of 0.3 and 0.34, standard deviations of 0.1, and sample sizes of 24 per group.
   A) 0.8
   B) 0.9
   C) 0.3
   D) 0.4

2. Calculate the equivalent Cohen's *d* for a correlation coefficient of $r = 0.7$ with a sample size of 10.
   A) 0.70
   B) 1.96
   C) 0.87
   D) 0.33

3. What is the odds ratio that corresponds to a t-test with a *p*-value of 0.01, and 30 participants per group?
   A) 3.48
   B) 1.25
   C) 0.69
   D) 4.19

4. What is the value of Hedge's *g* for an ANOVA with an F-ratio of 13.6 and 17 participants in each group?
   A) 1.26
   B) 0.53

C) 0.60

D) 1.24

5. What is the log odds ratio for comparing proportions of 0.7 and 0.6 with three participants per group?

A) 0.12

B) 1.56

C) 0.44

D) 0.24

6. Conduct a random effects meta-analysis using effect sizes of d = 0.1, 0.6, −0.2, 0.9, and 1.1, with standard deviations of 0.2, 0.3, 0.1, 0.4, and 0.5. What is the aggregate effect size?

A) 0.80

B) −0.01

C) 0.37

D) 0.20

7. What is the aggregate effect size using the values from question 6, but conducting a fixed effects analysis instead?

A) 0.80

B) −0.01

C) 0.37

D) 0.20

8. Produce a forest plot using the data from question 6 (assuming random effects). Is there a significant effect overall?

A) No, because the diamond overlaps the line of no effect

B) Yes, because the diamond overlaps the line of no effect

C) Yes, because most individual studies do not overlap the line of no effect

D) No, because one of the individual studies has a negative effect

9. Which of the following is **not** a plausible explanation for an asymmetrical funnel plot?

A) Small sample studies are more likely to produce significant effects

B) Random sampling

C) P-hacking

D) Publication bias

10. If 10 members of a treatment group of 500 recover from an illness, whereas only five members of a control group of 400 recover, what is the odds ratio?

A) 0.020

B) 0.013

C) 1.60

D) 1.61

Answers to all questions are provided in the answers to practice questions at the end of the book.

Mixed-effects models

Mixed-effects models (also sometimes referred to as linear mixed models, or hierarchical linear models) are a class of statistical tests that build on simpler tests that you may already be familiar with, such as regression, t-tests, and ANOVA. As readers of Chapter 4 will know, all of these tests are based on the *general linear model*, in that they all involve fitting straight lines to data to explain a portion of the total variance in a systematic way. The key difference with mixed-effects models is the inclusion of one or more *random effects*—variables where our observations are grouped into subcategories that have a systematic effect on the outcome. In many situations, a mixed-effects model could be used instead of one of the simpler tests. The main advantages are: (i) the mixed-effects approach models more of the variance, including item-level variance, (ii) mixed-effects models can incorporate group and individual differences, and (iii) mixed-effects models can cope well with missing data points and unequal group (cell) sizes. The aim of this chapter is to introduce the basic concepts of mixed-effects models, and illustrate them with examples and code.

It is important to be clear about the distinction between *fixed* and *random* effects. A fixed effect can be either a continuous or categorical independent variable, much as we might use in standard regression or ANOVA designs. We include these in our design because we anticipate that they may be able to explain a portion of the variance in our dependent variable (outcome measure), and we would like to know if this effect is statistically significant. A random effect is a grouping variable that we might also expect to have an effect on the dependent variable, but usually one that we are not interested in, and therefore want to control for. Examples of groups include different individuals (each of whom is tested multiple times), different testing sites (schools, countries, forests, oceans, farms, bee colonies, etc.), or any other plausible grouping variable such as species, family, breed (e.g. of dog), or genotype. Whereas the repeated measures ANOVA framework can cope with a single grouping variable, mixed-effects models can include multiple random effects.

In common with other versions of the general linear model (see Chapter 4), the general idea of mixed-effects models is to try to account for as much of the overall variance in the data set as possible, using our various predictors. We usually want to know if our fixed effects are able to account for a significant proportion of the variance, but we also want to account for the variance due to our random effects. Sometimes this is because random effects are 'nuisance' variables that we need to control for, but are not really interested in. Including the random effect in our model means that we can remove this variance, reducing the noise in our estimate of the fixed effects. Sometimes accounting for a random effect can reveal structure in a data set that is otherwise masked by group differences, as we will demonstrate with our first example.

Different types of mixed-effects model

In our first example of mixed-effects modelling, we will use a simulated data set with one continuous predictor variable (a fixed effect or independent variable), one outcome measure (dependent variable), and one random effect (grouping variable) with five levels. Figure 7.1 plots the outcome against the predictor—this just looks like a cloud of points without much overall structure. If we try conducting traditional linear regression, we get quite a flat line overall (dashed line), which does not differ much from a line that we constrain to be flat (solid grey line). In regression terminology, the solid line is an *intercept-only* model, which has its slope fixed at zero, and can only shift up or down. The dashed line has both an intercept and a slope that are free to vary. If there were an overall positive or negative relationship in this data set, linear regression would give us a fitted (dashed) line with a slope that differs from zero.

Using linear regression, we can ask statistically if the slope of the fitted line differs significantly from zero. In this case it does not ($p > 0.05$):

```
## Analysis of Variance Table
##
## Response: DV
##             Df Sum Sq Mean Sq F value Pr(>F)
## IV           1     28   28.34  0.0674 0.7953
## Residuals  498 209475  420.63
```

However, what if the data points came from five different groups (which is how they were actually simulated)? When we plot them this way (using different colours in Figure 7.2(a)) there is a much clearer structure. Within each group, the IV is quite strongly predictive of the DV. Mixed-effects models allow us to characterize these relationships in three different ways.

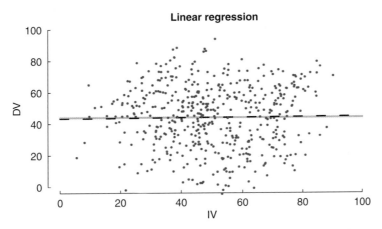

Linear regression

Figure 7.1 Simulated data showing the relationship between one independent variable (IV) and one dependent variable (DV). The solid grey line is an intercept-only regression, with slope constrained to be 0. The dashed black line is the best fitting regression line, with slope and intercept free to vary.

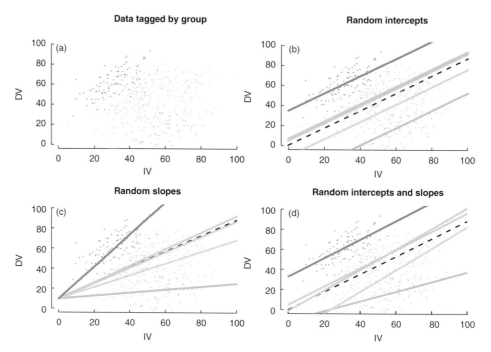

Figure 7.2 The same data as in Figure 7.1, but with groups tagged in different colours (a). Mixed-effects models with random intercepts (b), random slopes (c), and random intercepts and slopes (d), are shown by the regression lines.

The first approach is to allow each group to have its own *intercept*—this is referred to as a *random intercepts* model. This means we would fit a regression line across all five groups, which has a constant slope, but allow it to shift up and down for each group. This is what is represented by the individual coloured lines in Figure 7.2(b). The black dashed line is the grand average regression line. It is quite different from our traditional regression fit (in Figure 7.1), which was essentially flat. If we now perform a statistical comparison, we see that there is a highly significant effect of the IV:

```
## Type III Analysis of Variance Table with Satterthwaite's method
##     Sum Sq Mean Sq NumDF  DenDF F value    Pr(>F)
## IV  40802   40802      1 496.48  344.39 < 2.2e-16 ***
## ---
## Signif. codes:  0 '***' 0.001 '**' 0.01 '*' 0.05 '.' 0.1 ' ' 1
```

An alternative to the random intercepts model is the *random slopes* model, shown in Figure 7.2(c). In this model, the slopes are permitted to vary, but the intercept is fixed across all groups. Recall that the intercept is the value of y when x = 0 (in the regression equation $y = \beta_0 + \beta_1 x$), which is why all the lines in Figure 7.2(c) meet at this point. For our current data set, this model also does a reasonable job of describing the data.

Finally, we can allow both the intercepts and slopes to vary between groups, as shown in Figure 7.2(d). This captures the shallower slope of the bottom group (in purple), as well as the vertical offsets between groups. Overall this type of model has more degrees of freedom than the other two. We could alternatively have run five completely independent linear regressions (or used multiple regression) instead of our single mixed-effects model, but the mixed-effects approach additionally gives us the grand average regression line, which takes account of the sample size of each group and has greater overall power than for any individual group. In other words it tells us about the overall effect of the independent variable, rather than its effect only within each group. For models where either the slope or intercept is fixed, the mixed-effects framework lets us jointly estimate the value of the fixed parameter across all of our groups.

Mixed-effects regression models are enormously flexible, and we will learn about the syntax to implement them later in the chapter. The decision of whether to include random intercepts, random slopes, or both will depend heavily on the hypothesis you are trying to test. However it is quite rare to find a situation where only random slopes are required—most models either involve random intercepts, or allow both parameters to vary. Sometimes it might also be advisable to test more than one model, and use goodness of fit indicators such as R^2 to decide which model describes the data best (see 'Reporting and comparing mixed-effects models' later in the chapter for more detail). Note that because a random intercepts model requires fewer degrees of freedom than a model in which both parameters vary, it can sometimes be fitted to data sets with fewer observations. In the next section, we will run through an example of mixed-effects regression using data from the literature.

Mixed-effects regression example: lung function in bottlenose dolphins

A study by Fahlman et al. (2018) measured resting lung function in 32 bottlenose dolphins. The study used mixed-effects regression models to understand the relationships between several variables. The main dependent variable was the tidal volume (V_T), measured in litres, which is an index of lung capacity. Here we focus on the relationship between V_T and body mass (in kg), as shown in figure 2 of the original paper. The raw data are available online (**https://osf.io/6wjh8/**) and are replotted in Figure 7.3. (I have slightly tweaked the inspiration data by adding 1 to each data point for the purposes of this example.)

The traditional approach to analysing these data might be to conduct a single regression, incorporating all data points and fitting one straight line. However, this is not ideal in the current situation, for two reasons. First, each dolphin contributes between one and four observations (notice that there are 32 animals, but more than 32 data points of each condition). If we averaged across measurements for each animal, this would mean our measurement precision would change for different animals. If we included each data point, we would be double-counting (or triple- or quadruple-counting) some animals, meaning they would contribute disproportionately to estimating the regression line. Second, we have two measures here (expiration and inspiration—breathing out and in), and in principle the relationship with body mass might differ between them.

The mixed-effects approach allows us to deal with both of these issues in a coherent way. We treat body mass and breath direction (expiration/inspiration) as fixed effects, and animal as a random effect. This allows us to have separate regression lines for the two directions of breath (expiration and inspiration) and test if there is evidence overall for an effect of body

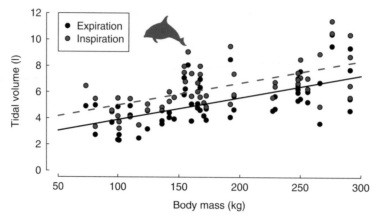

Figure 7.3 Tidal volume for bottlenose dolphins as a function of body mass. Modified and replotted from Fahlman et al. (2018).

mass. The two lines in Figure 7.3 show these fits, and the regression output tells us there is a significant effect of body mass, and also of breath direction:

```
## Type III Analysis of Variance Table with Satterthwaite's method
##
##          Sum Sq Mean Sq NumDF   DenDF F value     Pr(>F)
## bodymass 30.719  30.719     1  26.720  26.521 2.096e-05 ***
## direction 34.799 34.799     1  77.414  30.044 5.095e-07 ***
## ---
## Signif. codes:  0 '***' 0.001 '**' 0.01 '*' 0.05 '.' 0.1 ' ' 1
```

The significant effect of body mass indicates that the overall regression slope is significantly steeper than 0. The significant effect of direction tells us that the two breath directions involve different regression lines (though note that this is largely because I modified the data for this example).

To account for the multiple observations from some animals, we treat animal as a random effect. In the current context, this is similar to a repeated measures design for a t-test or ANOVA. But, crucially, those tests expect a balanced design, where each individual contributes the same number of observations. The mixed-effects approach relaxes this assumption, so it is more flexible when dealing with real data sets.

As you might recall from Chapter 4, the general linear model underlying regression can also be used in situations where the independent variable is categorical rather than continuous. The most familiar instances of this are t-tests and ANOVAs. We can bring the benefits of mixed-effects models to factorial experimental designs, and also include multiple random effects, as we will see in our next example.

Factorial mixed-effects example: lexical decision task for nouns and verbs

Mixed-effects models have become very popular the area of psycholinguistics—the study of how the brain processes language. In this field, it is necessary to have multiple different stimulus examples that are all drawn from a particular category—for example nouns or verbs. Some of these examples will be easier than others for a given task, and each example can be presented to the same participant only once, so that memory processes do not interfere with the task. In traditional analyses (e.g. ANOVAs), this item-level variance would be ignored, which reduces statistical power (Westfall, Kenny, and Judd 2014). Mixed-effects models allow us to treat such variance in a principled way, much as a repeated measures t-test or ANOVA deals with variance between individual participants.

For our next example, we consider a lexical decision task in which participants are presented with a string of characters, and must decide if they are a word or a non-word. Examples of non-words are often based on real words, but with some errors introduced, for example 'bekause', but they can also be nonsense strings of letters such as 'okjsdfj'. In our experiment, 20 participants each respond to 20 nouns and 20 verbs, and also 20 non-words based on the original nouns, and 20 non-words based on the original verbs. The dependent variable is the reaction time, measured in milliseconds. Our example stimuli are shown in Table 7.1.

For our first participant, we will show these stimuli in a random order, and measure reaction times for each decision (word vs non-word). Their data might look something like the values shown in Table 7.2.

In a traditional analysis, we would average across all of the examples in each category, and use the participant means (i.e. the numbers in the final row of Table 7.2). These would then be entered, along with the means of the other 19 participants, into a 2 x 2 repeated measures

Table 7.1 Example stimuli for a lexical decision task.

Example	Nouns	Non-nouns	Verbs	Non-verbs
1	pancake	poncake	speak	speek
2	eyeball	eyepall	break	bweak
3	doorway	doornay	drill	driil
4	computer	compuler	stretch	stredch
5	watermelon	wadermelon	manipulate	manipulake
6	husband	hubsand	pinch	pinsh
7	chicken	chisken	spell	srell
8	television	telerision	grow	mrow
9	flagpole	flagmole	watch	watsh
10	railway	rainway	grab	greb
11	tractor	tragtor	listen	listun
12	fertilizer	fertinizer	fly	flei
13	rectangle	rectangel	swim	swib
14	lettuce	lestuce	yawn	yorn
15	flower	flowor	scrape	scyape
16	rabbit	rabpit	wink	wimk
17	banana	banama	fight	finht
18	firefly	firafly	jump	fump
19	cheesecake	cheeseqake	punch	punxh
20	grape	grage	scratch	scratsh

I ask the forgiveness of any real psycholinguists reading this, who would no doubt have numerous objections to using these stimuli in an actual experiment.

Table 7.2 Example reaction times for a lexical decision task, for one participant (times in ms).

Example	Nouns	Non-nouns	Verbs	Non-verbs
1	453	634	495	509
2	494	547	614	557
3	477	553	473	611
4	414	587	567	602
5	476	537	530	614
6	441	597	480	600
7	496	550	516	599
8	533	652	519	585
9	477	590	526	582
10	457	473	498	556
11	446	616	544	624
12	497	593	556	584
13	477	510	603	600
14	474	586	553	539
15	456	620	600	623
16	546	591	512	640
17	403	638	477	579
18	452	566	476	616
19	450	585	514	526
20	517	663	504	541
Mean	472	584	528	584

ANOVA. The two factors for the ANOVA are word type (noun or verb) and word validity (word or non-word), and each participant would contribute a single mean reaction time for each of those four conditions (meaning 80 data points in total—20 participants $\times$ 4 conditions). The ANOVA results might look something like this:

```
##
## Error: subject
##            Df Sum Sq Mean Sq F value Pr(>F)
## Residuals  1   5470    5470

##
## Error: Within
##                    Df Sum Sq Mean Sq F value  Pr(>F)
## wordtype            1  20654   20654  18.010 6.21e-05 ***
## validity            1 103705  103705  90.431 1.61e-14 ***
## wordtype:validity  1  10878   10878   9.485 0.00289 **
## Residuals          75  86009    1147
## ---
## Signif. codes:  0 '***' 0.001 '**' 0.01 '*' 0.05 '.' 0.1 ' ' 1
```

We see significant effects for word type, validity, and their interaction. We can also plot the means for each condition, along with individual data points, as shown in Figure 7.4.

Figure 7.4 is a conventional plot, in which the error bars show the standard deviation across participants, and each point corresponds to a different individual person (N = 20). Note that

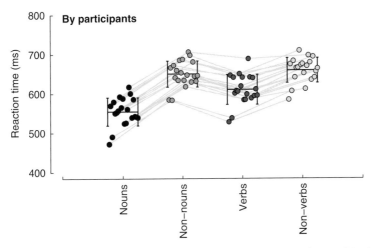

Figure 7.4 Graph showing condition means (black bars) and individual data points (symbols) for the four conditions. Error bars indicate the standard deviation across participants, and thin grey lines join points for an individual participant.

some individual participants are generally fast or generally slow, regardless of the condition. For example, you can see that the highest points stay near the top in each condition, as linked by the faint grey lines. It is this between-participant variance (the tendency for individuals to differ systematically across conditions) that the repeated measures design can discard, and which gives it greater statistical power relative to a between-participants design (where we test different individuals in each condition).

However, there is another way to think about the results of this experiment. Notice that some of the words in Table 7.1 are likely to be easier than others to identify. For example, in the non-nouns set, *poncake* might be quite a challenging word (because the *o* looks like an *a*, and *pancake* is a noun). On the other hand, *lestuce* might be an easier example to identify correctly as a non-word.

We can produce an alternative plot to Figure 7.4 by averaging reaction times across participants for each *item* (instead of averaging across items for each participant). Again, this will involve there being 80 observations: 20 items × 4 conditions. For the current example, we can see from Figure 7.5 that the 'By items' plot has a key similarity with the 'By participants' plot (Figure 7.4): the group means (horizontal black lines) are the same in both graphs. This has to be the case because these are the grand averages across both items and participants. However the group variances differ between the plots, as do the individual points.

Something to notice about the 'By items' plot (Figure 7.5) is that the standard deviations are much smaller than in the 'By participants' plot (Figure 7.4). This suggests that the responses to the different items are more similar to each other than are the responses of different individuals. Now, we could in principle do another ANOVA, this time treating *item* as the unit we average within, instead of *participant*. Note that item is repeated within the word class of noun or verb, but not across classes (see Table 7.1), which is why

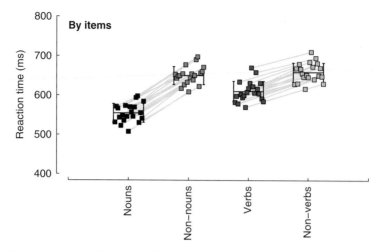

Figure 7.5 Graph showing condition means (black bars) and data points for each item (squares) for the four conditions. Error bars indicate the standard deviation across items, and thin grey lines join points for an individual stimulus item.

the faint grey lines only link within a word type category in Figure 7.5. The (mixed) ANOVA output looks like this:

```
##
## Error: item
##             Df Sum Sq Mean Sq
## wordtype   1  77105   77105
##
## Error: Within
##                     Df Sum Sq Mean Sq F value   Pr(>F)
## wordtype            1  37483   37483   67.41 4.75e-12 ***
## validity            1  12804   12804   23.02 7.95e-06 ***
## wordtype:validity   1  10878   10878   19.56 3.25e-05 ***
## Residuals          75  41707     556
## ---
## Signif. codes:  0 '***' 0.001 '**' 0.01 '*' 0.05 '.' 0.1 ' ' 1
```

This analysis produces larger F-ratios (sometimes referred to as F2, where the ANOVA based on participants is referred to as F1) and smaller p-values, because for our example the item-wise variance is lower than the participant-wise variance. However this feels somewhat uncomfortable—running the same analysis two ways on the same data set has a rather underhand feel. Most experimentalists would baulk at the idea of counting items as the unit of observation, because it divorces the number of observations from the number of participants. And indeed, there are more principled statistical objections to this practice (Rietveld and van Hout 2007).

Mixed-effects models offer an alternative. We can instead calculate a model that takes into account both the participant variance *and* the item-level variance, treating them both as *random effects*. Word type and validity—the independent variables from our ANOVAs— are included as fixed effects. Instead of averaging over items (or participants), we give the

model all our data points without any averaging. In this example, that consists of 20 participants × 20 items × 4 conditions = 1600 data points. The output of such a model might look
as follows:

```
## Type III Analysis of Variance Table with Satterthwaite's method
##                        Sum Sq Mean Sq NumDF   DenDF F value    Pr(>F)
## wordtype              180483  180483      1  100.96  115.15 < 2.2e-16 ***
## validity              806248  806248      1 1539.00  514.38 < 2.2e-16 ***
## wordtype:validity     217552  217552      1 1539.00  138.80 < 2.2e-16 ***
## ---
## Signif. codes:  0 '***' 0.001 '**' 0.01 '*' 0.05 '.' 0.1 ' ' 1
```

The mixed-effects approach is modelling each individual effect—each participant's change
in means across condition, and each item's change in means across condition. You will see
by comparing the output tables that this generally produces larger F-ratios, smaller p-values,
and that there are many more (denominator) degrees of freedom than in the ANOVAs. In the
ANOVAs, we had averaged across either items or participants to reduce the total variance.
Modelling the individual effects increases our statistical power, and also deals sensibly with
any correlations within participants or items.

How can I decide if an effect is fixed or random?

One common question in mixed-effects modelling is how to decide if an effect should be
considered as fixed or random. For example, in Figure 7.3 we treated breath direction as a
fixed effect with two levels—why wasn't this a random effect instead? This turns out to be
quite a complicated question, to which there is not always a definitive answer, and these
decisions are often left to the person doing the analysis. One good heuristic is to think about
whether you are interested in the effect in question: if you are, it should probably be a fixed
effect. In addition, there are two important factors that prevent a variable from being treated
as a random effect. First, if the variable is continuous, it cannot be used as a random effect;
only categorical variables can be treated in this way. This rules out variables like age and
weight from being treated as random factors, unless they are discretized into categories first.
Second, random effects should have at least five levels, as with fewer levels the estimate of the
standard deviation for that variable will be inaccurate. This is the reason why breath direction
needed to be a fixed effect in the dolphins example, and it also means that variables such
as sex/gender, handedness, blood group, and (in genetics) single-nucleotide polymorphism
must be either treated as fixed factors or ignored.

Dealing elegantly with missing data and unequal groups

An added bonus with mixed-effects models is that they cope well with missing values and
situations where we have groups of different sizes. Imagine that we decided to exclude trials
from our lexical decision task on which the participant got the task wrong (i.e. they indicated
that a stimulus was a word when it was a non-word, or vice versa). The excluded trials will
involve different words for each participant. This has the effect of making the experimental

design *unbalanced*. Depending on the research design, there could be many other reasons for missing observations, such as equipment malfunction, participants missing sessions in longitudinal work, and so on.

For traditional ANOVAs, we usually deal with missing data either by excluding participants (known as *listwise deletion*—this can substantially reduce power), or by averaging across only the data points we have (which means that some participants contribute more observations than others). In mixed-effects models, we are estimating the properties of an underlying regression line as best we can for each comparison. Critically, if some estimates are missing, we can still come up with a sensible parameter estimate. We can see how this might work by randomly removing some of the observations from our psycholinguistics data set. For example, if we remove 5% of the data points the analysis runs fine, and the summary table changes only very slightly:

```
## Type III Analysis of Variance Table with Satterthwaite's method
##                     Sum Sq Mean Sq NumDF   DenDF F value    Pr(>F)
## wordtype           187295  187295     1  103.89  120.63 < 2.2e-16 ***
## validity           801521  801521     1 1459.20  516.22 < 2.2e-16 ***
## wordtype:validity  225751  225751     1 1459.17  145.39 < 2.2e-16 ***
## ---
## Signif. codes:  0 '***' 0.001 '**' 0.01 '*' 0.05 '.' 0.1 ' ' 1
```

Another situation that causes problems for ANOVA is when different groups have very different sample sizes, also resulting in an unbalanced design. The main issues are that it is difficult to accurately test whether groups of very different sizes meet the homogeneity of variances assumption (i.e. that they have equal variances), and also that if sample size covaries with an independent variable, it can confound the main effect (in factorial designs). These issues can cause particular problems when conducting research on rare conditions and diseases, or in groups that comprise a minority of a population. Designs that aim to sample the population at random (such as polling research) will tend to select relatively few people in such categories, leading to highly unbalanced designs. Again, mixed-effects models are able to deal with this situation more appropriately than ANOVA, because they correctly account for the variance structure of the underlying data. This makes them a good choice for analysis of data that relate to equality and diversity of under-represented groups.

Reporting and comparing mixed-effects models

As we have seen in the examples we've looked at, the predictors in mixed-effects models can be assessed using F-tests, just like in regression and ANOVA. We can report the main effects and interactions in much the same way as we would report regression or ANOVA, along with *p*-values. However, it is critical that the design of the model is properly described, so that it is clear to the reader which variables are fixed effects and which are treated as random effects.

In some situations, it can also be helpful to demonstrate how well models with different assumptions (e.g. with a random effect added or removed) are able to describe the data. This is best achieved by fitting models in a systematic way, much like in multiple regression where additional predictors are added in sequence. Models can then be compared using a variety of statistics. The R^2 statistic tells us the proportion of the total variance that is explained by the model, just like in regression models, with larger values indicating a better model fit. The model with the largest *log-likelihood* statistic also indicates the best fit in absolute terms (regardless of the number of free parameters). Finally, the *Akaike information criterion* (AIC) and *Bayesian information criterion* (BIC) scores are based on the likelihood but also take the number of free parameters into account, the best model being the one with the *smallest* value.

Finally, it is important to display your data graphically in an intuitive way, so that the reader can visualize the important effects for themselves and understand what they mean. As well as showing main effects and interactions, by-items plots (e.g. Figure 7.5) are often useful, as are graphs that allow inspection of the residual variance that the model cannot explain (such as Q–Q plots). More detailed recommendations on reporting mixed-effects models are given by Meteyard and Davies (2020), though ultimately there are likely to be discipline-specific norms and study-specific priorities for what to report. A useful guide is to read relevant papers on the same topic that have used mixed-effects models, to see how the results have been reported.

Practical problems with model convergence

Values for the free parameters of mixed-effects models are determined using an optimization algorithm similar to those that we will discuss in Chapter 9. In this context, the free parameters are estimates of intercepts, slopes, and variances in the model. Sometimes the algorithm can fail to converge on the best parameter estimates, which will produce a convergence error. This is very similar to the problems with local minima that we will cover in 'Local minima' in Chapter 9. In such situations, the fitted parameters might not be the best possible estimates, though they will often be acceptable. There are several possible solutions, including running the fit again from different starting parameter values, or trying alternative optimization algorithms. It will sometimes also help to normalize any continuous fixed effects so that they have a mean of zero and a standard deviation of one (see 'Normalizing and rescaling data' in Chapter 3). This does not change the overall structure of your data, but it places the likely parameter values closer to the starting point for the algorithm.

Running mixed-effects models in *R* with *lmerTest*

This section will demonstrate how to build and run mixed-effects models using the *lmer* function in the *lmerTest* package (Kuznetsova, Brockhoff, and Christensen 2017). This builds on the older *lme4* package (Bates et al. 2015), and in particular adds tests for significance (i.e. *p*-values) that were previously not included. There are other *R* packages that can be used to run mixed-effects models, including the *nlme* package (which can also run non-linear mixed-effects models) and the *brms* package (Bürkner 2017, 2018) for running Bayesian versions of mixed-effects models. Outside of *R*, an extension to the *SAS* package called *PROC MIXED*, and bespoke software such as *MLwiN* can also be used.

Before we can run a mixed-effects model, we need to get our data into the correct format. The function expects data to be stored in a *data frame*, which we first introduced in Chapter 2. In brief, a data frame is a special type of data structure that can store numbers, strings, and factor variables, much like a spreadsheet. We can create one using the *data.frame* function to combine a series of vectors, or by reading data in from an external file such as a csv file (with the *read.csv* function).

The syntax for the *lmer* function is based on the linear modelling formula used in other *R* functions such as *lm* (linear model) and *aov* (to run ANOVAs), which we introduced in Chapter 4. These formulae have the general form: DV ~ IV, where DV represents the dependent variable, IV is the independent variable, and the tilde symbol (~) is read as 'is predicted by'. So, if we wanted to predict children's heights based on their age, we might use a command such as:

```
output <- lm(height ~ age, data=dataset)
```

This code would run a linear model using the data stored in a data frame called *dataset*, trying to predict values of the *height* column using the values in the *age* column. If we have additional independent variables, such as sex, we can include them either as single predictors (additive, as in regression formulae), or as factors that interact with the other independent variables (multiplicative, as for ANOVA formulae):

```
# regression notation (no interaction)
output <- lm(height ~ age + sex, data=dataset)
```

```
# ANOVA notation (with interaction)
output <- lm(height ~ age * sex, data=dataset)
```

Formulae for the *lmer* function follow similar rules, but there is an additional piece of syntax to consider. A *random effect* is always entered after the fixed effects (i.e. independent variables). It is entered in brackets, and after a vertical slash symbol. For example, if we wanted to include nationality as a grouping variable to predict height, we might do so as follows:

```
# mixed-effects model call with random intercepts
output <- lmer(height ~ age + (1|nationality), data=dataset)
```

The above line of code will run a mixed-effects model with *random intercepts* (see Figure 7.2(b)), using age as a predictor and nationality as a grouping variable (random effect). Alternatively, we can specify a *random slopes* model (see Figure 7.2(c)) by incorporating the independent variable into the random effects specification, and specifying (using a 0) that the intercept is *not* included as a random effect:

```
# mixed-effects model call with random slopes
output <- lmer(height ~ age + (0 + age|nationality), data=dataset)
```

Finally, we can specify a model with *random slopes and intercepts* (see Figure 7.2(d)) by allowing the intercepts to vary again:

```
# mixed-effects model call with random slopes and intercepts
output <- lmer(height ~ age + (1 + age|nationality), data=dataset)
```

A more generic specification for these three models is as follows:

```
# mixed-effects model call with random intercepts
output <- lmer(DV ~ IV + (1|group), data=dataset)

# mixed-effects model call with random slopes
output <- lmer(DV ~ IV + (0 + IV|group), data=dataset)

# mixed-effects model call with random slopes and intercepts
output <- lmer(DV ~ IV + (1 + IV|group), data=dataset)
```

Just as we can have more than one independent variable, random effects can be defined for multiple grouping variables, depending on the structure of our data set. For example, the call for our factorial mixed-effects model for the lexical decision task was:

```
model <- lmer(RT ~ wordtype * validity + (1|subject) + (1|item),
data=RTlmm)
```

This line of code specifies that reaction time (RT) is predicted by two factorially combined independent variables (word type and validity), with random intercepts on subject and item.

As with many *R* functions, the output of the model fit is stored in another data object (the one to the left of the <- assignment). This data structure contains a lot of information, and we can extract it in several ways. Simply inspecting the object (by typing its name) will give us some helpful numbers, such as the number of observations, but it is not generally very informative:

```
model
## Linear mixed model fit by REML ['lmerModLmerTest']
## Formula: RT ~ wordtype * validity + (1 | subject) + (1 | item)
##     Data: RTlmm
## REML criterion at convergence: 16473.13
## Random effects:
##    Groups   Name        Std.Dev.
##    item     (Intercept) 22.66
##    subject  (Intercept) 33.53
##    Residual             39.59
## Number of obs: 1600, groups:  item, 40; subject, 20
## Fixed Effects:
##     (Intercept)    wordtype    validity   wordtype:validity
##       357.22     102.10     141.97     -46.64
```

Alternatively, we can use the generic *summary* function to get some more useful numbers out:

```
summary(model)
## Linear mixed model fit by REML. t-tests use Satterthwaite's method [
## lmerModLmerTest]
## Formula: RT ~ wordtype * validity + (1 | subject) + (1 | item)
##     Data: RTlmm
##
## REML criterion at convergence: 16473.1
##
## Scaled residuals:
##    Min       1Q  Median      3Q     Max
## -3.2488 -0.6845 -0.0095  0.6634  3.1224
```

```
##
## Random effects:
##  Groups    Name    Variance Std.Dev.
##  item      (Intercept)  513.5   22.66
##  subject   (Intercept) 1124.1   33.53
##  Residual               1567.4  39.59
## Number of obs: 1600, groups:  item, 40; subject, 20
##
## Fixed effects:
##               Estimate Std. Error     df t value Pr(>|t|)
## (Intercept)     357.220    16.809  117.499   21.25  <2e-16 ***
## wordtype        102.099     9.515  100.961   10.73  <2e-16 ***
## validity        141.973     6.260 1539.000   22.68  <2e-16 ***
## wordtype:validity -46.643     3.959 1539.000  -11.78  <2e-16 ***
## ---
## Signif. codes:  0 '***' 0.001 '**' 0.01 '*' 0.05 '.' 0.1 ' ' 1
##
## Correlation of Fixed Effects:
##          (Intr) wrdtyp valdty
## wordtype    -0.849
## validity    -0.559  0.592
## wrdtyp:vldt  0.530 -0.624 -0.949
```

This provides a summary table, which includes coefficient estimates, t-statistics, and *p*-values for the fixed effects, as well as some helpful summary information about the residuals. It is more typical to report F-ratios for ANOVA-type designs, which we can obtain by using the *anova* function:

```
anova(model)
## Type III Analysis of Variance Table with Satterthwaite's method
##                    Sum Sq Mean Sq NumDF  DenDF F value   Pr(>F)
## wordtype           180483  180483     1 100.96  115.15 < 2.2e-16 ***
## validity           806248  806248     1 1539.00 514.38 < 2.2e-16 ***
## wordtype:validity  217552  217552     1 1539.00 138.80 < 2.2e-16 ***
## ---
## Signif. codes:  0 '***' 0.001 '**' 0.01 '*' 0.05 '.' 0.1 ' ' 1
```

Similarly, we can request a summary table for our random effects terms with the *ranova* function:

```
ranova(model)
## ANOVA-like table for random-effects: Single term deletions
##
## Model:
## RT ~ wordtype + validity + (1 | subject) + (1 | item) +
wordtype:validity
##                 npar  logLik   AIC    LRT Df Pr(>Chisq)
## <none>             7 -8236.6 16487
## (1 | subject)      6 -8611.1 17234 749.12  1  < 2.2e-16 ***
## (1 | item)         6 -8402.6 16817 332.16  1  < 2.2e-16 ***
## ---
## Signif. codes:  0 '***' 0.001 '**' 0.01 '*' 0.05 '.' 0.1 ' ' 1
```

The values included here are measures of how the model fit changes when a term is *removed*. So a significant *p*-value here implies that the random effects term is making a meaningful contribution to the fit of the model.

We may also wish to report R^2 values for the model fit. Calculating these requires a function from a different package—the *r.squaredGLMM* function from the *MuMIn* package—which we can provide with the model output as follows:

```
library(MuMIn)
r2 <- r.squaredGLMM(model)
r2
##              R2m       R2c
## [1,] 0.3454568 0.6798848
```

Two values are calculated and reported by this function. The first (R_m^2) is the *marginal* R^2 value, which represents the proportion of the variance explained by our fixed effects (i.e. traditional independent variables), excluding any random effects. The second (R_c^2) is the *conditional* R^2 value, which is the proportion of the variance explained by the full model, including both fixed and random effects. It is helpful to report both of these statistics for each model that you run.

If we want to compare two (or more) models, we can again use the *anova* function to produce a table of useful statistics, including AIC, BIC, and log-likelihood scores. The model with the smallest AIC and BIC scores, and the largest log-likelihood score, gives the best account of the data. Here is an example comparing the models from Figure 7.2(b)–(d):

```
anova(simmodel3,simmodel4,simmodel5)
## refitting model(s) with ML (instead of REML)
## Data: groupdata
## Models:
## simmodel3: DV ~ IV + (1 | group)
## simmodel4: DV ~ IV + (0 + IV | group)
## simmodel5: DV ~ IV + (1 + IV | group)
##           Df    AIC    BIC   logLik deviance   Chisq Chi Df Pr(>Chisq)
## simmodel3  4 3843.7 3860.5 -1917.8   3835.7
## simmodel4  4 3901.6 3918.4 -1946.8   3893.6   0.000      0          1
## simmodel5  6 3830.1 3855.4 -1909.0   3818.1  75.507      2     <2e-16 ***
## ---
## Signif. codes:  0 '***' 0.001 '**' 0.01 '*' 0.05 '.' 0.1 ' ' 1
```

It is sometimes necessary to find out the coefficients for our random effects groups (perhaps to plot individual regression lines as in Figures 7.2(b)–(d). The *coef* function allows us to extract these values (shown here for the model displayed in Figure 7.2(d), and stored in the *simmodel5* variable):

```
coef(simmodel5)
## $group
##    (Intercept)        IV
## 1    32.911341 0.9268825
## 2    -1.536704 1.0272352
## 3     4.649459 0.9156914
## 4   -11.237636 0.4863508
## 5   -25.522081 1.0810784
```

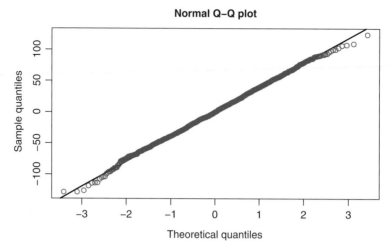

Figure 7.6 Q-Q plot showing the residuals.

```
##
## attr(,"class")
## [1] "coef.mer"
```

The first column contains the intercept values, and the second column the slope values.

Finally, we can inspect the residuals using a Q–Q plot, which we first encountered in Chapter 3. These graphs show the expected quantiles (based on a normal distribution) along the x-axis, and the actual residuals (from the data) along the y-axis. Substantial deviations from the major diagonal line (usually at the extremes) indicate that the normality of residuals assumption has been violated, and the model results should be treated with some caution. We can generate a Q–Q plot (see Figure 7.6) using the *qqnorm* function, after first extracting the residuals from the model object using the *resid* function:

```
modelresiduals <- resid(model) # extract the residuals
qqnorm(modelresiduals)  # create the plot
qqline(modelresiduals)  # add the diagonal line
```

Further resources for mixed-effects models

To learn more about mixed-effects models, many resources are available. The book *Linear Mixed-Effects Models Using R* (Galecki and Burzykowski 2013) is a comprehensive source that discusses a number of different designs with detailed examples. The documentation and paper (Bates et al. 2015) for the *lme4* package is helpful and contains example implementations. For item-level analysis specifically, the paper by Baayen, Davidson, and Bates (2008) is an influential source, Westfall, Kenny, and Judd (2014) discuss issues relating to statistical power, and Meteyard and Davies (2020) provide recommendations on reporting. Readers convinced by the arguments for Bayesian statistics (see Chapter 17) are advised to read about

Bayesian hierarchical models, which have similar properties (see e.g. Kruschke 2014). It is also worth identifying papers in your own area of research that use the mixed-effects approach, to find examples of common practice. Finally, there are many helpful blog posts and discussion board threads online that are well worth reading when troubleshooting specific issues.

Practice questions

1. In mixed-effects models, independent variables are called:
   A) Random effects
   B) Fixed effects
   C) Mediators
   D) Outcomes

2. A random effect is best described as:
   A) A variable that is noisier than other variables
   B) An effect that cannot be predicted based on group membership
   C) A variable where subgroups influence the outcome
   D) Choosing slope values with a random number generator

3. Which of the following is **not** an advantage of mixed-effects models?
   A) They can cope with situations where individuals contribute different numbers of observations
   B) They can handle group-level differences in the outcome
   C) They can deal with variation due to different stimulus items
   D) They are guaranteed to give significant results when ANOVAs do not

4. Mixed-effects models cope well with missing data because:
   A) They can still estimate parameters even when some values are missing
   B) They delete all participants with missing data
   C) They set all missing values to 0
   D) They delete all items with missing data

5. The formula DV ~ IV + (0 + IV|group) will fit a model with:
   A) Random slopes
   B) Random intercepts
   C) Random slopes and random intercepts
   D) An intercept fixed at 0

6. Which formula would fit a model with random slopes and intercepts?
   A) DV ~ IV + (0 + IV|group)
   B) DV ~ IV + (1 + IV|group)
   C) DV ~ IV + (1|group)
   D) DV ~ IV + (DV|group)

7. To check the assumption of normally distributed residuals, we usually plot a:

   A) Histogram
   B) Normal distribution
   C) Q–Q plot
   D) By-items plot

8. Which summary statistic tells us the proportion of the variance explained by a full mixed-effects model?

   A) R^2_c
   B) AIC
   C) Log-likelihood
   D) R^2_m

9. When comparing two models, which summary statistic will be smallest for the best fitting model?

   A) R^2_c
   B) AIC
   C) Log-likelihood
   D) R^2_m

10. Imagine a study in which participants respond to a sequence of 100 images of celebrity faces. Half of the participants are from a country in which the celebrities are well known, but the rest of the participants are from a country where they are not famous. What would be the most appropriate design to model these data?

   A) Both image and country are random effects
   B) Both image and country are fixed effects
   C) Image is a fixed effect and country is a random effect
   D) Country is a fixed effect and image is a random effect

Answers to all questions are provided in the answers to practice questions at the end of the book.

8 Stochastic methods

What does stochastic mean?

In lay terms, stochastic means *random*. The methods described in this chapter involve using random numbers in one way or another. For various historical reasons there are a number of more specific terms that are often used to refer to particular methods within this general area, including:

- Monte Carlo methods
- resampling
- bootstrapping
- jackknifing
- probabilistic simulations
- permutation testing
- plug-in principle

This plethora of terms hints at how flexible and widespread stochastic methods are. In this chapter, we will focus on running simulations with a stochastic component, and bootstrap resampling methods to calculate confidence intervals and test hypotheses.

I mostly find stochastic techniques useful because they offer a shortcut that allows one to avoid lengthy mathematical derivations, and they tend not to require parametric assumptions. My favourite quote that illustrates their utility comes from the *Numerical Recipes* books:

> Offered the choice between mastery of a five-foot shelf of analytical statistics books and middling ability at performing statistical Monte Carlo simulations, we would surely choose to have the latter skill.

(Press et al. 1986)

The quote pithily expresses the main advantage of stochastic methods—they allow us to work out useful stuff without requiring formal equations. Note that the methods involving random numbers we discuss in this chapter are distinct from the concept of a *random effect*, that we introduced in Chapters 6 and 7. Random effects are where individuals, groups, or studies differ in their means, whereas here we use random numbers for several other purposes.

Ways of generating random numbers

Almost everyone who needs random numbers for something will generate them using a computer. However computers are the opposite of random—they are *deterministic* machines. This means that for a given input, they should always produce the same output. For most applications, like calculating your tax return or running a statistical test, we want them to behave in this way. It would cause utter chaos if such calculations involved a stochastic component: you'd get a different answer every time! But when we do want random numbers, we need some way to create them.

The most common way, and the method used by *R*, is to use a complex algorithm to generate sequences of numbers that appear random to a first approximation. These are called *pseudo-random* numbers, and they are predictable if you run the same algorithm again with the same starting conditions. The algorithm takes as an input a number called a *seed*, and changing the value of the seed will dramatically affect the output. Even two consecutive seed values will produce completely different sequences of numbers. It is common practice to use the current time from the computer's clock as the seed, meaning that the same two sequences of pseudo-random numbers should never reoccur. But if we were to store the seed and use it again in the future, we would get exactly the same set of numbers out.

Pseudo-random numbers are absolutely fine for all the techniques we will discuss here. But they do have clear weaknesses for any applications where privacy or security is important. A good example is online gambling websites. If these used a pseudo-random number generator based on the computer's clock, and we knew the algorithm, we could predict in advance what cards would be drawn in an online poker game, or what numbers would be picked by a simulated roulette wheel, and perhaps fraudulently win large amounts of money. Apparently this has actually happened in the past, though gambling companies have since wised up and now mostly use something called a *true random number generator*.

Truly random numbers require some sort of input from the physical world. This might come from the weather, the decay of a radioactive material, temperature fluctuations in a piece of metal, and so on. Hardware true random number generators using this sort of information are commercially available, now usually as small USB devices. There are also online services, such as the website **www.random.org**, that will generate truly random numbers for you. However, these are not necessary for anything we will discuss here, and are only worth the bother and expense if you really need them.

The remainder of this chapter is divided into two parts. In Part 1, we will describe how stochastic methods can be used to model different situations, in order to gain insights into how a system, model, or experiment might behave. In Part 2, we will introduce the concept of resampling. This is a way of analysing data that can be used to estimate confidence intervals, and also to conduct statistical hypothesis testing.

Part 1: Using random numbers to find stuff out

In the spirit of the quote from Press et al. at the start of the chapter, let's do some simulations using random numbers to demonstrate their usefulness. I have included the *R* code here to show that there's no sleight of hand or funny business going on—you can run it yourself

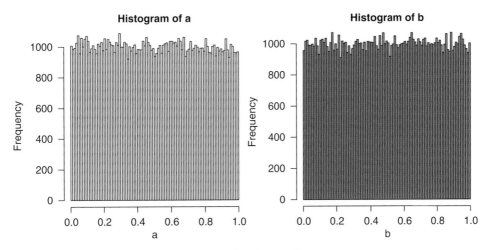

Figure 8.1 Two populations of uniformly distributed random numbers.

if you want to. Imagine that we have two sets of uniformly distributed random numbers (where each sample has an equal probability of taking on any value between 0 and 1), as shown in Figure 8.1:

```
par(mfrow=c(1,2), las=1)

a <- runif(100000)
b <- runif(100000)
hist(a, breaks = 100, col = 'white')
hist(b, breaks = 100, col = '#8783CF')
```

What would we expect the distribution to look like if we added these two samples together? Intuitively, we might guess that the distribution of the summed values should also be uniform. However this intuition would be incorrect. In fact, the summed distribution has a clear peak in the centre, as shown in Figure 8.2, generated by the following code:

```
hist(a+b, breaks = 100, col = '#CFCDEC')
```

Why does this happen? It is a consequence of something called *central limit theorem*, which is a mathematical theorem that states that the sum of several non-normal distributions will tend to approximate a normal distribution. The reason for this is that it is very unlikely that the two biggest numbers, or the two smallest numbers, of the two samples will be paired together (of course, this assumes that the two sets of numbers are not sorted before being added). It is much more likely that the summed numbers will be of middling value. If we kept on summing lots and lots of uniform distributions, we would eventually end up with a beautiful normal distribution, as shown in Figure 8.3, and generated using the following code:

```
bigsum <- runif(100000)
for (n in 1:99){bigsum <- bigsum + runif(100000)}
hist(bigsum, breaks = 100, col = 'grey')
```

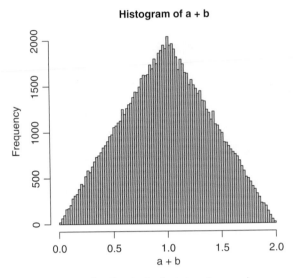

Figure 8.2 The sum of two populations of uniformly distributed random numbers.

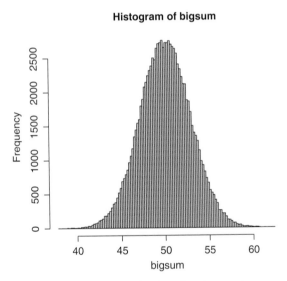

Figure 8.3 The sum of 100 populations of uniformly distributed random numbers.

With this simple example, using random numbers, we have demonstrated central limit theorem in action. Of course, there are a whole load of complex mathematical equations that explain how it works in detail (see Fischer 2011). But, in keeping with the spirit of the quote at the start of the chapter, we have been able to show that the theorem works without needing to think about the underlying mathematics.

Incidentally, central limit theorem explains why the assumption of normality holds for so many variables in scientific research. If a dependent variable is determined by a combination of underlying factors, it will tend to be normally distributed even if the underlying factors are not. Consider a physiological variable, like heart rate. For each individual, this will be determined by factors such as current arousal, fitness level, age, sex, and various genetic influences. But across a population of individuals, the distribution of heart rates should be approximately normal because central limit theorem says that all of these different influences will combine at random. The same is likely to be true of high-level psychological constructs (like IQ), brain activity measured using neuroimaging techniques, pollen yield of flowers, phenotypes dependent on many separate genes, and so on.

Stochastic simulations: models and synthetic data

If stochastic simulations can be used to find stuff out, what sorts of things might we use them for? One common application is to model systems which are themselves stochastic—a key example being the human brain. Many dynamic neural models incorporate a stochastic component, which often helps us to better understand how the brain works.

A recent example from my own research is a study in which my colleague Bruno Richard and I modelled perception in binocular rivalry (Baker and Richard 2019). Rivalry is a curious phenomenon where our perception of conflicting stimuli shown to the left and right eyes fluctuates over time in an unpredictable way. Since the process itself is stochastic, we needed to use a model containing a random component to understand it properly. In the paper, we were particularly concerned with working out the magnitude and characteristics of the neural noise (i.e. inside the participant's brain) that governs rivalry alternations. We approached this by adding different amounts of dynamic noise to the contrasts of the rivalling stimuli in a psychophysical experiment to provide a rich data set with many conditions. We then attempted to reproduce the pattern of human data using models with different types and amounts of internal noise. The key point is that this analysis could probably not have been done analytically (i.e. with noise-free equations)—we needed to be able to use stochastic methods to simulate human perception convincingly.

Another related and important use of stochastic methods is to simulate what we think might happen in an experiment before we collect the data. This is a form of modelling, but one that is prospective rather than retrospective. For simple designs, involving only one or two variables, it is not necessarily very informative. However for complex factorial designs with many interacting variables (see Chapter 7 on mixed-effects models and Chapter 12 on structural equation modelling), data simulation can reveal how interdependencies might affect the results. This can help researchers to develop their intuitions about how the results of an experiment might turn out. Power calculations (see Chapter 5) can also be done by simulation, affording greater flexibility (e.g. for complex or unbalanced designs) than analytic approximations (Colegrave and Ruxton 2020). An added advantage of simulating data is that one can construct an analysis pipeline in advance of running the experiment. This saves time later, is useful for clarifying and making explicit one's assumptions and expectations, and can also be included in preregistration materials.

Generating random numbers in *R* from different distributions

The *stats* package in *R* provides access to a number of different statistical distributions that can be used to generate random numbers. A full list is available by typing *help(Distributions)*.

Each distribution has four functions, which are shown in Figure 8.4 for a normal distribution. The following code chunk demonstrates their use:

```
par(mfrow=c(2,2), las=1) # divide the plot into four panels

hist(rnorm(1000, mean=0, sd=1), breaks = 50,
    main='rnorm',xlab='x',ylab='Frequency',xlim=c(-4,4))

plot(seq(-4,4,0.001),dnorm(seq(-4,4,0.001),mean=0,sd=1),
    type='l',lwd=3, main='dnorm',xlab='x',ylab='Density')

plot(seq(-4,4,0.001),pnorm(seq(-4,4,0.001),mean=0,sd=1),
    type='l',lwd=3, main='pnorm',xlab='x',ylab='Cumulative probability')

plot(seq(0,1,0.001),qnorm(seq(0,1,0.001),mean=0,sd=1),
    type='l',lwd=3, main='qnorm',xlab='Quantile',ylab='x')
```

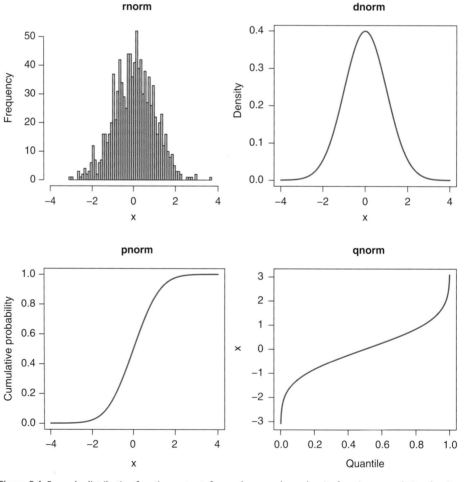

Figure 8.4 Example distribution function outputs for random numbers, density function, cumulative density, and quantiles, for a normal distribution with mean = 0 and sd = 1.

The top left panel of Figure 8.4 shows the output of the *rnorm* function, which generates a sequence of *n* random numbers drawn from a normal distribution, with mean and standard deviation defined by the function call (defaults are mean = 0 and sd = 1). The *rnorm* function is the most useful function for our current purposes, but for reference we will also describe the outputs of the other three related functions.

In the top right panel of Figure 8.4, the *dnorm* function produces a probability density plot for the same normal distribution. This gives the probability of drawing a number with value *x* from a normal distribution with the mean and standard deviation specified by the function call. Note that this function does not produce random numbers directly, but it has many uses, such as plotting smooth curves to summarize distributions.

In the lower left panel of Figure 8.4, the *pnorm* function provides the cumulative distribution function. A good way to understand this is to imagine that at each value of *x* on the curve, you are adding up the probabilities for every number between $-\infty$ and *x*. For any input value of *x*, this function will tell you the probability that a number drawn from a normal distribution will have a value smaller than *x*. Equivalently, subtracting the probability from 1 will tell you the probability of a number having a value larger than *x*. This is particularly important for calculating *p*-values in statistical testing. For example, if we run an ANOVA and calculate an F-ratio, we compare this to the (inverse) cumulative F-distribution (from the *pf* function) with an appropriate number of degrees of freedom. This provides the ubiquitous *p*-value that is used to determine whether a test is statistically significant.

Finally, in the lower right panel of Figure 8.4, the *qnorm* function provides quantiles from the normal distribution. This is the reverse of the cumulative distribution—notice that the *x* and *y* axes are switched between the lower two panels—so it can be used to reverse engineer a test statistic if we know the *p*-value. This is used in some of the effect size conversion tools we discussed in 'Converting between effect sizes' in Chapter 6.

Equivalent functions are available for other distributions with a consistent naming pattern. For example the *rgamma, dgamma, pgamma,* and *qgamma* functions generate a gamma distribution. This has a positive skew, and is sometimes used for modelling prior distributions in Bayesian statistics. Other particularly useful distributions include the uniform distribution (*runif, dunif, punif,* and *qunif*) that we encountered in the central limit theorem example above, the log-normal distribution (*rlognorm, dlognorm, plognorm,* and *qlognorm*), the F-distribution (*rf, df, pf,* and *qf*) used in ANOVA and related statistics, and the Poisson distribution (*rpois, dpois, ppois, qpois*) that is used to model event probabilities such as the spiking of neurons.

All of the above functions use the same underlying random number generator. We can set the seed to a specific (integer) value, and be confident that the sequence of pseudo-random numbers we generate will always be the same. For example, setting the seed to 100 and asking for five random numbers from a normal distribution produces the following output:

```
set.seed(100)
rnorm(5)
## [1] -0.50219235  0.13153117 -0.07891709  0.88678481  0.11697127
```

If we change the seed to a different value (99), we will get a different sequence:

```
set.seed(99)
rnorm(5)
## [1]  0.2139625  0.4796581  0.0878287  0.4438585 -0.3628379
```

Crucially, if we set the seed back to 100, we should get our first sequence of numbers out again:

```
set.seed(100)
rnorm(5)
## [1] -0.50219235  0.13153117 -0.07891709  0.88678481  0.11697127
```

If the user does not specify a seed, then one is generated automatically based on the computer clock the first time any functions involving random numbers are called. If it is important to reproduce the random seed again in future, it can be saved to a data object as follows:

```
seed <- .Random.seed
```

This needs to be done before any random numbers are actually generated, as each time we sample from the random number generator we change its state. The seed can then be restored by setting .*Random.seed <- seed*, which should permit full reproducibility of the original random sequence.

The *set.seed* function can also be used to specify the random number generator algorithm to be used (with the *kind* argument). The default is the exciting-sounding *Mersenne-Twister* (Matsumoto and Nishimura 1998), which is a widely used algorithm implemented in a number of programming languages and software packages (including SPSS). There are half a dozen alternatives with similarly exotic names, and also the option for users to specify their own algorithms if required.

Part 2: Resampling methods

A second widespread application of stochastic techniques is an approach known as *resampling*. This is a class of methods that is widely used to estimate confidence intervals and other measures of precision, and can also be used to test statistical hypotheses. Resampling is generally a non-parametric method that can be used on any type of data, and makes use of random sampling. The basic idea is that we take some data and perform a statistic or calculation on it. A simple example would be to calculate the mean, but any method that produces a summary statistic that we are interested in is suitable: other examples might include a t-statistic or correlation coefficient, or the parameters of a fitted model. Next, we repeatedly *resample* the original data by drawing randomly chosen values from it. With each set of resampled data we calculate the same test statistic. Over many iterations, we build up a population of resampled test statistics (e.g. resampled means, or t-values, or correlation coefficients).

This resampled distribution has several uses. One thing we can do is to take the *confidence intervals* of the population, usually at the upper and lower 2.5% of the distribution, as these points will provide 95% 'bootstrapped' confidence intervals. Confidence intervals can be used as error bars, giving an indication of the precision of our original estimate. We could also compare the distribution to a particular point, such as the mean of another condition, or a suitable benchmark (such as a value of 0). The technique is often called bootstrapping as a reference to the phrase 'pulling yourself up by your bootstraps', which means to improve your situation by your own efforts. In the context of data analysis, the idea is that we derive additional information directly from the data themselves without making further observations or assumptions.

The confidence intervals are a statement about the likelihood that the true population value falls within a particular range. When used with parametric data to calculate the confidence intervals of a mean, bootstrapping produces very similar estimates to analytic approximations (95% confidence intervals are well approximated by 1.96*SE when parametric assumptions hold). This is shown in Figure 8.5. The grey histogram bars show a sample of data derived from an underlying normal distribution (black curve). The white point and black error bars show the mean and analytic approximation of 95% confidence intervals. The blue shaded region shows the distribution of bootstrapped means, calculated by resampling the data 10000 times. The blue error whiskers give the 2.5% and 97.5% points of this distribution, which are the bootstrapped 95% confidence intervals of the mean. Evidently, the error bars on the two points are very similar, showing that for parametric data the two methods are equally useful.

The usefulness of bootstrapping becomes more apparent when we consider non-parametric data, such as the skewed data shown in Figure 8.5(b). Here the appropriate measure

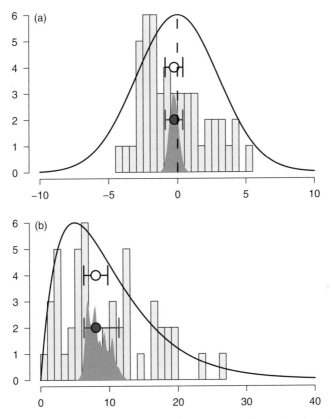

Figure 8.5 Examples of bootstrapping confidence intervals on measures of central tendency. Panel (a) shows a sample of data (histogram) from a normal distribution (black curve). The white point and error bars show the mean and analytically derived confidence intervals (1.96*SE). The blue point and error bars show the mean and 95% confidence intervals derived by bootstrapping. The blue shaded area is the distribution of bootstrapped means. Panel (b) has the same format, but the data are from a gamma distribution, and the measure of central tendency is the median rather than the mean.

of central tendency is the median, and approximating the confidence intervals from the standard error does not make sense (the error bars on the white point are symmetrical for a start, which must be wrong given the skewness of the data). By bootstrapping, we produce more plausible estimates of confidence about where the true (population) median might lie (error bars on the blue point). These are asymmetrical, which is consistent with the skew in the data.

Resampling with and without replacement

There are two varieties of resampling that can potentially be used to bootstrap. Resampling *without replacement* means that on any given bootstrapping iteration (repeat), each value from the original data set can be included in the resampled data set only once. This type of bootstrapping only makes sense if the resampled data set is smaller than the original data set—otherwise we would just end up with thousands of identical data sets! Reducing the sample size (known as subsampling) can cause problems in some situations because it reduces statistical power (see Chapter 5), for example if a statistical test is being conducted. However there are some situations where sampling without replacement is the right thing to do. Lottery draws are a good example: if we ended up drawing the same number twice it would cause a lot of confusion.

The second type of resampling, which is more widely used, is resampling *with replacement*. Here, each value from the original data set can be included in the resampled data set more than once. Assuming the resampled data set is the same size as the original data set, this also means that some values from the original set might not be included at all in the resampled data. I always think of bootstrapping using the analogy of a bag of ping-pong balls, each of which has a number from the original data set written on it. Resampling involves drawing a ball from the bag, and noting down the number. In resampling with replacement, the ball then goes back in the bag, meaning there is the possibility it will be pulled out again. In resampling without replacement, the ball stays out of the bag until the end of this bootstrap iteration.

Using resampling for hypothesis testing

It is also possible to use resampling to test statistical hypotheses. The basic idea is that we build up a population of resampled test statistics to represent the null hypothesis, and compare the test statistic from the original data to this distribution. Several specific variants have been proposed, but the general approach for a two-sample test is to randomize the group assignment of each data point on every iteration. This will generate a distribution of resampled test statistics that we might expect to observe if there were no true effect—in other words a *null distribution*. The null distribution is typically centred on 0, with a spread determined by the characteristics of the data. If the original test statistic lies close to the middle of this distribution then it is unlikely to indicate a real effect. On the other hand, if the original test statistic is in one of the tails of the distribution, it is more likely to indicate a true effect. This is because if the original test statistic is extreme, it means that group membership matters, and implies that there is a genuine difference between the groups. Figure 8.6 illustrates this procedure.

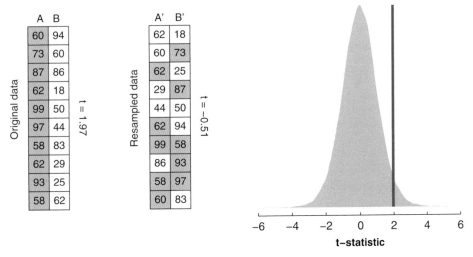

Figure 8.6 Illustration of the bootstrap test. The original data (left) consists of two groups, A and B (columns), which produce a t-statistic of 1.97. These data are resampled by randomly reshuffling the group allocations, and a new t-statistic is calculated (here −0.51) using the resampled groups, A' and B' (blue squares indicate values that originated in group A). The distribution of resampled t-statistics from 10000 such resampling iterations is shown in the plot on the right, along with the original t-statistic (blue line). For this example, 3.4% of the population lies to the right of the blue line, implying a one-sided p-value of 0.034, or a two-sided p-value of 0.069. This is close to the value from the original t-test of p = 0.067.

Once the null distribution has been generated, we can calculate a p-value by determining the proportion of resampled test statistics that are more extreme than the original test statistic. For a one-sided test this is the proportion of resampled statistics that are either smaller than or larger than the original statistic. For a two-sided test, the absolute values are used instead. Resampling approaches are inherently non-parametric, so can be used in situations where the assumptions of more traditional parametric statistics are not met. Crucially, this method works for any test statistic one might come up with, even if the expected distribution is unknown. Some variants include the bootstrap test, in which data are resampled with replacement, and the permutation test, in which all possible permutations (i.e. combinations of group ordering) of the data are included in the resampled distribution. A similar approach can also be taken with correlations by randomly reshuffling the pairings of the two dependent variables on each iteration (see below). For a more elaborate use of resampling methods, see 'Cluster correction for contiguous measurements' in Chapter 15, which describes a related method for controlling for multiple comparisons.

How to do resampling in *R*

Resampling is a sufficiently useful technique that it is now built into some commercial statistics packages (including SPSS) for some tests. In *R* there is a package and a function that are both called *boot* that can be used to bootstrap other functions. However in the interests of making the stages of bootstrapping explicit, the examples here use two built-in functions:

sample and *quantile*. The *sample* function resamples a data set. If we resample the values from 1 to 10 with replacement, we generally get a set of numbers that includes several duplicates, and also several missing values:

```
sample(1:10,10,replace=TRUE)
## [1]  7 10  5  3  5  8  9  6  5  9
```

If we resample without replacement, we get a random *permutation* of the numbers, which is useful in some situations (e.g. for randomizing the order of conditions in an experiment).

```
sample(1:10,10,replace=FALSE)
## [1] 10  3  9  4  6  5  2  7  1  8
```

Finally, we can also resample either with or without replacement but produce a smaller data set, by specifying how many values we need with the second argument to the *sample* function. This is known as subsampling:

```
sample(1:10,5,replace=FALSE)
## [1] 8 6 7 1 3
```

The *sample* function becomes particularly useful when it is embedded in a loop (see 'Loops' in Chapter 2) that repeats an operation many times on the resampled data. The following code resamples the mean of some data, and plots the distribution of resampled means in Figure 8.7:

```
# generate some random synthetic data
data <- rnorm(100, mean=1, sd=3)

# create an empty data object to store the resampled means
allmeans <- NULL
# repeat the resampling lots of times
for (n in 1:10000){
# use the sample function to resample the data, and store in rsdata
```

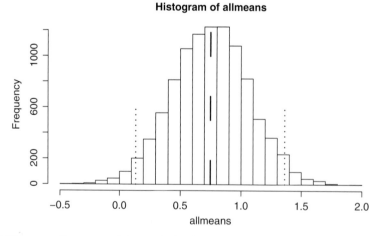

Figure 8.7 Histogram of bootstrapped means. The black dashed line is the true mean, and the dotted lines are the 95% confidence intervals.

```
rsdata <- sample(data,replace=TRUE)
# calculate the mean of the resampled data
allmeans[n] <- mean(rsdata)
}

# plot a histogram of the resampled means
b <- hist(allmeans,breaks=20)
# add a vertical line showing the true mean
lines(c(mean(data),mean(data)),c(0,max(b$counts)),col='black',lwd=8,
lty=2)
```

The data object *allmeans* now contains 10000 bootstrapped means. We can estimate the confidence intervals from this population using the *quantile* function. This function returns values at a specific proportion of a distribution. To get the 95% confidence intervals, we request proportions of 0.025 for the lower bound, and 0.975 for the upper bound, because 95% of the values will lie between these points.

```
# use the quantile function to get the confidence intervals
# from the population of bootstrapped means
CIs <- quantile(allmeans, c(0.025,0.975))
CIs
##     2.5%     97.5%
## 0.2493662 1.4879034
```

We can add the limits to our histogram (vertical dotted lines) to visualize them as follows:

```
# add vertical lines showing the confidence intervals
lines(CIs[c(1,1)],c(0,max(b$counts)/2),lty=3,lwd=4)
lines(CIs[c(2,2)],c(0,max(b$counts)/2),lty=3,lwd=4)
```

These upper and lower confidence intervals can then be used to plot error bars for the mean in other figures. Of course, we are not limited to bootstrapping the mean. We can bootstrap any test we are interested in, and obtain confidence intervals on the test statistic. For example, we could bootstrap confidence intervals on the t-statistic of a one-sample t-test using the same data (see Figure 8.8).

```
maint <- t.test(data,mu=0) # calculate a t-statistic instead of a mean

allT <- NULL
for (n in 1:10000){
  allT[n] <- t.test(sample(data,replace=TRUE),mu=0)$statistic}
CIs <- quantile(allT,c(0.025,0.975))

b <- hist(allT,breaks=20)
lines(c(maint$statistic,maint$statistic),c(0,max(b$counts)),lty=2,lwd=8)
lines(CIs[c(1,1)],c(0,max(b$counts)/2),lty=3,lwd=4)
lines(CIs[c(2,2)],c(0,max(b$counts)/2),lty=3,lwd=4)
```

These confidence intervals could be used to compare t-statistics across different data sets (e.g. in meta-analysis, see Chapter 6), or to compare the test statistic to a fixed point (e.g. to zero). One can also produce confidence intervals on *p*-values, effect size measures, and any other statistic one might be interested in. The key thing to remember is that we repeat whatever calculation we are interested in many times on sets of randomly resampled data.

Histogram of allT

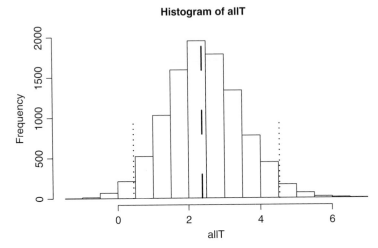

Figure 8.8 Distribution of resampled t-statistics, showing the true mean (dashed line) and 95% confidence intervals (dotted lines).

Finally, let's conduct a bootstrap test on some weakly correlated data. We'll generate these ourselves so that we have control over the extent of the correlation:

```
# generate a vector of 50 random values
var1 <- rnorm(50)
# generate a vector of 50 values that includes a fraction of var1
var2 <- rnorm(50) + 0.25*var1

# calculate the correlation coefficient for these two vectors
truecor <- cor(var1,var2)
truecor
## [1] 0.2186587
```

To generate a null distribution, we need to randomly reorder both of the vectors to destroy the correspondence between them (so that the value in row 1 of *var1* is unlikely to be paired with the value in row 1 of *var2*, and so on). We can do this using the sample function as follows:

```
nullR <- NULL
for(n in 1:10000){
  var1r <- sample(var1,50,replace=TRUE)
  var2r <- sample(var2,50,replace=TRUE)
  nullR[n] <- cor(var1r, var2r)
}
```

The data object *nullR* now contains the null distribution of 10000 resampled correlation coefficients. We can calculate a one-sided *p*-value by working out the proportion of this distribution that is larger than our original correlation coefficient (stored in the data object *truecor*):

```
length(which(nullR>truecor))/10000
## [1] 0.062
```

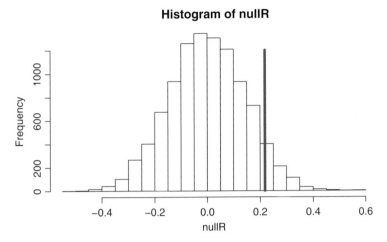

Figure 8.9 Null distribution from a bootstrap test of a correlation. Each correlation coefficient in the distribution is calculated using independently resampled data from each variable. The vertical blue line shows the correlation coefficient from the original data.

The *which* function here returns the indices of any entries in *nullR* that are larger than the value of *truecor*. Then the *length* function counts how many indices have been returned by the *which* function; this is converted to a proportion by dividing by the number of resampling iterations (10000). Finally, it is worth visualizing both the null distribution and the original correlation coefficient, as shown in Figure 8.9.

```
hist(nullR,breaks=20)
lines(c(truecor,truecor),c(0,1200),col='blue',lwd=6)
```

Notice that the number of resampling iterations determines the precision of the resulting *p*-value. Running 10000 iterations gives us a precision of 1/10000 = 0.0001. If we only ran 100 iterations, we would only have a precision of 1/100 = 0.01. Values smaller than this will default to 0.

Further reading

If you want to read more about stochastic methods, a very comprehensive source is the book called *An Introduction to the Bootstrap* by Efron and Tibshirani (1993). Bradley Efron essentially invented the technique of bootstrapping in the late 1970s, which had an enormous impact on the field, and this book is the definitive source on the method. Alternatively, for a tutorial-style paper on resampling methods with ecology-related examples in *R*, see Fieberg, Vitense, and Johnson (2020). For a general example of data simulation, the book on power simulations by Colegrave and Ruxton (2020) is a useful resource. If you need to simulate a particular type of data, there may well be specific *R* packages and online tutorials designed with this in mind.

Practice questions

1. Pseudo-random numbers can be generated by:
   A) Radioactive decay
   B) Weather patterns
   C) The computer's clock
   D) A mathematical algorithm

2. According to central limit theorem, the sum of many different distributions will be approximately:
   A) Normal
   B) Uniform
   C) Positively skewed
   D) Triangular

3. A sequence of pseudo-random numbers can be recreated if we know:
   A) The date the original numbers were generated
   B) The seed value
   C) The kernel number
   D) The first number in the sequence

4. Bootstrapping methods allow us to calculate:
   A) The mean or median value of a test statistic
   B) A t-statistic for any data set
   C) Confidence intervals on any statistic
   D) Whether or not a distribution is normal

5. Stochastic simulations are **not** useful for:
   A) Understanding dynamic systems such as the brain
   B) Modelling complex processes without using analytic equations
   C) Checking whether a sequence of numbers is truly random
   D) Testing our expectations about statistical power

6. Use the *rgamma* function to generate 100000 random numbers from a gamma distribution with shape and scale parameters of 2. The median is approximately:
   A) 3.35
   B) 2.85
   C) 4.52
   D) 1.11

7. Which R function is used to extract confidence intervals from a population of bootstrapped values?
   A) *qnorm*
   B) *hist*
   C) *mean*
   D) *quantile*

8. 95% confidence intervals of a population indicate:
   A) The points which 97.5% of values lie between
   B) The points which 95% of values lie between
   C) The points which 1.96% of values lie between
   D) The points which lie beyond an alpha level of 0.05

9. Which sequence of numbers could have been resampled without replacement from the following set? [5, 3, 7, 12, 15]
   A) 7 15 3 12 5
   B) 3 15 5 3 12
   C) 5 3 15 12 8
   D) 12 15 3 12 5

10. Use the *rpois* and *hist* functions to generate and plot 10000 samples from a Poisson distribution with a lambda value of 2. The histogram is best described as:
    A) Bimodal
    B) Negatively skewed
    C) Positively skewed
    D) Symmetrical

Answers to all questions are provided in the answers to practice questions at the end of the book.

Non-linear curve fitting

Fitting models to data

Real empirical data can be quite complicated. Using a model of some sort can often help us to make sense of our data. This might allow us to reduce the complexity of the data by summarizing it using one or more parameter estimates. Or it might help us to understand the processes that led to the data being produced. Actually, many familiar statistical tests are based on fitting linear models (straight lines) to data, including linear regression and ANOVA (see Chapter 4). But sometimes our data do not involve linear trends, and we need a more elaborate model to describe them. In this context, a model might be a mathematical equation that can predict values of a dependent variable for given levels of an independent variable. Such models will usually have some *parameters* (often called coefficients). These are numbers in the equation that control the model's behaviour, and which we can alter to try and improve the fit to the data. But how can we find appropriate values of these parameters that give the best description of our data?

The general problem here is called *parameter optimization*, and there is a whole class of computational techniques designed to solve it. This chapter will discuss some of the issues involved in fitting models to data, and introduce a well-established optimization algorithm called the *downhill simplex algorithm*. We will first outline the idea of linear and non-linear models, and how to calculate the error of a model fit. Next we will introduce the possible parameter space for a model, and discuss how this depends on the number of parameters in the model. Then we will introduce the simplex algorithm, and describe some problems it can encounter during optimization. Finally we will go through an illustrative example of function fitting in *R*.

Linear models

Most readers will be familiar with the idea of linear regression. In regression, we aim to fit a straight line to a data set, such that we can predict how our dependent variable (i.e. the thing we have measured) changes as a function of another variable. An intuitive example of this might be how height increases as a function of age during childhood, as shown in Figure 9.1.

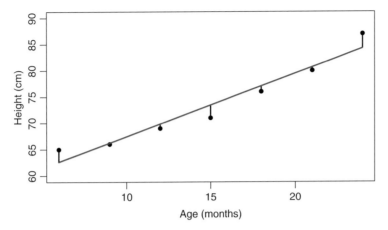

Figure 9.1 Straight line fit to age vs height data.

To the extent that height will continue to increase approximately linearly with age, we could use the fitted line to predict how tall a particular child might be in another three months. In regression notation, the equation of a straight line is:

$$y = \beta_0 + \beta_1 x \tag{9.1}$$

where the β_1 parameter determines the slope (gradient) of the line, and the β_0 parameter is a vertical offset that determines the value of y when $x = 0$ (often called the y-intercept). Performing regression involves finding the values of the two parameters (β_0 and β_1) that give the best description of the data. For the above example this turns out to be β_0 = 55.4 and β_1 = 1.2. The slope value is telling us that every month the child grows another 1.2 cm on average.

For real data, there will always be some amount of error between the straight line fit and the data points (shown by the thin vertical lines in Figure 9.1). One way of thinking about fitting is that we are trying to make this error as small as we possibly can. If the parameter estimates were completely wrong this would give a very poor fit. For example if the slope parameter were negative, the model (thick blue line) would predict that children should shrink as they age! It follows that the best fitting parameter values (of β_0 and β_1) are the ones that produce the smallest error between model and data.

Non-linear models

Sometimes data are not best described by a straight line. In principle we could fit some other equation to these data. For example, if the data followed a square law, we could adapt our equation as follows:

$$y = \beta_0 + \beta_1 x^2 \tag{9.2}$$

This transforms our straight line into a curve, as shown in Figure 9.2.

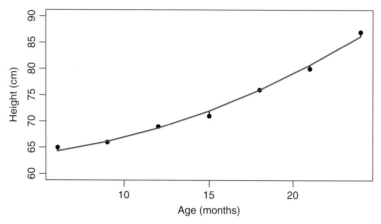

Figure 9.2 Quadratic model fit to age vs height data (a quadratic function is one that involves squaring).

The quadratic curve gives a slightly better fit to the data (the vertical lines are shorter than before). More generally, the exponent might not be exactly 2, and so its value could become another *free parameter* in the equation, much like β_0 and β_1:

$$y = \beta_0 + \beta_1 x^\gamma \tag{9.3}$$

This means that when we fit the model to our data, we would have three different parameters to adjust (β_0, β_1, and γ) instead of just two. In fact, we could in principle fit any equation to any set of data if we had reason to do so, and these equations would have as many parameters as they might need. As we will see later in the chapter, models with a large number of parameters quickly become very difficult to fit.

A practical example: exponential modelling of disease contagion

This book was mostly written during 2020, when the world experienced a pandemic outbreak of a novel coronavirus. At the time of writing, the news is full of graphs showing exponential increase in cases and deaths. A critical question during the early weeks was how quickly the disease would spread within different countries. One way in which this can be predicted is to fit non-linear growth curve models to the existing data for a country, and try to extrapolate forward into the future. Figure 9.3 shows some example data for the United Kingdom, with two different fitted curves. The dashed black curve is a fit to the first 30 days of data (indicated by the arrow). It provides a good fit to the data up to this point, but substantially overestimates the increase in cases afterwards. The curve shown in blue is a fit to all 44 days of data, which gives a much better overall description, despite overshooting slightly between days 20 and 30. The data on cases were aggregated by researchers at Johns Hopkins University (Dong, Du, and Gardner 2020), and were made publicly available online (**https://github.com/CSSEGISandData/COVID-19**).

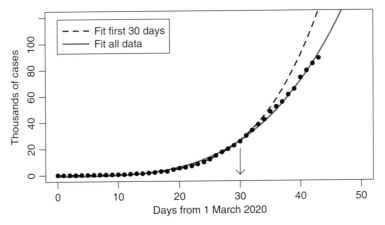

Figure 9.3 UK coronavirus cases for a six-week period in early 2020. The dashed black curve was fitted to the first 30 days of data, and predicts the following two weeks. The blue curve was fitted to the full data set.

Data source: Dong, Ensheng, Hongru Du, and Lauren Gardner. 2020. 'An Interactive Web-Based Dashboard to Track Covid-19 in Real Time'. *Lancet Infectious Diseases* 20 (5): 533–4. **https://doi.org/10.1016/S1473-3099(20)30120-1**.

This example illustrates two important points. First, fitting curves to data is an important research skill that can factor into critical life or death decision-making at the highest levels. Models of infection were used to guide government policy about how to control the virus. Second, extrapolating from curves fitted to limited data can be extremely misleading—the dashed curve in Figure 9.3 does not give accurate future predictions, and basing important decisions on it at the end of March would have been a bad idea. Of course most of the coronavirus modelling was rather more sophisticated than a three-parameter exponential function, but the same caveats apply no matter how elaborate the model.

Parameter spaces and the combinatorial explosion

When we fit a straight line with two free parameters, it is feasible to work out how good the fit is for all possible combinations of these parameters within a reasonable range (and with some specified resolution). For the example in Figure 9.4, we could work out how good the fit is for all intercept values from $\beta_0 = 0$ to $\beta_0 = 100$, and all slope values from $\beta_1 = -2$ to $\beta_1 = 2$. A useful measure of goodness of fit is the *root mean squared* (RMS) error. This involves taking the lengths of all the thin vertical lines in Figure 9.1, squaring them, and then taking the square root of their mean (average). We could then plot the error for each combination of β_0 and β_1 as shown in Figure 9.4.

This visualization tells us that the region of the *parameter space* that gives the best fit is somewhere around $\beta_1 = 1$ and $\beta_0 = 60$, which corresponds well to our original parameter estimates from the regression fit (shown by the black star). The parameters that give the best fit are those that produce the smallest error between the line and the data points, and so this is the lowest point in a virtual three-dimensional 'space' consisting of one dimension for each parameter (β_0 and β_1) and a third dimension for the value of the error (the height, indicated by the contours and shading). This error surface will be different for

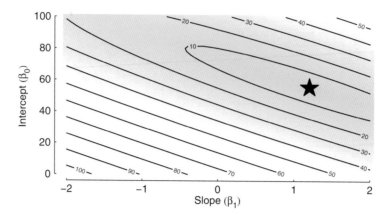

Figure 9.4 Parameter space for a linear model fit. The star indicates the best fitting parameters, which give the smallest error. Blue shading indicates the depth of the surface.

every data set, and for each model equation we might attempt to fit. Note that although in this chapter we plot several error surfaces, we would not typically visualize them, as the computational cost of evaluating the model for all possible parameter combinations is usually too great.

If we only have two free parameters, testing all plausible combinations of parameter values is possible on a modern computer (assuming some sensible level of sampling resolution). But as we add more free parameters, the amount of time required to do this will increase exponentially. This is known as the *combinatorial explosion*—the rapid growth in the complexity of a problem as more dimensions are added. Also, the error surface will have more and more dimensions (always $n + 1$, where n is the number of free parameters, and the extra dimension represents the error between the model and the data). Spaces with more than three dimensions are pretty much impossible to represent graphically or to imagine in our dimensionally limited brains. Clearly, we need an algorithm to do this for us.

Optimization algorithms

A class of computer algorithms exists that aim to find the lowest point on a multidimensional error surface without sampling every possible combination of parameters. These optimization algorithms are used in most areas of scientific research, as well as in many other fields, most notably economics and finance. Most algorithms involve *iterating* many times through some basic operations that are intended to improve the fit of the model. When the algorithm believes it has found the lowest point on the error surface, it returns the parameter values that give this best fit. Algorithms are often inspired by real-world processes, such as the cooling of metal, or the swarming of insects. Examples include:

- gradient descent (Curry 1944)
- simulated annealing (Pincus 1970)

- genetic algorithms (Holland 1992)
- particle swarm optimization (Kennedy and Eberhart 1995)
- artificial bee colony optimization (Karaboga 2005)

Different algorithms have various strengths and weaknesses on features such as computational efficiency, speed, accuracy, memory requirements, and the type of problem they are best suited to solving. It is therefore often the case that particular subdisciplines prefer a specific algorithm. In the next section, we will introduce a widely used generic optimization algorithm called the *downhill simplex* method. Because this is one of the oldest methods around, it has been implemented in many programming languages, including *R*.

The downhill simplex algorithm

The simplex algorithm, sometimes also called the *amoeba* method, was developed by Nelder and Mead (1965). It involves constructing a virtual shape (called the *simplex*) that navigates around the multidimensional error surface (e.g. Figure 9.4) in search of the lowest point. The shape has a number of vertices (corners), each of which consists of an estimate of all of the model parameters. When the algorithm is first started, the initial guess for these parameters is usually fed in by the programmer. On each iteration of the algorithm, the error between the model and the data (e.g. the height of the error surface) is calculated for the set of parameters at each vertex of the simplex.

Next, the simplex performs geometric operations to try and get closer to the surface minimum. These operations can include expansion, contraction, shrinkage, and reflection, or combinations of these operations. Expansion means that one corner of the simplex (usually the one giving the worst fit) moves away from the shape's centroid by some amount, so the shape gets larger. Contraction is the opposite—one corner moves towards the centroid, making the shape smaller. Shrinkage is similar, except that all points except for the one that is currently giving the best solution (i.e. the one at the lowest point on the surface) move in towards the best one. Reflection is a little more complicated, as it involves shifting one vertex to the opposite side of the mean of the other vertices. Note that reflection is the main way for the simplex to move around the surface, as this shifts the centroid of the shape quite substantially. An example of the progress of a simplex algorithm (every five iterations) across a three-dimensional error surface is shown by the blue triangles in Figure 9.5.

Over time, the simplex navigates around the error surface, and eventually returns a set of parameters corresponding to the best solution (i.e. lowest point) it has found. The algorithm decides when to stop based on a set of *termination criteria*—rules that tell it when it should finish. Several termination criteria are possible, including the total number of iterations of the algorithm, the total number of times the error function has been evaluated, and how close the vertices of the simplex can get before terminating. Usually all of these criteria can be adjusted by the programmer if the algorithm has either been taking too long or is not providing a very good fit. Once the set of best fitting parameters has been returned, these can be plugged back into the model equation to create a curve that fits the data points.

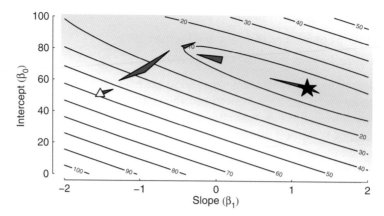

Figure 9.5 Path taken by the simplex, proceeding left to right from the starting position (white triangle), to the final solution (black star). The blue triangles are intermediate instances, sampled every five iterations.

Local minima

One common problem with the simplex algorithm (and other optimization methods) is that it can sometimes get stuck in a *local* minimum. This is a region of the error surface that is lower than all surrounding points, but is not the overall lowest possible point (the *global minimum*). The algorithm gets stuck because once it is in the local minimum, any direction it tries to move in makes the model fit worse. So it stays put, and cannot find the global minimum. A good example of a surface with multiple minima is the Himmelblau function, shown by the contour plot in Figure 9.6.

The black star at around ($x = 3$, $y = -2$) indicates the true global minimum, but often an optimization algorithm will get stuck in one of the two local minima on the left-hand side of the plot, and return parameter values from one of these two locations instead. These are also

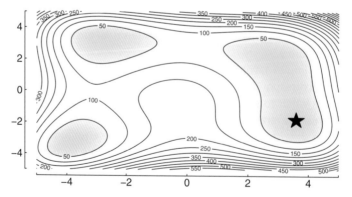

Figure 9.6 The Himmelblau function. This is a mathematical function that has four minima of slightly different depths (shaded blue), which is often used for benchmarking optimization algorithms.

good solutions, and sometimes they will be sufficient for whatever purpose we have in fitting our model (the model curve will likely follow the data quite closely). However they are not quite as good as the global minimum, which we would ideally like to find.

There are two main ways to fix the local minima problem. The first is to restart the simplex algorithm from many random starting points, in the hope that one version finds the global minimum. The other is to alter one of the fitted parameter values by a large amount, and then restart the algorithm from this new location. This method, referred to as *casting the stone*, assumes that if the original solution is the global minimum, the algorithm will not find a better solution in the new region of the search space to which it has been 'cast', and will reconverge to the original solution.

Some practical considerations

Complex models can often take a long time to fit. One way to speed things up is to optimize the code that calculates the model predictions as much as possible. This can involve removing extraneous commands, replacing loops with matrix operations, pre-allocating memory, compiling code, and making use of parallel processing capabilities. Because the model code will be called hundreds or thousands of times during optimization, even small increases in efficiency can often translate to time savings of several hours.

Another way to speed up fitting is to constrain the range of values that one or more parameters can take. Often this can be determined on practical grounds. In the child height example from Figure 9.1, we could constrain the slope value to always be positive (because babies shrinking as they get older doesn't make sense). When model parameters represent real-world properties of a physical system, it is sometimes reasonable to constrain them to lie within a sensible range. For example, a model parameter representing body temperature in live humans could be constrained to lie between $10°$ and $50°$C because temperatures outside this range would be fatal.

Models that contain a stochastic (random) element (see Chapter 8) are not typically suitable for use with optimization algorithms. This is because the model itself returns a different solution each time it is run, even with a fixed set of parameters. This means that the error surface changes on every iteration, causing obvious problems for fitting. One solution to this is to use the same *seed* value for the random number generator on each iteration. This freezes the surface and allows the minimum to be found. However, it will be important to rerun the fitting with different values of the random seed to check that similar parameter values are found each time.

Real data sets often contain some data points that are more reliable than others. This might be because more observations were made in some conditions than in others. In such situations, it can be useful to weight the data points by some measure of their reliability when calculating the error of the fit. Doing this might prevent very noisy data points from having an undue influence on the fit; the precise values used for the weights will depend on the type of data you are fitting. The general idea of weighting was introduced in the section 'Weighted averaging' in Chapter 6. As with many aspects of computational modelling, the precise details of the implementation are left to the modeller, and as you gain more experience you will usually develop heuristics that work for the type of data you are interested in.

Two philosophical approaches in model fitting

There are many reasons why it is useful to fit models to data. One advantage is that complex data can often be summarized into one or more parameter estimates. These parameters can be taken forward to a further stage of analysis—for example, if a model is fitted to the data for each individual in an experiment and then the model parameters are compared between groups or conditions. In such cases, the model can be essentially descriptive, perhaps involving some convenient mathematical function (e.g. an exponential decay function). This can be a valuable thing to do, and is a widespread and perfectly legitimate use of model fitting.

However, another type of model is one that attempts to represent, on some level, the underlying processes involved in producing the data. This type of model is called a *functional* (or sometimes *generative*) model, and can often be more informative than the descriptive models outlined above. For example, by comparing different functional models, it is often possible to infer the processes involved in the system under study. Models that do not fit the data well can be rejected, whereas more accurate models are preferred. Furthermore, if a model is more than just a description, it can often generate predictions for new conditions and situations, which can subsequently be tested empirically. This might be particularly useful for predicting future events, e.g. for climate conditions, stock market activity, or neural responses to a novel stimulus. Thinking about the type of model you wish to construct is an important first step in undertaking any form of computational modelling.

To illustrate the distinction between descriptive and functional models, let's think again about the coronavirus case data shown in Figure 9.3. We fitted a descriptive model here, using an exponential function with three free parameters. As we saw, the model was not very good at predicting the future when we fitted it to the first 30 days of case data. And of course it is just a simple equation, so the model has no concept of things like typical transmission rates, legal restrictions, vaccination, and mask wearing, that might affect the numbers of cases. If instead we had built a functional model, it might have parameters representing these sorts of factors, and would therefore be more robust in its future predictions (for an example see Friston et al. 2020). We could also see what effect changing these parameters would be likely to have, perhaps letting us predict the impact of different public health policy interventions.

Tools to help with model development

Given the above, it is reasonable to ask how one should decide on the model one is trying to fit. This is a very domain-specific question, and so it is not sensible to try to be prescriptive here. However, there are some general rules of thumb that may be of use. First, there is no substitute for visualizing the data you are trying to model (see Chapters 3 and 18 for some guidance on plotting). If the data have a clear form, for example a Gaussian-like distribution, this might suggest—or rule out—particular mathematical functions. Second, the simpler a model is, the easier its behaviour will be to understand. One way to simplify a model is to reduce the number of free parameters as far as possible. Statistics have been proposed to mathematically compare the performance of models with different numbers of parameters. The Akaike information criterion (AIC; Akaike 1974) is a widely used example, and contains a penalty term that increases with the number of free parameters. This can help to avoid 'over-fitting', by excluding parts of a model that may not be necessary to provide an acceptable fit. In general, though, reading existing studies on a similar topic is the best way to get a feel for the types of models that might be suitable for a particular data set. Some more detailed practical suggestions for model development are proposed by Blohm, Kording, and Schrater (2020)–although the authors focus on neuroscience, the points they make are generally applicable to modelling in other domains.

How to fit curves to data in *R*

The *R* language contains many packages that implement different optimization algorithms. We will focus on the *pracma* package, which contains a version of the Nelder–Mead downhill simplex algorithm that we described earlier in the chapter (Nelder and Mead 1965). The package has dozens of useful functions, but the key function for our present purpose is called *nelder_mead*. However this function requires as an input the name of the function that we want to optimize, which we will need to write ourselves. The function we optimize must:

- take a vector of parameters as its input
- return the error between the model and data as its output

Let's fit something that's a bit more interesting than a straight line. A Gaussian function can be described with two free parameters as follows:

$$f(x) = e^{\frac{-(x-a)^2}{2\sigma^2}} \tag{9.4}$$

where a and σ are free parameters, and x is the value along the x-axis. The parameter a controls the horizontal offset of the Gaussian, and σ controls the spread (width). Gaussian functions can be used to characterize many biological processes, such as the tuning functions of neurons. We can implement the equation as the first line of an *R* function (see 'Functions' in Chapter 2 for a refresher on how function definitions work in R). I have named the function *errorfit*, because it calculates the error between the model and some data (in other words, the error of the fit):

```
errorfit <- function(p){     # define a new function called 'errorfit'
    # equation of a gaussian, with parameters from the input, p
    gaus    <- exp(-((x-p[1])^2)/(2*p[2]^2))
    # root mean square error between model and data
    rms <- sqrt(sum((gaus-ydata)^2)/length(ydata))
  return(rms)}     # the function returns the rms error
```

The second line of the function calculates the root mean square (RMS) error by taking the differences between the model (stored in the *gaus* data object) and the data (stored in the *ydata* data object), squaring the differences, and calculating the square root of the mean. We will generate some synthetic data for the model to fit, where we know the true values of the free parameters, and see how well the simplex algorithm can recover them. Let's set the values to be $a = 2$ and $\sigma = 3$, and generate data using the Gaussian function for a range of x-values, adding a bit of noise (to simulate measurement error):

```
x <<- seq(-10,15,1) # sequence of x-values from -10 to 15
p <- c(2,3)      # true parameter values used to simulate some data
ydata <- exp(-((x-p[1])^2)/(2*p[2]^2))   # simulated data from gaussian
ydata <<- ydata + 0.05*rnorm(length(x)) # add noise to simulated data
plot(x,ydata,type='p')   # plot simulated data
```

This code generates the graph shown in Figure 9.7. One small point to notice: when we define the data objects *x* and *ydata*, we use the double arrow assignment (<<-) to specify that they are *global variables*. This means they are available from within the *errorfit* function, so we do not need to explicitly pass them to it as inputs.

If we tried to guess the model parameters without knowing them in advance, we might estimate that the middle of the function was around 5, and the spread was around 1. This would produce the (very poor) fit shown in Figure 9.8, and given by the following code:

```
p <- c(5,1)# a guess at some possible parameter values
pred <- exp(-((x-p[1])^2)/(2*p[2]^2)) # model prediction using these
parameters
plot(x,ydata,type='p')   # plot the data again
lines(x,pred,lwd=2,col='blue')   # add the model prediction
```

Notice that we store the parameters in a data object called *p*, which we could pass to the *errorfit* function to get a numerical estimate of how good (or otherwise) the fit is:

```
# calculate the error of the fit with our first guess parameter values
errorfit(p)
## [1] 0.3870659
```

The fit is obviously poor, and hopefully by optimizing our parameters we will be able to improve on the RMS error of 0.4. We will do this using the downhill simplex algorithm, which is called using the *nelder_mead* function. We provide the function with the name of the *errorfit* function and a starting 'guess' for what the parameter values might be. It returns a data object,

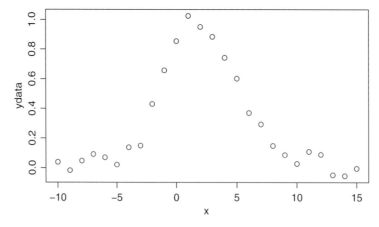

Figure 9.7 Simulated data from a Gaussian function.

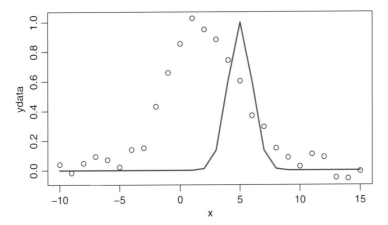

Figure 9.8 Simulated data with a poorly fitting Gaussian function.

which contains the estimated parameters. We can then plug those parameters into our equation, and generate a curve that fits the data well (see Figure 9.9).

```
library(pracma) # load the pracma package

sout <- nelder_mead(errorfit, c(1,1)) # fit the model to the data
p <- sout$xmin    # extract the parameter estimates

# get the model predictions for these parameters
pred <- exp(-((x-p[1])^2)/(2*p[2]^2))

plot(x,ydata,type='p')
lines(x,pred,lwd=2,col='blue')   # plot the model fit as a line
```

The data object produced by the *nelder_mead* function (the *sout* object) contains several pieces of information besides the final parameter estimates. For example it includes the value

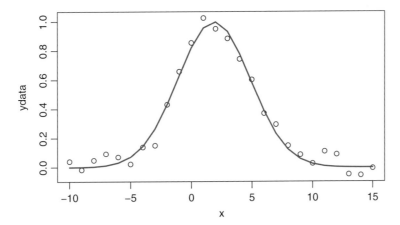

Figure 9.9 Simulated data with the best fitting Gaussian function (curve).

of the function, and the number of iterations that were run. We don't need to look at this information now, but it is there if you ever need it.

The estimated parameter values should be close to the original values used to generate the data, and as you can see (from Figure 9.9) the curve provides a good fit to the data points. We have extracted the estimated parameters from *sout$xmin* and stored them in the data object *p*:

```
p
## [1] 2.080850 2.970019
```

Notice that our estimated parameter values are close to the original values (2 and 3) that we used to generate the data in the first place. If we want to know how good the fit is numerically, we can call the *errorfit* function ourselves, and it will tell us the RMS error:

```
errorfit(p)
## [1] 0.04361436
```

The RMS error is much smaller than for the non-optimized best guess parameters that we started with. This means that the simplex algorithm has done a good job of fitting the model and finding some good parameter values. Something we could potentially do next is to bootstrap this whole process (see Chapter 8) to obtain confidence intervals on our parameter values.

This example of function fitting contains all of the same steps that we would go through for a more sophisticated model fit. We need to create an *R* function that calculates the model predictions for a given set of parameters, and calculates the error between the model predictions and the data. We then pass this function to the simplex algorithm (*nelder_mead* function), along with an initial guess about the parameters. This initial guess will often influence the end result, so it is sensible to repeat the fitting process many times using random starting parameters, and choose the model parameters with the best overall fit. For the example above, such a procedure might look something like this:

```
# first initialize data objects to store the best error value and pa-
rameters
bestrms <- 10000
bestp <- c(0,0)

# now loop through some number of repetitions
for (n in 1:100){

# run the simplex with some random starting values
sout <- nelder_mead(errorfit, 10*rnorm(2))
p <- sout$xmin     # extract the fitted parameters
thiserror <- errorfit(p)     # work out the error of the fit

# if this is the best fit we've found so far, store the parameters
if (thiserror<bestrms){
  bestrms <- thiserror
  bestp <- p}
}
```

This general approach can be used to fit models of arbitrary complexity to any type of data, and it is an enormously flexible and useful scientific tool. Of course, more complex models will take longer to fit, and might require additional lines of code (or extra functions) to specify. You could use different measures of error, and specify some additional options in the simplex fit, such as the maximum number of iterations or evaluations, as described in the help files for *nelder_mead*. Alternatively, it is also possible to use a Bayesian approach (see Chapter 17) to fitting models. This involves sampling a version of the error surface using a stochastic process (see Kruschke 2014). Happy fitting!

Practice questions

1. If a model has 12 free parameters, how many dimensions does the error surface have?
    A) 3
    B) 12
    C) 13
    D) 24

2. The 'height' of the error surface represents:
    A) The value of the parameters
    B) The error between model and data
    C) The number of degrees of freedom
    D) The number of free parameters

3. Which of the following is not an operation performed by a simplex?
    A) Circulation
    B) Expansion
    C) Reflection
    D) Contraction

4. A good solution that is not the global minimum is often called:
    A) An iteration
    B) The RMS error
    C) An error surface
    D) A local minimum

5. One straightforward way to speed up model fitting is to:
    A) Optimize the code that calculates the model predictions
    B) Delete all the data objects in the *Environment*
    C) Start the simplex from a global minimum
    D) Collect more data before fitting your model

6. Which of the following would be the most sensible constraint on model parameters?
    A) Constraining a date parameter to always be in the future
    B) Constraining a parameter that represents blood pressure to be negative
    C) Constraining a parameter that represents heart rate to be positive
    D) Keeping a parameter value as large as possible

7. The *nelder_mead* function must operate on:
   A) A function that has fewer than three free parameters
   B) A function that is built into the core *R* toolbox
   C) A function that returns the best fitting parameters
   D) A function that calculates the error between model and data

8. A useful tool for comparing models with different numbers of free parameters is:
   A) The downhill simplex algorithm
   B) Akaike's information criterion (AIC)
   C) Cohen's *d*
   D) The *p*-value of a t-test

9. The best fitting model for a particular data set should have:
   A) A bigger RMS error value than other models
   B) A smaller RMS error value than other models
   C) The largest number of free parameters
   D) The smallest number of free parameters

10. A functional model attempts to:
    A) Model the underlying processes that produced the data
    B) Describe the data using a convenient mathematical equation
    C) Incorporate as many free parameters as possible
    D) Predict all possible states of the system under study

Answers to all questions are provided in the answers to practice questions at the end of the book.

10 Fourier analysis

Fourier analysis is a technique that allows us to determine the frequency content of waveforms. It is used in a variety of signal processing situations, and is a core analysis method in many areas of science and engineering. In this chapter we will explain the theory behind Fourier analysis in one and two dimensions, and illustrate its usefulness with several examples. After finishing the chapter, you should have an understanding of how to conduct Fourier analysis in R, and how to interpret and manipulate the frequency content of signals.

Joseph Fourier: polymath

Fourier analysis takes its name from a nineteenth-century French mathematician called Joseph Fourier (see Figure 10.1). Fourier was a polymath: an expert on many topics. Of particular note, he is generally credited as being the first person to describe the greenhouse

Figure 10.1 *Portrait of Joseph Fourier. By Louis-Léopold Boilly,* https://commons.wikimedia.org/w/index.php?curid=3308441

effect—the process by which carbon dioxide traps heat near the surface of a planet and causes global temperatures to rise.

The basic idea behind the technique that bears his name is that any waveform can be decomposed into a bunch of sine waves of different frequencies (you may have encountered the sine and cosine functions when learning trigonometry). Of course, in Fourier's era there were no computers, meaning that this procedure had to be carried out by hand. This was a prohibitively slow process, and so Fourier analysis was not widely used until long after his death. However, modern computers make the calculations straightforward and very fast, and Fourier analysis has been used in a wide variety of signal processing applications, as we will describe below.

Example applications

So what's the point in breaking waveforms down into sine waves? The main purpose is to describe the waveform in terms of the *frequencies* it contains. In the time domain, frequency refers to how often a signal repeats in a given period of time. The unit for this is the Hertz (Hz)—1 Hz means one repetition per second, 10 Hz means 10 repetitions per second (see Figure 10.2). Describing a sound recording in terms of frequencies might allow us to identify any interference in a particular frequency range, or perhaps alter the signal to remove certain frequencies. Indeed, most methods for compressing audio files (such as MP3 compression) involve removing frequencies that are inaudible to human ears.

Many biological processes are also periodic or semi-periodic, and can usefully be summarized using Fourier analysis. Examples include heart rate and breathing, as well as slower processes like the circadian rhythm and the cell cycle. Furthermore, many modern scientific methods, such as crystallography and magnetic resonance imaging, crucially depend on Fourier analysis to work.

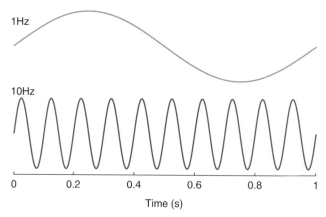

Figure 10.2 Example sine waves of different frequencies. The 1 Hz wave at the top goes through a single cycle (it increases, then decreases, then returns to baseline) during the one second of time depicted. The 10 Hz wave below it goes through ten cycles in the same period of time.

In psychology and neuroscience, an important application of Fourier analysis is to analyse time-varying brain signals, of the type recorded using electroencephalography (EEG) and magnetoencephalography (MEG). These 'brainwaves' often have activity at characteristic frequencies that are associated with specific mental states or cognitive operations. For example, when participants in an experiment become tired, they produce more activity in the alpha band of frequencies from 8–12 Hz. Synchronization and suppression between brain areas involves oscillations at higher frequencies in the gamma band, above 30 Hz. To quantify this type of activity, it is typical to use Fourier analysis to decompose the signal into its component frequencies.

Finally, as we will see later, it is also possible to apply Fourier analysis to two-dimensional images (or even movies). This has numerous applications in image processing, filtering, and synthesis that are useful for designing and understanding experimental stimuli.

Terminology

A key first step in understanding Fourier analysis is to get your head around some important terms. If we take a waveform and calculate the *Fourier transform*, this will break the waveform down into its component frequencies. The result is referred to as the *Fourier spectrum*, and has two parts, as we will describe in a moment. If we want to convert from the spectrum back to the waveform, we perform the *inverse* Fourier transform (see Figure 10.3 for an example). These operations are also referred to as *Fourier analysis* and *Fourier synthesis* respectively, and the underlying mathematics are known as the *Fourier theorem*. Understanding the maths is not necessary to use these methods, which are implemented as core functions in most computer programming languages.

Fourier analysis is an invertible linear process, which means that no information is lost when converting between the waveform and its spectrum. A consequence is that you can

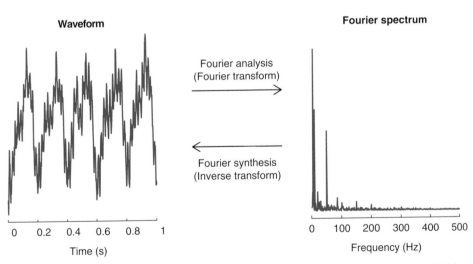

Figure 10.3 Illustration of the Fourier transform between a waveform (left) and its Fourier spectrum (right), for one second of brain activity measured using electroencephalography (EEG).

perform these operations as many times as you like without distorting your original signal. I think of the Fourier spectrum as an alternative way of representing the same information. Graphically, this is most often represented by the amplitude spectrum.

The amplitude spectrum

The amplitude spectrum is typically plotted as a graph, in which the x-axis shows the frequency, and the y-axis shows the amplitude. The frequency, as mentioned above (see Figure 10.2), is the number of cycles per unit of time. The amplitude is the vertical difference between the peaks and the troughs of the sine wave (see Figure 10.4). A low amplitude means a very small change, and a high amplitude means a large change.

Figure 10.5 shows an example amplitude spectrum. This is based on the same data as in Figure 10.3, but here we have zoomed in on the portion of the spectrum from 0 to 30 Hz. This is where most of the action is in human brain activity (because of the intrinsic timescales at which neurons operate), and so is a worthwhile frequency range to focus on. The highest amplitude is at 5 Hz, which is consistent with the clearly periodic nature of the waveform in Figure 10.3, showing five peaks and five troughs in one second. The amplitude spectrum therefore has a direct mapping to features of the waveform it represents. The units of amplitude are the same as the units used to measure the original waveform. For the EEG data used in the examples here, these are microvolts (μV), but the units will correspond to whatever dependent variable you have chosen to measure.

The phase spectrum

The amplitude spectrum contains only half of the information produced by the Fourier transform. The remainder is found in the phase spectrum. The property of phase corresponds to the horizontal alignment of a waveform. It is always a relative measure, and is usually

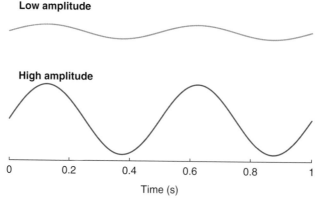

Figure 10.4 Example sine waves of different amplitudes. The sine wave at the top has a low amplitude, and the one below it has a high amplitude.

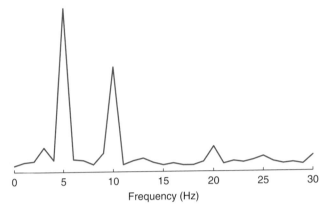

Figure 10.5 Example amplitude spectrum, zoomed in from 0 to 30 Hz.

expressed with reference to some meaningful event, such as the onset of a stimulus or the start of a recording. Phase is a circular (periodic) term, with phase angles being measured in degrees (or sometimes radians, which are an alternative angular unit). Figure 10.6 shows two example sine waves with different phases. The upper waveform has a phase angle of 0° relative to the vertical dashed line. This is referred to as being in *sine phase*. The lower waveform has a phase angle of 90° relative to the vertical dashed line. This is referred to as being in *cosine phase*. Two other common phases are *negative sine* (180°) and *negative cosine* (270°) phase, which correspond to a vertical flip (peaks become troughs and vice versa) of the two waveforms shown in Figure 10.6.

The Fourier *component* at each frequency in the amplitude spectrum has an accompanying phase term. This is difficult to show graphically, and so is not generally represented, at least for

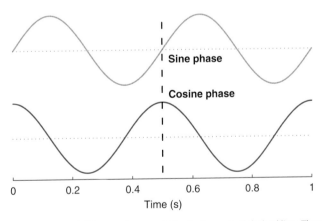

Figure 10.6 Example sine waves of different phases, relative to the vertical dashed line. The waveform at the top is in sine phase with the line (i.e. the line is midway through a cycle). The waveform below is in cosine phase with the line (i.e. the line is at a peak).

many frequencies at once. However the phase information is critical, as without it the original waveform cannot be reproduced. Indeed, *phase scrambled* stimuli are often used in experiments as a control (or baseline) condition because recognizable information (e.g. a speech signal) is destroyed by the scrambling process.

Limitations on sampling: the Nyquist limit and frequency resolution

Fourier analysis is not magic—it has two limitations. The first limitation is the maximum frequency that can be included in the spectrum, which is determined by the sample rate of the original waveform. The sample rate is how often the dependent measure is taken, which is usually determined by the hardware used to record the data. Consumer video cameras usually record around 25 or 30 frames per second, giving a sample rate of 25–30 Hz. Microphones and other audio hardware often record at a very high sample rate, around 44100 Hz, so that they can capture high-frequency sounds. EEG data is often sampled at a rate of 1000 Hz (e.g. a thousand measurements per second). Functional magnetic resonance imaging (fMRI) data are sampled much more slowly, at around one sample every three seconds (i.e. 1/3 Hz). Eye-tracking hardware usually operates somewhere between 30 Hz and 1000 Hz, depending on the make and model of equipment.

The maximum frequency that can be resolved is known as the *Nyquist limit*, and it is always exactly half the sample rate. So a signal recorded at 1000 Hz has a Nyquist limit of 500 Hz. This is because a frequency of half the sample rate has sufficient resolution for one high sample (e.g. a peak) and one low sample (e.g. a trough).

The second limitation is the frequency resolution—e.g. the granularity of steps along the frequency axis of the amplitude spectrum. Surprisingly, this is determined by the duration of the sample, and not the sample rate. A 1-second sample has a frequency resolution of 1 Hz. That means the spectrum will contain information at 1 Hz, 2 Hz, 3 Hz, and so on, but not at any intermediate frequencies (1.5 Hz for example). A 10-second sample has a frequency resolution of $\frac{1}{10}$ Hz, meaning that there are intermediate frequency bins between the integer frequencies in steps of 0.1 Hz. The frequency resolution is therefore given by 1/(duration in seconds), *regardless* of the sample rate. Only the Nyquist limit is determined by the sample rate.

Example: bat species identification by frequency

A good applied example of how Fourier analysis can be used is in the identification of bat echolocation signals. Different species of bats produce calls at distinct frequencies, which are generally well above the limits of human hearing. By combining the Fourier spectra with classification techniques (see Chapter 14), it is possible to identify individual species by their calls. For example, waveforms and Fourier spectra for two bat species are shown in Figure 10.7. The waveforms look broadly similar—the offset along the x-axis is arbitrary, and determined only by when in the recording the call began. However the amplitude spectra in the lower plot have peaks at very different frequencies. The common pipistrelle's call (black) peaks at around 5000 Hz, whereas the noctule's call (blue) peaks at around 2000 Hz.

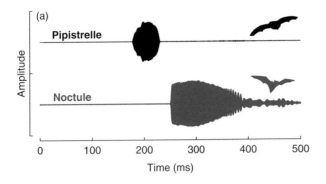

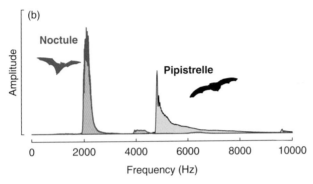

Figure 10.7 Waveforms (a) and Fourier spectra (b) for example calls from two bat species: *Pipistrellus pipistrellus* (black), and *Nyctalus noctula* (blue).

There have been many different classification systems proposed that use Fourier-transformed echolocation signals to identify bat species. For example, Walters et al. (2012) trained an artificial neural network to discriminate between 34 different bat species. The calls were first assigned to one of five different groups; this classification had an extremely high accuracy of around 98%. Calls were subsequently assigned to individual species, which had a slightly lower accuracy of around 84% (but still far above chance performance of 100/34 = 2.9%). There are online tools to classify bat calls, and mobile phone applications and dedicated handheld devices are now available that can perform classification in real time out in the field. These tools are all based on Fourier analysis.

Fourier analysis in two dimensions

One potential use of Fourier analysis (that is sometimes surprising when first encountered) is that it can be applied in more than one dimension. This means that images (which we can think of as two-dimensional signals) and movies (three-dimensional signals) can also be analysed in this way. The mathematics turn out to be equivalent to the one-dimensional case, albeit slightly more complex. Because Fourier analysis uses the sine wave as its basis function, a consequence is that we can consider *all* images to be reducible to a combination of sine waves of various *spatial frequencies* and *orientations*. See Weisstein (1980) for an excellent and detailed tutorial on two-dimensional Fourier analysis.

So what is spatial frequency? For the waveforms we have considered so far, the signal (usually a measure like voltage, or sound pressure level) changes as a function of time, and the frequency units are Hertz (cycles per second). Because the signal is changing over time, we call this property temporal frequency. For an image, the 'signal' is the change in luminance as a function of space, and we call the rate of this change the spatial frequency. The units of spatial frequency are cycles per degree of visual angle, or sometimes (for simplicity) cycles per image. Figure 10.8 shows example sine wave grating images, with different spatial frequencies and orientations. By applying Fourier analysis in two dimensions, it becomes apparent that all photographs can be described as the sum of multiple sine waves of different spatial frequencies, orientations, phases, and amplitudes.

Just as we took the Fourier transform of a 1D waveform, so we can Fourier-transform a 2D image. The Fourier spectrum will then also be two-dimensional, and is referred to as *Fourier space* (or sometimes the *Fourier domain*). The left panel of Figure 10.9 shows an image of a bug hotel. Although it is hard to think of a photograph as a 'waveform', consider that an image is made up of pixels that vary in intensity across space, and that if we plot those values for one row of the image (superimposed in blue), they look very much like a waveform. In the right panel is the Fourier spectrum of the image. This has been zoomed into the central low-spatial frequency portion, where most of the energy resides.

The two-dimensional Fourier spectrum is hard to interpret without some guidance. Figure 10.10 uses small patches of grating to illustrate the layout graphically. The lowest spatial frequencies are represented in the centre of the plot, with higher spatial frequencies towards the edges. Orientation is represented by the angle from vertical. However, because of a somewhat confusing convention, horizontal orientations are traditionally represented along the vertical axis, and vertical orientations are represented along the horizontal axis. Oblique (diagonal) orientations are represented in between. The spectrum is symmetrical (mirrored) about the midpoint (illustrated by the vertical gratings along the horizontal axis).

Figure 10.8 Example sine wave gratings of different spatial frequencies and orientations. The left grating has a low spatial frequency (3 cycles per image), while the middle grating has a higher spatial frequency (10 cycles per image). The right grating has an oblique orientation.

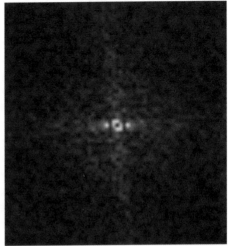

Figure 10.9 Greyscale image of a bug hotel (left), and its Fourier spectrum (right).

Of course in a real Fourier spectrum, the small grating icons are not shown. Instead the value (brightness) at each point (i.e. each x,y coordinate) represents the amplitude at that particular combination of orientation and spatial frequency. The right-hand panel of Figure 10.9 shows an example in which most of the energy is concentrated at low spatial frequencies (as is typical for natural images), with dominant vertical energy (along the horizontal axis), caused by the vertical contours (wooden poles) in the original image.

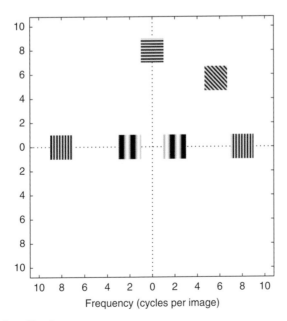

Figure 10.10 Illustration of Fourier space.

Example: using 2D Fourier analysis to measure goosebumps

If you are cold, tiny hairs on your arms and down your back will stand on end and you will get goosebumps—small raised lumps at the base of each hair. The same response, called the *piloerection* response, can also be triggered by other experiences, such as being frightened, or listening to a particular piece of music that gives you 'chills'. A group of researchers in Germany and Austria developed an elegant technique to quantify the piloerection response using Fourier analysis. Benedek et al. (2010) built a device that involved attaching a small video camera to a patch of skin, to continuously film a close-up view of the hair follicles. An LED light source illuminates the patch of skin from one direction. Most of the time no goosebumps are visible, and the output of the camera might look something like the image in Figure 10.11(a). However, when a participant experiences goosebumps, they cast shadows in the opposite direction to the light source, which are clearly visible in the images (see Figure 10.11(b)).

Hair follicles are distributed with approximately even spacing across the skin. That means the pattern of shadows will have a consistent spatial frequency, which turns out to be around 0.4 cycles per mm (in other words, you have a hair follicle about every 2.5 mm). If we look at the ratio of the Fourier spectra of the two images, there is a clear peak corresponding to the goosebumps (see Figure 10.11(c)). Benedek et al. (2010) created software (available at

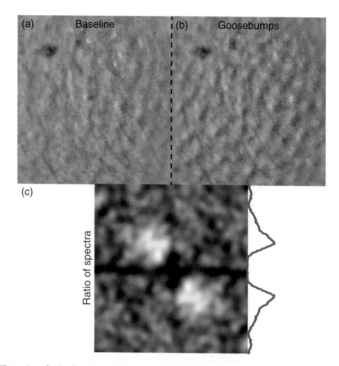

Figure 10.11 Illustration of using Fourier analysis to quantify the piloerection response. The images on the top row show a small patch of skin (a) without and (b) with goosebumps. Panel (c) shows the ratio of the Fourier spectra of the two images (smoothed), which exhibits a strong peak corresponding to the presence of goosebumps. The blue trace at the right hand margin shows the polar average of the ratio. Image credit: Rémi de Fleurian.

www.goosecam.de) that takes the Fourier transform of the camera images, and sums the amount of signal around this frequency to give a continuous, objective estimate of the pilo-erection response. This method has many potential applications in several areas of research, particularly in understanding our experience of music and film. Amazingly, the researchers also found an individual who has direct control over their own piloerection response, and could give themselves goosebumps on demand!

Filtering: altering signals in the Fourier domain

One common reason for using Fourier analysis is that signals can often be manipulated in the Fourier domain in various ways that would be challenging to achieve using only the original signal. A simple example is to remove high-frequency 'noise' from a signal. This is achieved by *filtering*: multiplying the Fourier spectrum by a *filter* constructed to include some frequencies and exclude others, and then taking the inverse transform. An example of this for removing noise from an EEG waveform is shown in Figure 10.12. The filter here is referred to as a *low pass* filter, because it *passes* (allows through) low frequencies (those within the blue shaded region), but blocks higher frequencies (those outside that region). Because high frequencies here contain mostly noise, this has the effect of smoothing the waveform (shown by the black curve in the right panel of Figure 10.12).

We can also apply filters in two dimensions. Figure 10.13 shows low pass and high pass filters, and their effect on the bug hotel image. The low pass filtered image looks blurry, as the fine detail is stored at the higher spatial frequencies which have been removed by the filter. The high pass filtered image lacks extended light and dark regions (represented by the lower

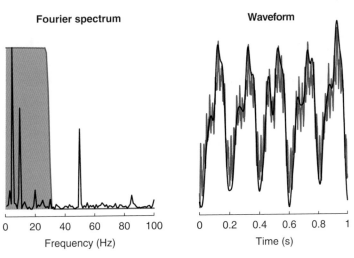

Figure 10.12 Illustration of low pass filtering. The left plot shows the Fourier spectrum, with superimposed low pass filter (blue), which excludes the high-frequency components outside of the shaded region. The right panel shows the original waveform (blue) and the filtered waveform (black) which lacks the high-frequency noise and is therefore visibly smoother.

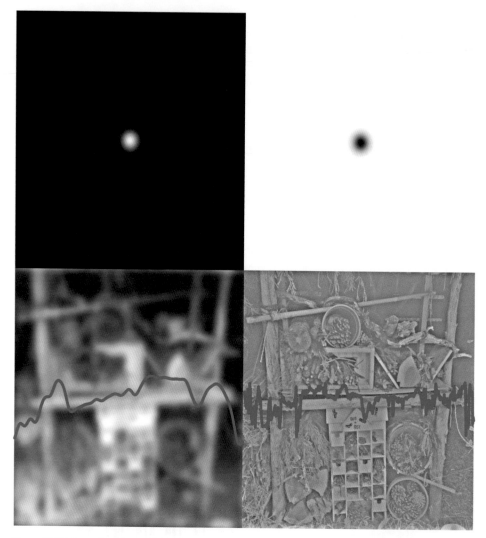

Figure 10.13 Example of low and high pass filtering on the bughouse image. The top row shows low pass and high pass filters, in which frequencies in the lighter regions pass the filter, but frequencies in the darker regions are attenuated. The lower row shows the resulting filtered images: low pass filtering produces a blurred image; high pass filtering produces a sharp-looking image but without coarse changes in light and dark.

spatial frequencies), and retains only edges at higher frequencies. Note also how the overlaid pixel intensities for the central row (shown in dark blue) are smooth in the low pass filtered version, and jagged in the high pass filtered version.

Finally, we can filter orientation information. Figure 10.14 shows filters and the resulting images in which either horizontal (left) or vertical (right) information is removed, leaving information at the orthogonal orientation. Notice how in the image where horizontal information is removed (left) we can still clearly see the vertical poles at either side of the image, and the vertical white parts of the little drawer unit in the centre at the bottom. On the other

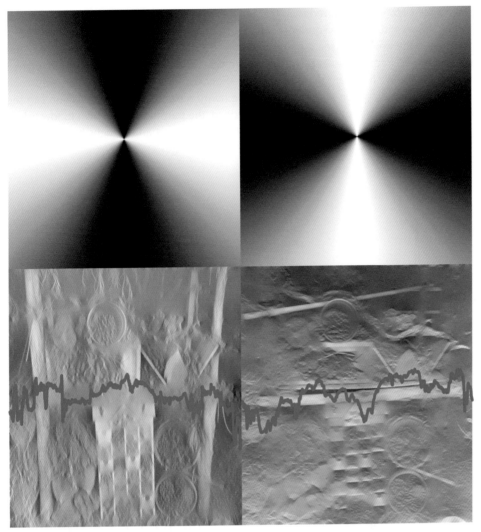

Figure 10.14 Example of filtering orientation information. The left column shows a filter that blocks horizontal information, but retains vertical information. The right column shows the opposite filter.

hand, in the image where vertical information is removed (right) these features are missing, but a central horizontal bar, and the horizontal parts of the drawer unit are visible.

Stimulus construction in the Fourier domain

As well as altering existing stimuli in Fourier space, we can also construct stimuli from scratch in this way. The classic example of this is to synthesize a square wave from the *Fourier series* of odd harmonic sine waves. We begin with a sine wave at the lowest frequency, which for this example is 1 Hz. This is referred to as the *fundamental*, often denoted 1F. It is generated by setting the amplitude of the Fourier spectrum at 1 Hz to 1, and taking the inverse transform to visualize the waveform (see top row of Figure 10.15). We then add energy at each of the odd harmonic frequencies in turn—the harmonics are the multiples of the fundamental frequency, and we just want the odd numbered ones: 3F, 5F, 7F, and so on. The amplitude of each harmonic is given by $\frac{1}{h}$, where h is the harmonic (e.g. 3, 5, 7 etc.). As you can see from the example, as more components are added, the transitions from high to low values in the synthesized waveform become sharper, and the peaks and troughs become flatter. Eventually, we have a complete square wave with sharp edges.

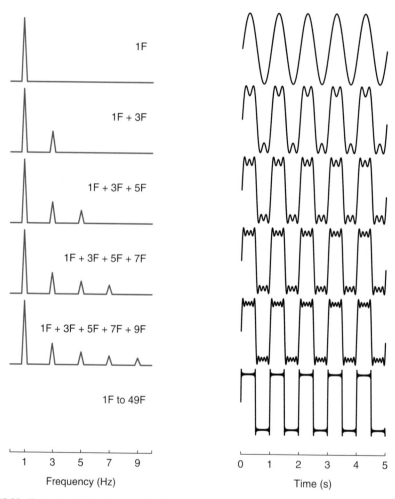

Figure 10.15 Illustration of Fourier synthesis of a square wave. The left column shows the Fourier spectrum, and the right column the synthesized waveform created by inverse transforming the spectrum. Successive rows add additional components, at odd harmonics of the fundamental frequency. By the final row, with 25 harmonics, the square wave is well defined.

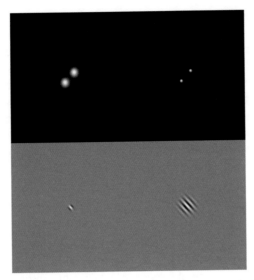

Figure 10.16 Gabor stimuli synthesized in Fourier space. The upper row shows the Fourier spectra, and the lower row the spatial transforms.

The same approach can be taken to generate images. For example, we can create sinusoidal stimuli with very tightly defined properties (specified bandwidths) in the Fourier domain. A popular stimulus in computer vision research is the Gabor pattern, which is a spatially localized sine wave grating. We can generate these in the Fourier domain by shifting a two-dimensional Gaussian blob (like the low pass filter in Figure 10.13) away from the origin of Fourier space. This will produce a Gabor pattern in the spatial domain (see Figure 10.16 for examples). An interesting observation is that patterns with a small footprint in the Fourier domain have a large spatial extent in the spatial domain, and vice versa. This means that small patches of grating have a broader frequency *bandwidth* than large ones, and so their orientation and spatial frequency are less clearly defined.

Doing Fourier analysis in *R*

So far in this chapter, we have seen that we can:

- represent waveforms and images by their frequencies
- filter the Fourier spectrum to change the original signal
- construct new signals by synthesizing a spectrum

In this section we will discuss how to implement these operations in *R*. Most of the examples use base *R* functions, though we will also use some custom functions, and a function from the *signal* library. The waveform in Figure 10.3 is loaded in from an external data file as follows:

```
load('data/EEGdata.RData')
thiswave <- allwaves[1,]
```

The key function is the *fft* (Fast Fourier Transform) function. This takes a vector or matrix as its input, and returns a complex-valued Fourier spectrum of the same dimensions. Complex numbers are a mathematical convenience, and contain 'real' and 'imaginary' components. It is not necessary to fully grasp the mathematics of complex numbers, but in contemporary implementations of Fourier analysis, the amplitude and phase information are represented in Cartesian coordinates by the real and imaginary components of the number. An optional argument to the *fft* function, *inverse = TRUE*, will request the inverse transform. By convention, we also scale the output of the function by the length of its input. The following lines of code perform the Fourier transform on the waveform and confirm (using the *is.complex* function) that we have a complex-valued output:

```
output <- fft(thiswave)/length(thiswave)
is.complex(output)
## [1] TRUE
```

We can determine the frequencies for plotting the amplitude spectrum if we know the duration of the signal (here it was 1 second) and the sample rate (here 1000 Hz). The amplitudes can then be plotted as a function of frequency by taking the absolute values of the Fourier spectrum (e.g. forcing any negative values to be positive using the *abs* function), as follows:

```
samplerate <- 1000
duration <- 1
frequencies <- ((1:(samplerate*duration))-1)/duration
plot(frequencies[2:500],abs(output[2:500]),type='l',lwd=2)
```

Note that in the above code (see Figure 10.17 for the output), we plot values only up to the Nyquist limit of 1000/2 = 500 Hz. The spectrum is mirrored about its midpoint, so the values from an index of 501 onwards are a reflection of the spectrum plotted in Figure 10.17. Notice also that we begin plotting at the second index of the vectors containing the frequency and Fourier spectrum data. This is because the first entry in the spectrum, known as the *DC component* (by analogy to direct current), often has a much larger amplitude than the other

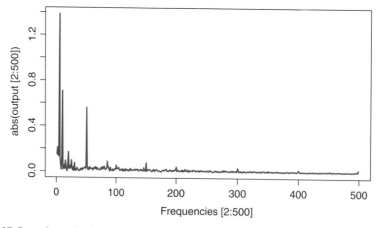

Figure 10.17 Example amplitude spectrum from 0 to 500 Hz.

frequencies. The DC component has a frequency of 0 Hz, which corresponds to the vertical offset of the waveform (a bit like the intercept term in regression and ANOVA). Since it is often uninteresting, we have omitted it from the plot above, but it can be included if required.

To apply a filter to the amplitude spectrum, we can construct one using a function from the *signal* package. The *fir1* function produces a type of filter called a *finite impulse response* (or FIR) filter. This is a commonly used type of filter in signal processing, which has several convenient properties such as being symmetrical and also very stable. We can create and plot a FIR filter in the temporal (e.g. non-Fourier) domain as follows (see Figure 10.18):

```
cutfrequency <- 15
filter1 <- fir1(samplerate-1,2*cutfrequency/(samplerate/2),type='
low')
plot(filter1,type='l',lwd=2)
```

Rendering the filter in the temporal domain is sometimes useful, as it can be used for *convolution* with a signal. Convolution is equivalent to multiplying the filter with the signal at each consecutive time point. However this can be quite a slow and inefficient process, especially for long signals. A more efficient approach is to take the Fourier transform of the filter, and multiply this by the Fourier transform of the signal. This produces the same result, because **convolution in the temporal domain is the same as multiplication in the Fourier domain**. So we can apply the filter in the Fourier domain, and then take the inverse transform to view the filtered signal as follows:

```
# multiply the fourier spectra of the waveform and filter
filteredspectrum <- output*abs(fft(filter1))
# inverse transform and take the Real values
filteredwave <- Re(fft(filteredspectrum,inverse=TRUE))
plot(1:1000,filteredwave,type='l',lwd=2)
```

The filtered waveform in Figure 10.19 is much smoother than the original, shown in Figure 10.3.

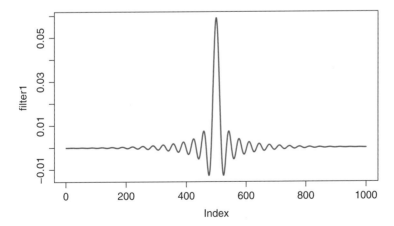

Figure 10.18 Waveform of a finite impulse response filter.

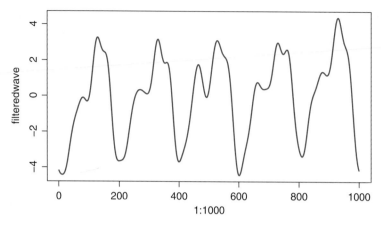

Figure 10.19 The filtered waveform.

Next we will demonstrate Fourier analysis in two dimensions. However there is a small issue that needs dealing with first. Recall that the 1D amplitude spectrum is mirrored about its centre, with the lowest frequencies at the extremes, and the highest frequencies in the centre. Well, this is the opposite of how 2D amplitude spectra are conventionally plotted— with the lowest frequencies in the centre (see Figure 10.9). To represent the 2D spectrum in the conventional way, we need to perform an operation called the *quadrant shift*. In brief, this involves switching the top left and bottom right quadrants of the spectrum, and the top right and bottom left quadrants. This means that the low spatial frequencies previously represented in the corners of the spectrum are now represented in the centre. In many programming languages there is a built-in function to implement the quadrant shift, but in *R* we need to define the following single line function:

```
fftshift <- function(im) {im * (-1)^(row(im) + col(im))}
```

My grasp of imaginary numbers is not sufficient to understand exactly what this function is doing, so I defer to the mathematical wizards who came up with it and trust that it does the job.

We first load the image in from a file using the *readJPEG* function from the *jpeg* package. The image is stored as a 512x512x3 matrix. The 512x512 part is the size of the image in pixels (in the x and y directions), and the third dimension contains three colour channels: red, green, and blue. We will just use the information in the red colour channel and discard the others, so that our image is black and white.

```
library(jpeg)
bughouse <- readJPEG('images/bughouse.jpg')
bughouse <- bughouse[,,1]
```

The image is now stored as a 512x512 matrix of pixel intensities. We can take the Fourier transform, applying the quadrant shift, as follows:

```
bugspectrum <- fft(fftshift(bughouse))
```

Now that we have Fourier-transformed the image, we can do some more aggressive filtering. Perhaps we could include only oblique orientations within a narrow range of spatial frequencies, using an oriented bandpass filter like those in Figure 10.16. These are created in the Fourier domain using a short function called *offsetgaus* as follows:

```
offsetgaus <- function(n,std,x,y){
i <- matrix(data = (1-(n/2)):(n/2), nrow=n, ncol=n)
j <- t(apply(i,2,rev))
h <- exp(-(((i+x)^2) / (2 * std^2)) - (((j+y)^2) / (2 * std^2)))
return(h)}

# create a Gabor filter using two Gaussian functions, offset from the
origin
g <- offsetgaus(512,8,20,20) + offsetgaus(512,8,-20,-20)
```

The filter and its spatial transform will look very similar to those shown in Figure 10.16. We then multiply the filter by the Fourier spectrum of the image, and take the inverse transform, with a bit of quadrant shifting sleight of hand. Finally, we rescale the luminances to between 0 and 1, and then plot the resulting image (see Figure 10.20).

```
# apply the filter and inverse transform
filteredimage <- Re(fftshift(fft((bugspectrum*g), inverse=TRUE)))
filteredimage <- filteredimage - min(filteredimage)   # scale the
luminances
filteredimage <- filteredimage/max(filteredimage)   # to between 0 and 1

# plot the filtered and original image side by side
plot(x=NULL,y=NULL,xlim=c(0,4.5),ylim=c(-1,1),axes=FALSE, ann=FALSE,
lwd=2)
rasterImage(filteredimage,0,-1,2,1)
rasterImage(bughouse,2.5,-1,4.5,1)
points(3.75,0.25,pch=1,col='blue',cex=8,lwd=8)
```

The clearest feature in the filtered image is a diagonal plank of wood, which has the most left-oblique energy. This is circled in blue in the original image (right panel of Figure 10.20).

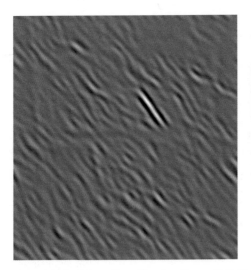

Figure 10.20 Oblique filtered bug hotel image (left). The strongest feature corresponds to a diagonal plank of wood, circled in blue in the original image (right).

This section has provided example code for performing Fourier analysis and filtering in both one and two dimensions. The practice questions below test your understanding with further examples.

Practice questions

1. What operation will convert the Fourier spectrum to a waveform?
   A) The Fourier transform
   B) The quadrant shift
   C) The inverse Fourier transform
   D) The phase spectrum

2. What determines the Nyquist limit of a signal?
   A) The sample rate
   B) The frequency with the largest amplitude
   C) The signal duration
   D) The sample rate multiplied by the duration

3. What determines the frequency resolution of the Fourier spectrum?
   A) The sample rate
   B) The number of samples in the signal
   C) The signal duration
   D) The sample rate multiplied by the duration

4. What units do we use to measure Fourier phase?
   A) Hertz
   B) The units the dependent variable is measured in
   C) Cycles per degree
   D) Degrees

5. Which type of filter would we use to remove only high frequencies from a signal?
   A) A high pass filter
   B) A low pass filter
   C) A bandpass filter
   D) A notch filter

6. Which line of code will convert the data object *waveform* to its Fourier amplitude spectrum?
   A) angle(fft(waveform))
   B) abs(fft(waveform,inverse=TRUE))
   C) abs(fft(waveform))
   D) angle(fftshift(waveform))

7. Which pair of operations are equivalent?
    A) Convolution in the temporal domain and division in the Fourier domain
    B) Squaring in the temporal domain and subtraction in the Fourier domain
    C) Addition in the temporal domain and convolution in the Fourier domain
    D) Convolution in the temporal domain and multiplication in the Fourier domain

8. In Fourier space, the highest spatial frequencies are traditionally represented:
    A) In the corners
    B) In the upper half
    C) In the centre
    D) In the lower half

9. What will the line of code *angle(fft(waveform))* do?
    A) Return the amplitude spectrum
    B) Return the phase spectrum
    C) Return the full Fourier spectrum
    D) Return a smoothed waveform

10. Which line of code will return a filtered version of the data object *signal*?
    A) abs(fft(fft(signal)*filter,inverse=TRUE))
    B) Re(fft(fft(signal)*filter,inverse=TRUE))
    C) Re(fft(fft(signal,inverse=TRUE)*filter))
    D) abs(fft(signal*filter,inverse=TRUE))

Answers to all questions are provided in the answers to practice questions at the end of the book.

Multivariate t-tests

Many widely used statistics are *univariate* in nature, in that they involve a single dependent variable (outcome measure). If you have more than one dependent variable, a number of alternative statistical tests are available that can deal with all of the dependent variables at once, rather than running a series of univariate tests. The next four chapters will introduce a selection of these methods, which are referred to as *multivariate* techniques.

Multivariate statistics have some advantages over their univariate cousins. In particular, because they consider more than one dependent variable at a time, they will typically have greater statistical *power* for detecting an effect (see Chapter 5 for an explanation of power). For some research designs, using multivariate methods also means that a single (omnibus) test can be conducted, rather than a series of univariate tests, one for each dependent variable. This helps to avoid issues with multiple comparisons, and the required corrections, which we will discuss in more detail in Chapter 15.

In this chapter we will consider a multivariate extension of the t-test, first introduced by Hotelling (1931). Known as *Hotelling's* T^2, this statistic allows us to compare a set of multivariate observations to a particular value, or to compare two sets of multivariate observations (e.g. from two groups). We will also consider a variant of the T^2 statistic proposed by Victor and Mast (1991), called T^2_{circ}, and discuss situations where this might be used. Because visualizing multivariate data can be challenging, we will introduce the T^2 statistic using bivariate data (i.e. data with two dependent variables), but the mathematics of T^2 extends to any number of dependent variables.

Thinking about and visualizing bivariate data

We will begin by considering some possible forms that bivariate data might take. For these examples, we will refer to our two dependent variables as *x* and *y*, but in principle these can be any two things that we might care to measure. They could be weight and height, for example, or reaction times and accuracy, or any other two dependent measures. They can be in any units, and there is no requirement that the units of the two measures be the same. One key situation that we will discuss later in the chapter is when the components are the real and imaginary terms from a *Fourier transform* (see Chapter 10).

Figure 11.1(a) shows some example bivariate data that are normally distributed about zero in both the *x* and *y* dimensions, and show no correlation between the two measures. These data do not differ significantly from the reference point of 0 (at the origin), and we would ideally like a statistical test that can indicate this. In Figure 11.1(b), the data are displaced in the *x* direction, but still have a mean of 0 in the *y* direction, and in Figure 11.1(c) they are similarly

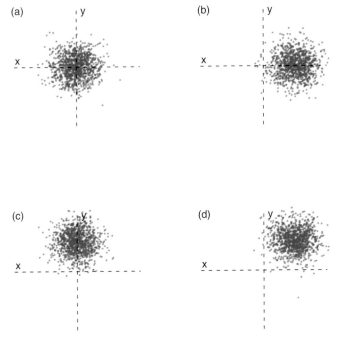

Figure 11.1 Example scatterplots of 1000 samples of uncorrelated bivariate data, with different offsets in the x and y directions.

displaced in the y direction. In principle, we could analyse the data in these panels using a univariate test, simply by discarding the uninformative variable. But this is not ideal, as we have no principled way of knowing in advance which variable (x or y) is going to be the informative one (if we knew, we probably wouldn't bother to measure both!). Finally, in Figure 11.1(d) the data points are displaced in both directions. One can envisage a situation where this offset does not quite reach statistical significance in either the x or y direction when assessed using a univariate test, but if we could somehow take both variables into account we might be more likely to detect an effect.

A further complication is illustrated in Figure 11.2. Here, we see similar arrangements as before, but this time with a positive correlation between the two variables (it could equally well be negative). We will need our statistical test to be able to take any such correlations into account when assessing statistical significance, as they account for a proportion of the variance of each measure. Put another way, when two variables are correlated, we can partly predict one variable from scores on the other. This effectively reduces the number of *degrees of freedom*, which needs to be considered when estimating statistical significance.

The one-sample and paired Hotelling's T^2 statistic

A one-sample Student's t-test calculates the difference between the sample mean and a fixed point (often zero), scaled by the sample variance. Hotelling's T^2 takes a similar approach but with multivariate data. The sample mean is the average across the x and y (and any additional)

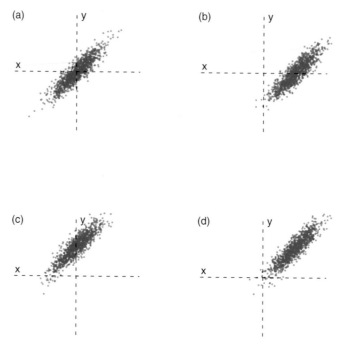

Figure 11.2 Example scatterplots of 1000 samples of positively correlated bivariate data, with different offsets in the x and y directions.

dimensions, also known as the *centroid*. This is shown by the large point in Figure 11.3. The centroid is compared with some other point in the space, for example the origin ($x = 0, y = 0$) in this example. The distance between the two points is the length of the vector that joins them, shown by the black line. The variance term is calculated from the lengths of the residuals. These are the thin grey lines that join the mean to each data point. Also included in the variance term is the covariance between the two variables, which is best thought of conceptually as the correlation between them.

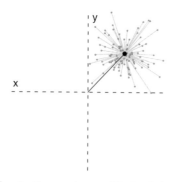

Figure 11.3 Example scatterplot showing the sample mean (black point), vector line between the sample mean and the origin (black line), and residual lines joining each data point to the sample mean (grey lines).

The equation for calculating T^2 is:

$$T^2 = N(\bar{x} - \mu)'C^{-1}(\bar{x} - \mu) \tag{11.1}$$

where N is the sample size, $(\bar{x} - \mu)$ is a vector of differences between the sample mean ($\bar{x}$) and the point we are comparing it to (μ; i.e. the black point and the origin in Figure 11.3), and C is the covariance matrix (and C^{-1} its inverse). The tick symbol (') indicates transposition of the vector. Calculating the inverse covariance matrix is impractical by hand, so it is always done by computer. However I have included the equation here so that you can see the role the covariance matrix plays in calculating the test statistic.

As a reminder, covariance matrices have the following structure:

```
##              x          y
## x         0.904     -0.924
## y        -0.924      1.221
```

The values on the diagonal of the matrix (x,x and y,y) give the variance for each of the two variables (which must always be positive). The off-diagonal values (x,y and y,x) give the covariance between the two variables (note that both these values are identical, and may be negative as in the above example). All of the values are in the original units of measurement—if the matrix is standardized, it becomes a correlation matrix. The covariance matrix fully describes the variance and covariance of a multivariate data set.

By including the inverse covariance matrix in the calculation, the T^2 statistic is effectively *decorrelating* the dependent variables, and rescaling them to have equal variance. This means that increases or decreases in the amount of correlation between the variables do not affect the test's statistical power (all else being equal). It also means that the test is assessing only the difference between the centroid and the comparison point—it is not telling you whether the variables are significantly correlated. Of course you could find this out by running a standard correlation test if you need to know if your variables are correlated.

To determine statistical significance, the T^2 statistic is converted to an equivalent F-statistic by multiplying by a scaling factor based on the number of dependent variables (m) and the sample size (N):

$$F = \frac{N-m}{m(N-1)} T^2 \tag{11.2}$$

The expected F-distribution then has m and $N-m$ degrees of freedom. A p-value can be estimated by comparing the calculated F-statistic to the expected F-distribution, in much the same way as for ANOVA.

For repeated measures designs (where the same participants complete two different conditions), a paired samples version of T^2 is achieved by subtracting each participant's scores across the two conditions, and performing the one-sample test comparing to zero. (It is not always appreciated that for univariate t-tests, a paired samples test is identical to a one-sample test conducted on the differences between the conditions.) Furthermore, the same approach works with an arbitrary number of dependent variables ($m > 2$), making T^2 a multivariate (rather than a bivariate) statistic.

Example: multivariate analysis of periodic EEG data

To demonstrate how to use the T^2 test on real data, we will reanalyse two conditions from a data set reported by Vilidaite et al. (2018). This is an EEG study that used the steady-state visually evoked potential (SSVEP) method. In this paradigm a sensory stimulus oscillates (flickers) at a fixed frequency, and neurons responsive to the stimulus modulate their firing at the same frequency. These modulations can be detected as electrical fluctuations at the scalp using electroencephalography (EEG). Steady-state methods are widely used in research into visual and auditory processing, and have the advantage that they do not require participants to make responses, and so can be used in infants, animals, and patients who are non-verbal.

The typical approach to analysing steady-state data is to take the Fourier transform (see Chapter 10) of the EEG waveform, and look at activity at the stimulus flicker frequency. The signal is represented using complex numbers, which can be plotted as x (real) and y (imaginary) coordinates in a Cartesian space, and analysed using multivariate statistics. Data for 100 participants at two different stimulus levels are shown in Figure 11.4. In panel (a) the stimulus level was 0 (the baseline condition), so we do not expect to see a signal. In panel (b) the stimulus level was 32% contrast, so we anticipate a measurable signal.

We will conduct a one-sample T^2 test on each data set, and then a paired samples T^2 test comparing the two conditions. For the baseline condition there was no significant effect ($T^2 = 1.12$, F(2,98) = 0.56, $p = 0.575$). For the 32% contrast condition there was a highly significant effect ($T^2 = 40.19$, F(2,98) = 19.89, $p < 0.001$). We can also conduct a paired samples T^2 test to compare these two conditions, which again produces a highly significant difference ($T^2 = 42.29$, F(2,98) = 20.93, $p < 0.001$).

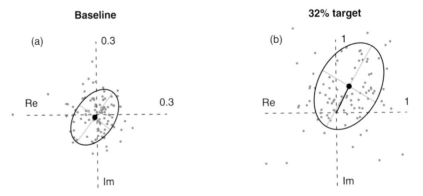

Figure 11.4 Example SSVEP data. Each blue point is an individual participant (N = 100), the black points are the group means, and the orthogonal lines show the eigenvectors of the bounding ellipse. Panel (a) shows data from the baseline condition where no stimulus was shown, and panel (b) shows data from a condition where 32% contrast sine wave grating patches flickered at 7 Hz. Both data sets are from the 7 Hz frequency bin of the Fourier spectrum of the EEG data recorded at the occipital pole. The x-axis represents the real component, and the y-axis the imaginary component of the complex number. Data are from Vilidaite et al. (2018), available online at: **http://doi.org/10.17605/OSF.IO/Y4N5K**

The two-sample (independent) Hotelling's T^2 statistic

An independent two-sample version of the T^2 statistic is also possible, using a slightly modified formula. For this version of the test, our two groups can have different numbers of observations and should be comprised of different individuals, though they must always involve the same dependent variables. The equation for the two-sample T^2 statistic is defined as:

$$T^2 = \frac{N_1 * N_2}{N_1 + N_2} \left(\overline{x_1} - \overline{x_2} \right)' C^{-1} \left(\overline{x_1} - \overline{x_2} \right) \tag{11.3}$$

where $\overline{x_1}$ and $\overline{x_2}$ are the vectors of sample means for the two groups, and N_1 and N_2 are the sample sizes. The covariance matrix (C) is the pooled covariance matrix across the two samples, taking sample size into account:

$$C = \frac{(N_1 - 1)C_1 + (N_2 - 1)C_2}{N_1 + N_2 - m} \tag{11.4}$$

where m is the number of dependent variables, and C_1 and C_2 are the covariance matrices of the two groups. For the two-sample version, the F-ratio is calculated as:

$$F = \frac{N_1 + N_2 - m - 1}{m(N_1 + N_2 - 2)} T^2 \tag{11.5}$$

and the degrees of freedom are given by $df1 = m$ and $df2 = N_1 + N_2 - m - 1$.

Conceptually, this test determines the distance between the centroids of the two groups, taking into account their variances and covariances. It therefore allows pairs of conditions to be compared statistically. If you have more than two conditions, it is possible to run a MANOVA (multivariate analysis of variance)—the multivariate extension of ANOVA. This method is covered in many introductory statistics texts, so we will not discuss it further here. However if you have conducted a MANOVA, the T^2 test can be used for *post hoc* tests, assuming appropriate correction for multiple comparisons is applied (see Chapter 15).

Example: visual motor responses in zebrafish larvae

A study by Liu et al. (2015) measured the locomotor responses of zebrafish larvae to a sudden onset or offset of light. They used an infrared camera to record arrays of 96 larvae simultaneously, between three and nine days after fertilization, and for different wild-type genetic strains. The visual motor response is a widely used assay of neural function that can be used to study development, or assess the effects of different drugs or genetic mutations on the nervous system. The data collected are rich and high dimensional, and can be analysed in many different ways. In the Liu et al. study, the authors calculated a burst duration index for each one second of time, which summarized the proportion of that time window that the animal was moving. They then conducted two-sample Hotelling's T^2 tests using the burst duration index for 30 1-second time windows to compare different developmental time points and different genetic strains (note that the use of 30 time points means that there were 30 dependent variables for most of their tests).

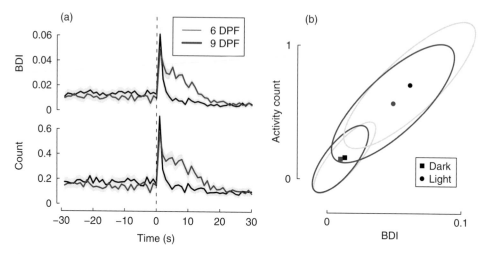

Figure 11.5 Zebrafish larvae visual motor reflex data from Liu et al. (2015). Panel (a) shows the time course for the burst duration index (BDI; upper) and the average activity count (lower), for larvae six (black) and nine (blue) days post-fertilization (DPF). Shaded regions indicate 95% confidence intervals across 192 individuals, and the vertical dashed line indicates light onset. Panel (b) shows the bivariate means across both variables (BDI and activity count) for the 1-second period before (dark, squares) and after (light, circles) the light stimulus onset, again for six (black) and nine (blue) days post-fertilization.

Figure 11.5 shows a reanalysis of a subset of the zebrafish data, which are publicly available at: **https://doi.org/10.7910/DVN/HTXXKW**. The left panel shows the time course for two developmental stages (six and nine days post-fertilization), for two measurement indices—the burst duration index (top) and the burst count (bottom). The right panel illustrates the two measures plotted against each other at two time points (1 second before or after light onset). Two-sample T^2 tests indicate no difference at the time point immediately before stimulus onset (squares; $T^2 = 3.45$, F(2,381) = 1.72, $p = 0.18$), but a significant effect 1 second after the light was presented (circles; $T^2 = 8.85$, F(2,381) = 4.41, $p = 0.01$). This suggests that older larvae have a slightly weaker initial response to light, though it is clear from Figure 11.5(a) that movement persists for longer in the 9-day-old larvae. Overall, the Liu et al. (2015) study is a good example of how multivariate statistics can be used to analyse complex data sets.

The T^2_{circ} statistic

Victor and Mast (1991) proposed a variant of the T^2 statistic called T^2_{circ} (the *circ* is short for circular). This was intended specifically for analysing complex Fourier components like those we encountered in Figure 11.4. The test has some additional assumptions—specifically that the units of the dependent variables have equal variance, and that there is no correlation between them. In other words, the data should conform to a circular cloud of points (as in Figure 11.1) and not an ellipsoidal one (as in Figure 11.2). If these conditions are met, the one-sample version of the statistic is calculated as:

$$T_{circ}^2 = (N-1)\frac{|\bar{x}-\mu|^2}{\Sigma|x_j-\bar{x}|^2}$$

(11.6)

where N is the sample size, $\bar{x}$ is the sample mean, μ is the point of comparison, and x_j represents individual observations. The vertical slash symbols ($||$) denote the absolute value of the numbers inside (i.e. the vector lengths). In words, this equation takes the squared length of the line joining the sample mean to the comparison point (i.e. the black line in Figure 11.3), and divides by the sum of the squared residuals (i.e. the grey lines in Figure 11.3).

Note that, crucially, there is no covariance term in this equation, which makes it substantially simpler to calculate. As with the original T^2 statistic, statistical significance is estimated by comparison with an F-distribution, which for two dependent variables has 2 and $2N-2$ degrees of freedom for $F = NT_{circ}^2$. Repeated measures and two-sample versions are also possible.

Victor and Mast (1991) demonstrate that the T_{circ}^2 statistic can be more sensitive (i.e. have greater power) than Hotelling's T^2 when its assumptions are met. However there is an issue with the false positive rate when the assumptions are violated (i.e. when the variables are correlated or have different variances). I recently proposed a method for testing the assumptions that involves comparing the *condition index* of a data set to that expected by chance (Baker 2021). The condition index is the square root of the ratio of eigenvector lengths (eigenvectors are the axes of a bounding ellipse, and eigenvalues are their lengths: see examples given by the grey lines in Figure 11.4). This functions like other assumption tests, in that a significant result means that the T_{circ}^2 should not be used, and Hotelling's T^2 is a safer alternative.

Mahalanobis distance as an effect size measure for multivariate statistics

Back in Chapter 3, we encountered a statistic called the Mahalanobis distance (Mahalanobis 1936). This was like a multivariate version of a z-score, in that it told us the distance between a single data point and the sample mean, taking into account the sample variance and covariance. For the one-sample situation, we can again use this statistic to calculate the distance between the multivariate sample mean (centroid) and a comparison point (such as the origin). This gives us a standardized measure of distance that is a multivariate generalization of the Cohen's d statistic (see 'Effect size' in Chapter 5). It is therefore an appropriate effect size measure to include when reporting the results of the T^2 and T_{circ}^2 statistics.

For the two-sample situation, there is a variant of the Mahalanobis distance that can be applied for two independent groups, often called the *pairwise Mahalanobis distance*. Just as the two-sample T^2 test combines the variances from the two groups, we must do the same for the pairwise Mahalanobis distance. Indeed, the equation is closely related to the two-sample T^2 equation (equation 11.3), and is given by:

$$D = \sqrt{(\overline{x_1-x_2})'C^{-1}(\overline{x_1-x_2})}$$

(11.7)

where all terms are as defined previously, and C is the pooled covariance matrix calculated using equation 11.4. Note that some implementations of the Mahalanobis distance actually return D^2, which can be converted back to D by taking the square root (as in equation 11.7). As with Cohen's d, the D statistic is standardized so it can be compared across different data sets, studies, and dependent variables, and could in principle be used as an effect size for meta-analysis (see Chapter 6). I strongly recommend reporting it alongside the results of any T^2 or T^2_{circ} test.

Calculating Hotelling's T^2 in R

It is surprisingly hard to find a working implementation of a one-sample Hotelling's T^2 test in R. Several packages exist that contain one, including the *ICSNP* and *MVTests* packages, however both of these are deprecated at the time of writing and do not work with recent versions of R. As part of a paper on analysing periodic data using multivariate statistics (Baker 2021), I created my own package that implements all of the tests we have discussed in this chapter.

The *FourierStats* package is hosted on the code repository *GitHub*, at **https://github.com/bakerdh/FourierStats** (for an overview of GitHub, see 'Version control of analysis scripts' in Chapter 19). Because it is on GitHub rather than the CRAN repository, we need to install it in a slightly different way from normal, using a function called *install_github* from the *devtools* package as follows:

```
install.packages('devtools')
library(devtools)
install_github("bakerdh/FourierStats")
library(FourierStats)
```

The package contains a function called *tsqh.test* that can calculate one-sample, two-sample, and repeated measures versions of Hotelling's T^2 test. Let's assume that our first data set is stored in an $N \times 2$ array called *data*:

```
head(data)
##                 [,1]         [,2]
## [1,]   0.9796330   1.83289050
## [2,]   0.5984391  -0.65855164
## [3,]   2.3306366   1.44780388
## [4,]  -0.2141754  -0.06080895
## [5,]  -0.1746285  -0.28229951
## [6,]  -0.7105413  -0.68882363
```

We can conduct a one-sample T^2 test using the *tsqh.test* function as follows:

```
tsqh.test(data)
##           tsq    Fratio df1 df2          pval                    method
## 1 40.18827 19.89116    2  98 5.618178e-08 One-sample T-squared test
```

If we wish to compare to a specific point in the two-dimensional space, we can define this by adding the optional argument *mu*:

```
tsqh.test(data,mu=c(0.25,0.25))
##           tsq   Fratio df1 df2          pval                    method
## 1 7.933029 3.926449    2  98 0.02288985 One-sample T-squared test
```

For the one-sample case, we can calculate the Mahalanobis distance using the built-in *mahalanobis* function from the *stats* package:

```
D2 <- mahalanobis(c(0,0),center=colMeans(data),cov=cov(data))
sqrt(D2)
## [1] 0.6339422
```

We are passing to the function the two points we wish to compare—the data centroid (calculated using the *colMeans* function), and the comparison point (0,0). We also provide the covariance matrix from the data (*cov(data)*). Note that the function returns the squared distance, so we must take the square root to find *D*. If we want to compare to a different point, we can change the input to the first argument, for example:

```
D2 <- mahalanobis(c(0.25,0.25),center=colMeans(data),cov=cov(data))
sqrt(D2)
## [1] 0.2816563
```

To compare two groups, we can again use the *tsqh.test* function, providing it with both data sets, and specifying either a paired or unpaired test:

```
tsqh.test(data,y=baseline,paired=TRUE)
##         tsq   Fratio df1 df2         pval              method
## 1 42.28931 20.93107   2  98 2.696304e-08 Paired T-squared test
tsqh.test(data,y=baseline,paired=FALSE)
##         tsq   Fratio df1 df2         pval              method
## 1 41.21264 20.50225   2 197 8.152384e-09 Independent samples T-
squared test
```

Calculating the Mahalanobis distance for the paired samples case again uses the *mahalanobis* function, but this time on the difference between the data sets, comparing to (0,0) (recall that a paired test is identical to a one-sample test on the differences):

```
diff <- data-baseline
D2 <- mahalanobis(c(0,0),center=colMeans(diff),cov=cov(diff))
sqrt(D2)
## [1] 0.6503023
```

For the independent samples (unpaired) case, we instead use the *pairwisemahal* function from the *FourierStats* package. The function expects the data to be stored in a single matrix, with an additional grouping variable to identify which group each observation belongs to. We can combine our two data objects using the *rbind* function, and generate the group indices with the *rep* function:

```
# combine both data sets into a single 200x2 matrix
alldata <- rbind(data,baseline)
# create group labels of 100 1s and 100 2s
grouplabels <- rep(1:2,each=nrow(data))
```

Then both of these new data objects are passed to the *pairwisemahal* function:

```
pairwisemahal(alldata,grouplabels)
##           1         2
## 1 0.0000000 0.9078837
## 2 0.9078837 0.0000000
```

Note that this function returns *D* (like it should!) and not D^2, so there is no need to take the square root. It returns a data object that is structured like a correlation matrix, showing the pairwise distance between each pair of groups. This allows you to pass in any number of groups, and obtain a full matrix of distances.

The *FourierStats* package also contains a function called *tsqc.test*, that implements the T^2_{circ} test. The syntax is identical to that for *tsqh.test*, so these functions can be used interchangeably (though note that *tsqc.test* only works for bivariate data, whereas *tsqh.test* can cope with any number of dependent variables). However, in order to justify running a T^2_{circ} test, we should first test the condition index of each data set. The function *CI.test* runs the condition index test as follows:

```
CI.test(data)
##          CI   N criticalCI        pval
## 1 1.484294 100       1.282 0.0005631189
```

A full explanation of how this test works is given by Baker (2021). However you can think of it as being similar to other assumption tests you might be familiar with (see 'Transforming data and testing assumptions' in Chapter 3), such as Mauchly's test of sphericity, which is used to test the assumptions of repeated measures ANOVA, or Levene's test of homogeneity of variances. Just like these other assumption tests, if the condition index test is significant at $p < 0.05$ (as it is above), then the assumptions of T^2_{circ} are violated, and we should instead run the T^2 test.

These are the basics of how to calculate the T^2 and T^2_{circ} statistics, and the Mahalanobis distance in R. They are quite rarely used tests, and my hope is that by including them here more people will know about and use them in the future. Readers interested in the implementation of the tests are welcome to inspect the code underlying the *FourierStats* package for further insights.

Practice questions

1. Multivariate tests are necessary when you have:
   A) A single independent variable
   B) A single dependent variable
   C) More than one independent variable
   D) More than one dependent variable

2. When is it appropriate to use Hotelling's T^2 statistic?
   A) With two or more dependent variables in any units
   B) With only two dependent variables in any units
   C) With two or more dependent variables, which must have the same units
   D) With only two dependent variables, which must have the same units

3. Hotelling's T^2 statistic takes into account the:
   A) Means and covariances
   B) Means, variances, and covariances
   C) Means and variances
   D) Variances and covariances

4. The significance of a T^2 statistic is determined using:
   A) A normal distribution
   B) An F-distribution

C) A t-distribution

D) A Poisson distribution

5. The degrees of freedom for Hotelling's T^2 depend on:

A) The sample size and number of groups

B) The sample size only

C) The sample size and number of dependent variables

D) The number of groups and the number of dependent variables

6. The T^2_{circ} statistic assumes that the dependent variables:

A) Are uncorrelated and have equal variance

B) Are correlated and have different variances

C) Are uncorrelated but have different variances

D) Are correlated and have equal variance

7. The lengths of the major and minor axes of an ellipse are called the:

A) Eigenvectors

B) Eigenvalues

C) Condition index

D) Eigenmatrices

8. Which effect size would be the most appropriate for summarizing the difference in means between two independent groups of multivariate data?

A) Cohen's d

B) The Mahalanobis distance

C) The pairwise Mahalanobis distance

D) The z-score

9. When running a two-sample Hotelling's T^2 test, how is the covariance matrix calculated?

A) It is the covariance matrix of the first sample

B) It is the covariance matrix of the second sample

C) It is the covariance matrix of the difference between the samples

D) It is the pooled covariance matrix across both samples

10. If the dependent variables are correlated, what effect does this have on the power of the Hotelling's T^2 test?

A) There is no effect on the power

B) The power will generally increase

C) The power will generally decrease

D) It will depend on the specific data set

Answers to all questions are provided in the answers to practice questions at the end of the book.

12 Structural equation modelling

Structural equation modelling (SEM) is a technique that allows us to make sense of the relationships between different variables. This is achieved by creating a model that specifies how the variables are connected to each other, and how they relate to hidden internal constructs (called *latent variables*) that we cannot measure directly (without any latent variables, the method is called *path analysis*). Examples of latent variables are things like intelligence, belief, political leanings, personality, quality of life, and nationality or other group affiliation. These are concepts that we might be interested in understanding, but can only infer by measuring more explicit observable variables—things like responses on a test or questionnaire, income, health records, or voting behaviour. Structural equation modelling is often what researchers really want to do when they run multiple correlations between several different pairs of variables. It allows us to explicitly compare different models that might reflect competing theories or hypotheses about a situation. Because structural equation modelling is used in situations with many different measures (e.g. multiple dependent variables), it is an example of a *multivariate* technique. It is best introduced using a concrete example.

How are different mental abilities related?

Our example uses a classic data set from the literature on the testing of mental abilities. The Holzinger and Swineford (1939) data set contains test scores from around 300 teenagers at two different schools, each of whom completed a series of tests measuring performance for different tasks. There were 26 different tests measuring different facets of mental ability, including visual ability, literacy, and performance under timed conditions. For our example, we will consider a subset of nine of the tests from the full study. Here is a snippet of the data set:

```
##   id sex ageyr agemo  school grade       x1   x2    x3       x4
## 1  1   1    13     1 Pasteur     7 3.333333 7.75 0.375 2.333333
## 2  2   2    13     7 Pasteur     7 5.333333 5.25 2.125 1.666667
## 3  3   2    13     1 Pasteur     7 4.500000 5.25 1.875 1.000000
## 4  4   1    13     2 Pasteur     7 5.333333 7.75 3.000 2.666667
## 5  5   2    12     2 Pasteur     7 4.833333 4.75 0.875 2.666667
## 6  6   2    14     1 Pasteur     7 5.333333 5.00 2.250 1.000000
```

```
##       x5        x6        x7    x8        x9
## 1 5.75 1.2857143 3.391304 5.75 6.361111
## 2 3.00 1.2857143 3.782609 6.25 7.916667
## 3 1.75 0.4285714 3.260870 3.90 4.416667
## 4 4.50 2.4285714 3.000000 5.30 4.861111
## 5 4.00 2.5714286 3.695652 6.30 5.916667
## 6 3.00 0.8571429 4.347826 6.65 7.500000
```

In the above output, the first six columns give demographic data about the participants, including age, sex, school year, and school attended. These are not of particular interest for the analysis we have in mind. The remaining columns contain the nine dependent measures, which correspond to the tests described in Table 12.1.

The nine tests probe different aspects of mental ability, from basic perception through to numerical and linguistic functions. We can summarize the relationships between the variables by generating a covariance matrix:

```
round(cov(HolzingerSwineford1939[,7:15]),digits=2)
##       x1    x2    x3    x4    x5    x6    x7    x8    x9
## x1 1.36  0.41  0.58  0.51  0.44  0.46  0.09  0.26  0.46
## x2 0.41  1.39  0.45  0.21  0.21  0.25 -0.10  0.11  0.24
## x3 0.58  0.45  1.28  0.21  0.11  0.24  0.09  0.21  0.38
## x4 0.51  0.21  0.21  1.36  1.10  0.90  0.22  0.13  0.24
## x5 0.44  0.21  0.11  1.10  1.67  1.02  0.14  0.18  0.30
## x6 0.46  0.25  0.24  0.90  1.02  1.20  0.14  0.17  0.24
## x7 0.09 -0.10  0.09  0.22  0.14  0.14  1.19  0.54  0.37
## x8 0.26  0.11  0.21  0.13  0.18  0.17  0.54  1.03  0.46
## x9 0.46  0.24  0.38  0.24  0.30  0.24  0.37  0.46  1.02
```

The covariance matrix quantifies the relationships between the dependent variables. Covariance measures joint variability, or the tendency for two variables to increase or decrease together. Actually, the covariance matrix is often hard to interpret, because the values are not

Table 12.1 Summary of variables in the Holzinger and Swineford data set.

Variable name	Test content
x1	Visual perception
x2	Cubes
x3	Lozenges
x4	Paragraph comprehension
x5	Sentence completion
x6	Word meaning
x7	Speeded addition
x8	Speeded counting
x9	Speeded discrimination

standardized and so they depend on the units of each dependent variable. Instead, it can be more helpful to look at the correlation matrix, which is a standardized version of the covariance matrix:

```
round(cor(HolzingerSwineford1939[,7:15]),digits=2)
##        x1      x2    x3    x4    x5    x6     x7    x8    x9
## x1  1.00   0.30  0.44  0.37  0.29  0.36   0.07  0.22  0.39
## x2  0.30   1.00  0.34  0.15  0.14  0.19  -0.08  0.09  0.21
## x3  0.44   0.34  1.00  0.16  0.08  0.20   0.07  0.19  0.33
## x4  0.37   0.15  0.16  1.00  0.73  0.70   0.17  0.11  0.21
## x5  0.29   0.14  0.08  0.73  1.00  0.72   0.10  0.14  0.23
## x6  0.36   0.19  0.20  0.70  0.72  1.00   0.12  0.15  0.21
## x7  0.07  -0.08  0.07  0.17  0.10  0.12   1.00  0.49  0.34
## x8  0.22   0.09  0.19  0.11  0.14  0.15   0.49  1.00  0.45
## x9  0.39   0.21  0.33  0.21  0.23  0.21   0.34  0.45  1.00
```

Figure 12.1 shows the same correlation matrix in a graphical format. The matrix shows generally positive correlations between different combinations of variables. The strongest of these ($r = 0.73$) is between x4 and x5—the paragraph comprehension and sentence completion tasks—and there appears to be a cluster of high correlations involving x4, x5, and x6 in the centre of the matrix. But even so, just from inspecting the correlation matrix it is rather hard to understand the structure of the data set.

An alternative approach is to construct a hypothetical model of the potential relationships. One very simple model is that a single underlying factor determines performance on all tasks.

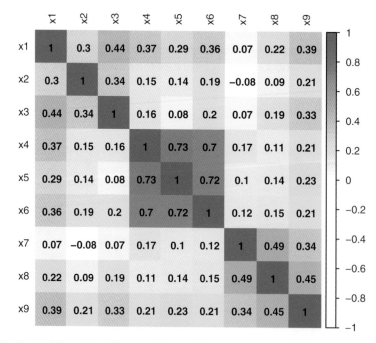

Figure 12.1 Graphical illustration of the correlation coefficients in the Holzinger and Swineford data set.

This general intelligence, or *g*, factor is widely discussed in the literature on human cognitive ability (Spearman 1904). It is the classic example of a *latent variable*—a construct that we hypothesize might exist, but we cannot measure directly. This model can be expressed diagrammatically, as shown in Figure 12.2.

The path diagram shown in Figure 12.2 has several key features. The nine dependent variables from the Holzinger and Swineford data set are shown in square boxes. In the centre is the latent variable g, shown in a circle. These shapes are the accepted conventions in SEM—squares or rectangles contain measured variables, and circles or ovals contain latent variables. The arrows joining the variables indicate the relationships between them, such as factor loadings, or covariances. Notice that the arrow connecting g to x1 is dashed—this indicates that the model weights are standardized relative to this covariance. Finally, the double-headed arrows that loop round on each variable represent the residual error that cannot be explained by the other relationships in the model.

An alternative model might be to propose that there are several latent variables, which map onto specific abilities that are probed by more than one test. For example, we might propose a latent variable for the visual tasks (x1–x3), another for the literacy tasks (x4–x6) and a final one for the timed tasks (x7–x9). We could allow interdependencies (i.e. correlations) between these three latent variables, and represent the model with the diagram in Figure 12.3.

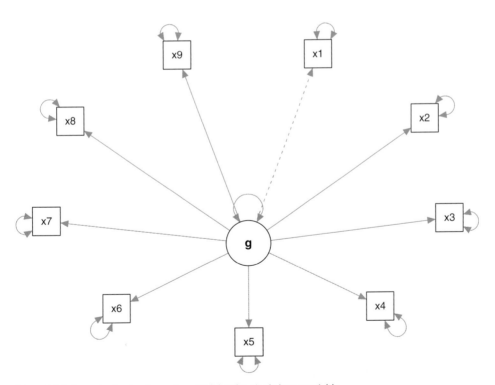

Figure 12.2 Example structural equation model with a single latent variable.

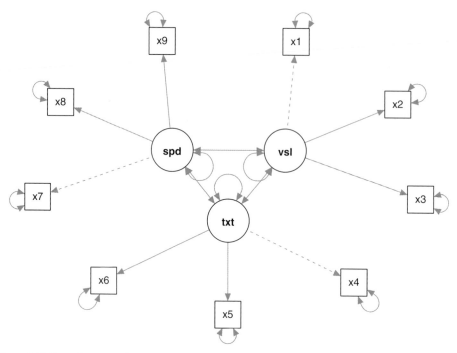

Figure 12.3 Example structural equation model with three latent variables.

Testing hypotheses using data with structural equation modelling

We can think of the two model diagrams shown in Figures 12.2 and 12.3 as explicit hypotheses about how different variables might be related to each other. Structural equation modelling is a statistical technique that allows us to test how well such models explain a given data set. This is a very powerful framework that can uncover (and quantify) the underlying relationships between different measures. We could, for example, see which of the above models gives the best quantitative description of the data set. There might also be a case for altering the connections between different nodes in a model to obtain a better fit; that could change our views on how different variables are related. The following sections will go through four stages involved in SEM, before discussing some general issues worth being aware of when conducting this type of analysis.

SEM stage 1: model specification

The model specification stage is broadly what we have just done. We consider the possible relationships between the variables we have measured (or are planning to measure), and one or more hypothesized latent variables. Note that at this stage we do not say anything about the magnitude of any relationships—these are estimated later. It can often be helpful to create diagrams like the ones in Figures 12.2 and 12.3 when designing a model, as a guide to thinking about possible relationships. This is one of the real strengths of structural equation modelling, as it allows us to make theories and hypotheses explicit by instantiating them in a model, which we then go on to test empirically.

The models we create are limited only by our imagination and our theoretical understanding of the topic (though it is important to have a large enough sample size, and to fulfil some other constraints on model design that we will describe later). It might seem that there are many possible degrees of freedom when designing a model like this. However, usually we will be guided by previous studies, and our intuitions about how different variables might be related. If we have designed the study that generated the data set being analysed, it is likely that we included measures because we had some sort of expectation about how they would be related. If we really have no idea about how to design a model, there is a technique called *exploratory factor analysis* that can try to derive the relationships for us. However this is beyond the scope of this chapter, and is perhaps less well suited to hypothesis-driven research.

SEM stage 2: model identification

Once we have specified our model, we then check that it is suitable for conducting SEM. This involves a process called *model identification*, where we check that the degrees of freedom in the model (known as the number of *free parameters*) does not exceed the degrees of freedom in the data set (known as the number of *data points*). If there are more free parameters than data points, the model is *under-identified*. This is a problem, because we cannot calculate a unique numerical solution for each of the model parameters, and SEM cannot proceed. If there are more data points than free parameters, the model is *over-identified*: this is what we are hoping for, and so we can proceed to the next stage. If the two numbers are equal, the model is referred to as *just identified*. This means that the model is describing the data set, but not simplifying it at all, and so may be less able to generalize beyond the current data set. It is also generally recommended that there are at least three measured variables for each latent variable (for a detailed explanation, see chapter 9 of Kline 2015).

The number of data points can be determined from the number of measured variables in our data set, according to: $N = m(m+1)/2$, where m is the number of variables. So, for our example data set we have nine variables (x1–x9), and $N = 9*(9+1)/2 = 45$. This turns out also to be the number of unique entries in the *covariance matrix* (the matrix that calculates the covariance between each pair of variables, which we inspected earlier).

The number of free parameters in a model is the sum of:

- the total number of latent variables
- the number of error terms on the measured variables
- any covariances between measured variables

It does **not** include links between the latent variables and the measures, or the error terms on the latent variables. Our single latent variable model (Figure 12.2) has one latent variable, and nine error terms on the measured variables, so it has 10 free parameters. The three-latent variable model (Figure 12.3) has three latent variables, and nine error terms, so it has 12 free parameters. We could potentially specify further covariances between data points, which would add to the number of free parameters. But for now, both of these models have far more data points than free parameters, so both are safely over-identified (as will typically be the case for data sets with many measures). Notice that model identification does not depend on the number of cases (i.e. participants) included in the data set, only on the structure of the data set and the model.

SEM stage 3: model evaluation

Once the model has been specified and identified, it is fitted to the data set. This process involves estimating values of the parameters (variances and covariances) that give the best description of the data set, and is done automatically using computer software. Formally, we are looking for values that produce a model covariance matrix that is as close as possible to the empirical covariance matrix (calculated from the data). This involves conceptually very similar procedures to those described in Chapter 9 on function optimization techniques.

There are several different methods of parameter optimization. The most common is called *maximum likelihood estimation*, which involves finding the parameter values that are most likely (in a probabilistic sense) to have resulted in the observed data. This approach generally rests on the parametric assumption that the data are normally distributed. There are other methods, including least squares fitting, asymptotically distribution-free (which has fewer assumptions), and various other scaled and corrected measures.

Once fitted, the model is traditionally assessed using a chi-square statistic. Somewhat counterintuitively, a *non-significant* chi-square statistic indicates a good fit, because the statistic is comparing model and data, so the null hypothesis (that they do not differ) indicates a good fit. However, as with other significance tests, this turns out to be highly dependent on sample size (see Chapter 5), and with large samples (e.g. N > 400) will often be statistically significant even when the model fit is actually quite good. To address this, several alternative fit indices have been developed. Some of these (for example, the Bentler–Bonett index, Comparative fit index, Bollen index, and McDonald index) indicate a good fit when they have values near 1. Conventions about exactly what values are considered 'good' will differ across disciplines, but 0.9 is often acceptable. For other fit estimates, such as the root mean square error or measures of residual variance, a low value near 0 indicates a good fit. Direct comparisons of different fit indices are discussed by Cangur and Ercan (2015). When reporting the outcome of SEM, it is typical to report several fit indices to give a complete picture of the model's performance.

A summary of SEM output might look something like this:

```
## lavaan 0.6-5 ended normally after 35 iterations
##
##    Estimator                                         ML
##    Optimization method                           NLMINB
##    Number of free parameters                         21
##
##    Number of observations                           301
##
```

```
## Model Test User Model:
##
##    Test statistic                              85.306
##    Degrees of freedom                              24
##    P-value (Chi-square)                         0.000
##
## Model Test Baseline Model:
##
##    Test statistic                             918.852
##    Degrees of freedom                              36
##    P-value 0.000
##
## User Model versus Baseline Model:
##
##    Comparative Fit Index (CFI)                  0.931
##    Tucker-Lewis Index (TLI)                     0.896
##
## Loglikelihood and Information Criteria:
##
##    Loglikelihood user model (H0)            -3737.745
##    Loglikelihood unrestricted model (H1)    -3695.092
##
##    Akaike (AIC)    7517.490
##    Bayesian (BIC) 7595.339
##    Sample-size adjusted Bayesian (BIC)       7528.739
##
## Root Mean Square Error of Approximation:
##
##    RMSEA    0.092
##    90 Percent confidence interval - lower       0.071
##    90 Percent confidence interval - upper       0.114
##    P-value RMSEA <= 0.05                        0.001
##
## Standardized Root Mean Square Residual:
##
##    SRMR                                         0.065
##
## Parameter Estimates:
##
##    Information                               Expected
##    Information saturated (h1) model        Structured
##    Standard errors                          Standard
##
## Latent Variables:
##                      Estimate  Std.Err  z-value  P(>|z|)
##    visual =~
##      x1              1.000
##      x2              0.554    0.100    5.554    0.000
##      x3              0.729    0.109    6.685    0.000
##    textual =~
##      x4              1.000
```

```
##    x5                      1.113    0.065    17.014    0.000
##    x6                      0.926    0.055    16.703    0.000
##  speed =~
##    x7                      1.000
##    x8                      1.180    0.165     7.152    0.000
##    x9                      1.082    0.151     7.155    0.000
##
## Covariances:
##                        Estimate  Std.Err  z-value  P(>|z|)
##    visual ~~
##      textual             0.408    0.074     5.552    0.000
##      speed               0.262    0.056     4.660    0.000
##    textual ~~
##      speed               0.173    0.049     3.518    0.000
##
## Variances:
##                        Estimate  Std.Err  z-value  P(>|z|)
##    .x1                   0.549    0.114     4.833    0.000
##    .x2                   1.134    0.102    11.146    0.000
##    .x3                   0.844    0.091     9.317    0.000
##    .x4                   0.371    0.048     7.779    0.000
##    .x5                   0.446    0.058     7.642    0.000
##    .x6                   0.356    0.043     8.277    0.000
##    .x7                   0.799    0.081     9.823    0.000
##    .x8                   0.488    0.074     6.573    0.000
##    .x9                   0.566    0.071     8.003    0.000
##    visual                0.809    0.145     5.564    0.000
##    textual               0.979    0.112     8.737    0.000
##    speed                 0.384    0.086     4.451    0.000
```

There is quite a lot of information in the above output, so we will go through it one section at a time. The first line tells us the version of the *lavaan* software that we are using (*lavaan* is an R package we will describe later in the chapter), and how the fitting proceeded, here with 35 iterations to optimize the model (see Chapter 9 for more details on model fitting). The next section tells us that a maximum likelihood (ML) estimator was used, along with the *nlminb* optimization method. We also learn that the model had 21 free parameters, and there were 301 observations (i.e. participants in the data set).

Next, the sections headed *Model Test User Model* and *Model Test Baseline Model* give us the results of chi-square tests for the model fit and for a baseline (null) model in which co-variances are all fixed at 0. We should expect the model we designed to do better than the baseline model, and indeed we see that it has a smaller chi-square test statistic, indicating a closer fit to the data. For this example both tests are significant; recall that a significant chi-square test can indicate a poor fit to the data, but that as discussed above this is hard to evalu-ate because of the confounding effect of sample size on significance. The following section of the output compares the model to the baseline using the Comparative Fit Index and the Tucker–Lewis Index. Both of these values are quite high, around 0.9, indicating that the model we designed gives a better fit than the baseline model.

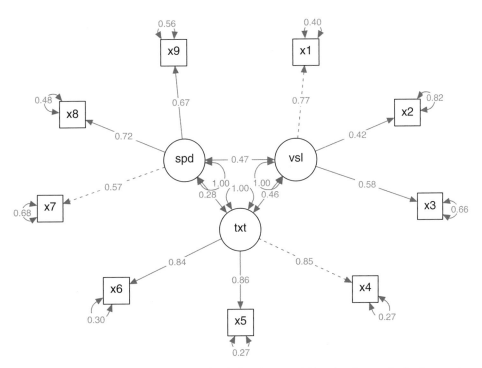

Figure 12.4 Example structural equation model with three latent variables, showing standardized parameter estimates.

The three subsequent sections of the output report additional measures of goodness of fit, including the log-likelihood, the Akaike information criterion (AIC), the Bayesian information criterion (BIC), and the root mean square (RMS) error. These values are particularly useful for comparing between different possible models, as we will describe in more detail later in this chapter.

The final sections of the output show parameter estimates for the latent variables, covariances, and variances. These are somewhat difficult to interpret in table format, so we can add the parameter estimates to the path diagram to give a numerical indication of the strength of the links between variables (see Figure 12.4). This can be done using standardized or unstandardized values. In general, standardized values are more useful, as the values are then similar to correlation coefficients. The fitted parameters show high loading of individual measures onto the three latent variables (coefficients between 0.42 and 0.86), and somewhat smaller correlations between the latent variables (0.28 to 0.47).

SEM stage 4: model modification

The final stage of SEM is to consider possible modifications to the model that might improve its description of the data. To do this, parameters can be added or removed (or both). The change in fit when parameters are added is assessed by the *Lagrange Multiplier test* (sometimes called the *score test*). This is based on evaluating the gradient of the likelihood function (the function that maps between model parameter values and the probability of the data), and if the test is significant for a particular parameter we should consider adding it. Be aware that this can cause problems for theory testing and interpretation if the new parameter does not make sense. The *Wald test* does something similar, but to check if any parameters can be removed from the model without significantly reducing the fit quality (again, the Wald test is based on the likelihood function, but evaluates its gradient slightly differently from the Lagrange Multiplier).

These procedures are conceptually similar to stepwise and backward stepwise entry of predictors in multiple regression, and come with similar pitfalls. Adding many new parameters at once is not advisable, as the parameters may be highly correlated (and therefore not very informative). The order in which parameters are added and removed can also affect the outcome, so care is advised when attempting changes to the model.

Comparing different models

At the start of this chapter we designed two possible models, with different numbers of latent variables (see Figures 12.2 and 12.3). It is natural to ask which of these models gives the best description of the data. We can do this by comparing some of the fit indices between the models. The AIC (Akaike 1974) and BIC (Schwarz 1978) scores take into account the fit quality and the number of free parameters, so the model with the smallest score on these statistics gives the best overall fit, given the number of degrees of freedom. For our example, the model with a single latent variable has an AIC score of 7738, and the model with three latent variables has an AIC score of 7517 (the BIC statistic behaves similarly). This suggests that the model with three latent variables explains the data better. We can also calculate a chi-square difference test to assess whether the difference in model fits is significant. For our two example models, the difference statistic is $\chi^2 = 227$, with a p-value of $p < 0.001$.

Cross-validation on fresh data sets

One concern in SEM is to ensure that the model generalizes beyond the data set it was constructed to explain. This can be confirmed by cross-validating the model on fresh data. If the data set is large enough, we might split the data in two (male and female is sometimes used, assuming these are expected to involve the same relationships between variables). If the data set is not sufficiently large to allow this, collecting additional data might be required. To cross-validate, one approach is to fit the model to the two data sets separately, and then compare the model parameters (coefficient estimates) to check that they are similar. Another option is to fit the model to one data set, and then assess how well the fitted model (with all parameters fixed) describes the other data set.

Power and SEM

SEM is a large sample size technique, and parameter estimates will only be stable with N > 200 participants. A further standard recommendation is to test at least five (and ideally 10) participants per measured variable (while still requiring at least 200 observations). This is because the parameters we are estimating are effectively correlation coefficients, and these are very hard to estimate precisely with small sample sizes. It is possible to conduct power analyses (see Chapter 5) for study designs, often using stochastic simulation (see Satorra and Saris 1985; Wolf et al. 2013). However often these decisions will be limited more by practical concerns such as the resources and time available. Versions of SEM that perform better with small sample sizes have also been developed (Bollen et al. 2007). For a detailed recent treatment of power in SEM see Wang and Rhemtulla (2021).

Dealing with missing data

One issue that can dramatically reduce power occurrs when observations are missing from a data set. If we excluded all participants with at least one missing data point, for some data sets this would substantially decrease the overall sample size. To avoid this situation, it is common practice to replace missing data points with an estimated value. This maintains the sample size and keeps the model as robust as possible. A simple method to do this is to replace a missing data point with the mean score for that variable (referred to as *unconditional mean imputation*). More sophisticated approaches have also been proposed to *impute* (i.e. estimate) the missing values, using methods such as regression or a technique called *expectation maximization* that tries to calculate the most likely estimates of the missing data points (Allison 2003). A popular option is *full information maximum likelihood*, which uses all the available data to estimate the most likely values of missing data points. It assumes that the dependent variables are continuous and normally distributed, and performs well when these assumptions are met (see e.g. Cham et al. 2017).

Doing SEM in *R* using the *lavaan* package

Several packages exist for conducting SEM in *R*, but we will focus on the *lavaan* package (Rosseel 2012), which is an acronym of **la**tent **va**riable **an**alysis. Lavaan contains tools for specifying models, fitting them to data, and assessing the fit. It is very well documented, and there is a detailed tutorial available at: **http://lavaan.ugent.be/tutorial/index.html**.

To specify a model, we must decide on the relationships between the variables, and define any latent variables, using a special syntax. To define a latent variable, we use the =~ operator. For example, defining the *visual* latent variable from the Holzinger and Swineford example as being based on the first three dependent variables looks like this:

```
visual =~ x1 + x2 + x3
```

If two variables are correlated, we can define this using the double tilde:

```
x1 ~~ x2
```

It is also possible to specify intercepts (~1) and regression (~) if desired. The model definition is stored in a single text string. So the three-factor model syntax for the model shown in Figure 12.3 is defined as follows:

```
library(lavaan)
HS.model2 <- 'visual   =~ x1 + x2 + x3
 textual =~ x4 + x5 + x6
 speed   =~ x7 + x8 + x9'

HS.model2
## [1] "visual   =~ x1 + x2 + x3\n textual =~ x4 + x5 + x6\n speed   =~ x7
+ x8 + x9"
```

The function that parses these model definitions is relatively insensitive to spacing, but note in the output that the carriage returns (i.e. new lines) have been replaced by backslash-*n*. Each new definition should appear on a separate line, and the variable names need to correspond to the variable names in the data set (i.e. the column names of a data frame). The *lavaan* syntax is sufficiently flexible that almost any conceivable structural equation model can potentially be expressed. It is not necessary to specify relationships between latent variables, or to include the error terms—these are added automatically.

Note that the default model specification in *lavaan* is to standardize the factor loading of the first dependent variable for each latent variable to 1 (known as the *marker method*). It is possible to change this behaviour in the model definition, but this is only advised for advanced users and we will not consider it further here.

The model can then be fitted to the data using the *cfa* (confirmatory factor analysis) function, provided that the data set is stored in a data frame. For our example data set, this is achieved as follows:

```
fit <- cfa(HS.model2, data = HolzingerSwineford1939)
```

The *fit* object stores the model definition, a summary of the fitting process, and all the various indices and test statistics. We can request a summary like the example earlier in the chapter, using the generic *summary* function, and specifying that we want to see the fit indices, as follows (I have suppressed the output of this command in order to save space, but it is identical to that shown previously):

```
summary(fit, fit.measures = TRUE)
```

From the output, we can extract the various statistics we might want to report. If we want to compare the fits of two models statistically, we can use the *anova* function as follows:

```
anova(fit,fitG)
## Chi-Squared Difference Test
##
##       Df    AIC    BIC   Chisq Chisq diff Df diff Pr(>Chisq)
## fit   24 7517.5 7595.3  85.305
## fitG  27 7738.4 7805.2 312.264     226.96     3   < 2.2e-16 ***
## ---
## Signif. codes:  0 '***' 0.001 '**' 0.01 '*' 0.05 '.' 0.1 ' ' 1
```

The output of the *anova* function includes the AIC and BIC scores, and the chi-square difference statistic with its accompanying *p*-value (here *fit* is the model with three latent variables, and *fitG* is the model with a single latent variable).

If we want to view the path diagram for the model, we can use an automated plotting function. We need to pass the fit object into the *semPaths* function, which is part of a separate package called *semPlot*. Again, I have suppressed the output, which is identical to that shown in Figure 12.4.

```
library(semPlot)
semPaths(fit,layout="circle",whatLabels="stand",edge.label.cex=1)
```

There are numerous plotting options, explained in the help file for the *semPaths* function. These can be used to change the layout and style of the plot. In these examples I have used the *circle* layout, as this shows the latent variables in the middle of the diagram. Other options include *tree* and *spring*—it is worth checking several of these alternatives to find the most natural and appropriate way to present a given model. For more general discussion of producing attractive and informative figures, see Chapter 18.

Model modification can then be conducted. We first calculate *modification indices* for the factor loadings, which will tell us the effect of removing one parameter on the other parameters in the model. The *modindices* function calculates this information for all possible operators. Since our model does not have any covariances between dependent variables, we will only inspect the links to latent variables (though this does not mean that covariances between dependent variables do not exist—we are just not considering them here).

```
mi <- modindices(fit)
mi[mi$op == "=~",1:4] # display only the indices involving latent
variables
##          lhs op rhs     mi
## 25   visual =~  x4  1.211
## 26   visual =~  x5  7.441
## 27   visual =~  x6  2.843
## 28   visual =~  x7 18.631
## 29   visual =~  x8  4.295
## 30   visual =~  x9 36.411
## 31  textual =~  x1  8.903
## 32  textual =~  x2  0.017
## 33  textual =~  x3  9.151
## 34  textual =~  x7  0.098
## 35  textual =~  x8  3.359
## 36  textual =~  x9  4.796
## 37    speed =~  x1  0.014
## 38    speed =~  x2  1.580
## 39    speed =~  x3  0.716
## 40    speed =~  x4  0.003
## 41    speed =~  x5  0.201
## 42    speed =~  x6  0.273
```

The largest modification index (in the *mi* column) is 36.4, and corresponds to the link between the visual latent variable and the speeded discrimination task. This isn't part of our original model, but we could consider an updated model that includes such a link (see Figure 12.5):

```
HS.model3 <- ' visual  =~ x1 + x2 + x3 + x9
               textual =~ x4 + x5 + x6
               speed   =~ x7 + x8 + x9 '

fit3 <- cfa(HS.model3, data = HolzingerSwineford1939)

semPaths(fit3,layout="circle",whatLabels="stand",edge.label.cex=1)
```

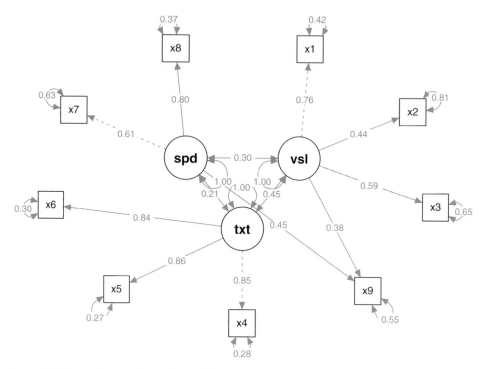

Figure 12.5 Updated structural equation model with an additional link between variable x9 and the visual latent variable.

Note that the new link between visual (*vsl*) and *x9* is now included, and has a substantial coefficient (0.38). We can assess the improvement in fit statistically using the Lagrange Multiplier test in the *lavTestScore* function as follows:

```
a <- lavTestScore(fit, add = 'visual =~ x9')
a$uni
##
## univariate score tests:
##
##         lhs op rhs    X2 df p.value
## 1 visual=~x9 ==   0 36.411  1    0
```

Note that we have passed in our original fit (the *fit* object), and not our updated fit (the *fit3* object). This test gives us a very small *p*-value, suggesting that the modification has significantly improved the model fit. We can also compare the root mean square error values of the two models:

```
fitmeasures(fit,'rmsea')
## rmsea
## 0.092
fitmeasures(fit3,'rmsea')
## rmsea
## 0.065
```

These statistics show us that the root mean square error value is smallest for the updated model (*fit3*), indicating a better fit to the data.

A similar approach can be taken for removing parameters using the Wald test (*lavTestWald* function). This time, let's remove the link with the lowest standardized coefficient—the one between the visual latent variable and *x2*. We achieve this by introducing a weight term onto this parameter in the model definition, and then checking what happens when the weight is set to zero:

```
HS.model4 <- ' visual  =~ x1 + b1*x2 + x3
               textual =~ x4 + x5 + x6
               speed   =~ x7 + x8 + x9 '

fit4 <- cfa(HS.model4, data = HolzingerSwineford1939)
lavTestWald(fit4, constraints = 'b1 == 0')
## $stat
## [1] 30.84248
##
## $df
## [1] 1
##
## $p.value
## [1] 2.79844e-08
##
## $se
## [1] "standard"
```

The Wald test also produces a significant *p*-value, suggesting this change to the model should be investigated more thoroughly. However, on further inspection, it actually produces a larger RMS error (and therefore a worse fit) than our original model:

```
fitmeasures(fit,'rmsea')
## rmsea
## 0.092
HS.model5 <- ' visual  =~ x1 + x3
               textual =~ x4 + x5 + x6
               speed   =~ x7 + x8 + x9 '

fit5 <- cfa(HS.model5, data = HolzingerSwineford1939)
fitmeasures(fit5,'rmsea')
## rmsea
## 0.099
```

The examples in this section provide a basic introduction to the capabilities of structural equation modelling. Of course, as with most of the techniques in this book, there is much more to learn, and many excellent resources are available to help. The book *Principles and Practice of Structural Equation Modelling* by Kline (2015) is an authoritative but readable text that goes into much more detail than we have had space for in this chapter. Another useful resource is the journal *Structural Equation Modeling*, which publishes technical papers on this topic. It is also worthwhile reading some empirical papers that use the methods, to see how they are implemented and reported in your area of interest. Outside of the *R* ecosystem, there are several commercial software packages designed for structural equation modelling, including *LISREL*, *Stata*, *Mplus*, and the *Amos* extension to IBM's SPSS.

Practice questions

1. A latent variable is:
   A) Something we measure in an experiment
   B) Something we manipulate in an experiment
   C) A hypothetical construct that we cannot directly observe
   D) A measured variable that mediates the relationship between other variables

2. When we fit a structural equation model, we are trying to:
   A) Reproduce the precise values of each measurement
   B) Model the covariances between the variables
   C) Find the largest correlation coefficient in the data
   D) Maximize the error between model and data

3. In a path diagram, squares and rectangles represent:
   A) Latent variables
   B) Measured variables
   C) Independent variables
   D) Error terms

4. In a path diagram, a double-headed arrow with both heads pointing to the same measure indicates:
   A) Covariance
   B) Correlation
   C) Regression
   D) Residual error

5. How many data points are there in a data set with five measures?
   A) 5
   B) 13
   C) 15
   D) 25.5

6. In the *lavaan* syntax, the operator ~~ indicates:
   A) A correlation between two dependent variables
   B) A latent variable definition
   C) An intercept
   D) Regression

7. Consider a data set with four measured variables. We attempt to model this using a single latent variable, and significant covariance between two pairs of variables. The model is:
   A) Over-identified
   B) Under-identified
   C) Just identified
   D) It is impossible to say without seeing the data

8. Which of the following fit indices indicates a good fit when it has a value near zero?

A) Bentler–Bonett

B) Chi-square

C) RMSEA

D) McDonald

9. To assess whether a parameter can be removed from a model, we should use the:

A) Chi-square test

B) Lagrange Multiplier test

C) Comparative fit index

D) Wald test

10. Structural equation modelling is typically unstable with sample sizes less than:

A) $N = 200$

B) $N = 300$

C) $N = 400$

D) $N = 1000$

Answers to all questions are provided in the answers to practice questions at the end of the book.

Multidimensional scaling and *k*-means clustering

In this chapter we will discuss two multivariate statistical techniques called *k-means cluster-ing*, and *multidimensional scaling* (MDS). The purpose of *k*-means clustering is to partition multivariate data into a number of clusters, such that observations within a given cluster are more similar to each other than to observations from other clusters. The purpose of multidimensional scaling is to reduce complex multivariate data sets to a smaller num-ber of dimensions (usually two or three) to facilitate graphical representation. These two methods are included in the same chapter because they can often be used together—for example MDS can be used to visualize the results of a multidimensional *k*-means clustering analysis. Both methods are *unsupervised* machine learning techniques, meaning that the algorithms involved try to find the best solution without knowing any ground truth group or category labels.

To give an example of how these methods might be used together, let's imagine that we discover some new varieties of insect in an underground cave. The insects are all about 10 mm long, but vary in their colouring from grey to blue, and in the thickness and angle of the characteristic stripes that cover their backs (see Figure 13.1(a)). You suspect that there might be three distinct species of insect, but how might we test this hypothesis? One option might be to measure all of the key variables from the insects (stripe thickness and angle, colour) and use *k*-means clustering to try to determine the underlying structure of the data set. Because it is challenging to visualize multivariate data with more than two dimensions (i.e. variables), we could then use multidimensional scaling to collapse the data into a two-dimensional space for plotting. The end result might look something like the graph shown in Figure 13.1(b). There is evidence of three primary clusters, for which the example insects in Figure 13.1(a) are prototypical examples.

The *k*-means clustering algorithm

To conduct *k*-means clustering, we must decide in advance how many clusters we are dividing our data into (i.e. the value of *k*). This will depend heavily on the data set we are working with, and what our aims are. The standard algorithm then defines *k cen-troids* (midpoints of a cluster) using *k* randomly selected observations from the data set (coloured points in Figure 13.2(a)). The Euclidean distance (i.e. shortest straight line) between a data point and each centroid is calculated, and the data point is allocated to its nearest cluster (Figure 13.2(b)). Once all data points have been allocated to a cluster,

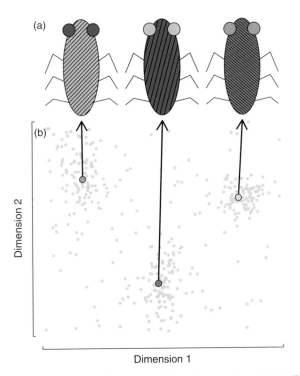

Figure 13.1 Illustration of fictitious insects (a), and a multidimensional scaling of their different traits (b) that suggests three distinct clusters. Note that the dimensions on the x- and y-axes are produced by the multidimensional scaling algorithm, and do not necessarily map directly onto the specific traits of the insects, such as colour and stripe thickness/orientation.

the midpoint of each cluster is then recalculated, using the mean location of all points within the cluster (Figure 13.2(c)). The cluster allocation process repeats iteratively to find a solution with the smallest within-cluster variance, which you can think of as the summed squared error between data points and centroids. Figure 13.2(d) shows the cluster centres on each iteration, as they move from the start points to their eventual locations.

Figure 13.3(a) shows some more complex simulated data (see Chapter 8) generated from five two-dimensional Gaussian distributions. The colours of the points indicate the true groupings, and you can see that there is some overlap between the groups in either the *x* or *y* direction. Figure 13.3(b) shows the *k*-means solution with *k* = 5, where each black point indicates a cluster centroid. The algorithm has identified sensible clusters, though you can see that some data points have been grouped with other points that come from a different generating distribution (i.e. true group). The lines are the residuals that are used to calculate the distance between each data point and its centroid. We can also see what happens if we choose different values of *k*. Figure 13.3(c) shows clustering with *k* = 2, and Figure 13.3(d) shows clustering with *k* = 10. These do produce plausible clusterings, though the original (generating) groupings are not preserved.

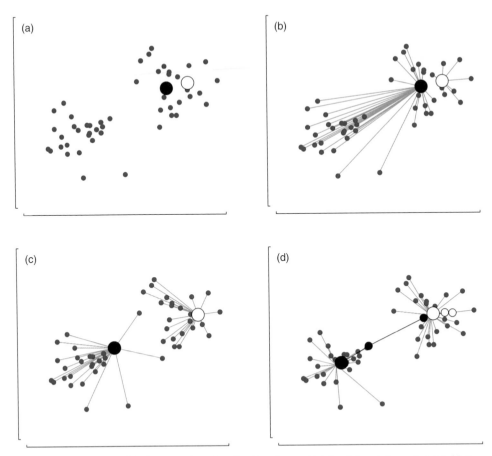

Figure 13.2 Illustration of the *k*-means clustering algorithm. In panel (a), the data points are shown in blue, with the initial centroid estimates in black and white. Panel (b) shows the initial cluster assignments, and residual vectors (lines). Panel (c) shows the revised centroid locations and cluster assignments on the second iteration of the algorithm. Panel (d) shows the path of each centroid across four iterations of the algorithm, with data points assigned to their final clusters.

Comparing different numbers of clusters

What can we do if we don't know how many clusters there should be in our data set? One option is to repeat the clustering for a range of different values of *k*, and pick the one that best describes the data. There are various figures of merit we can use for this. One possibility is to calculate R^2, which tells us the proportion of the total variance explained by cluster membership (in much the same way as is done with regression or ANOVA). However this type of statistic does not take into account the number of degrees of freedom, and so the 'best' fit will be when $N = k$, and each data point is its own cluster!

A better alternative is to compute a statistic that penalizes model complexity (here, the number of clusters). The Akaike information criterion (AIC; Akaike 1974) and Bayesian information criterion (BIC; Schwarz 1978) are two such statistics. Both of them take an error term

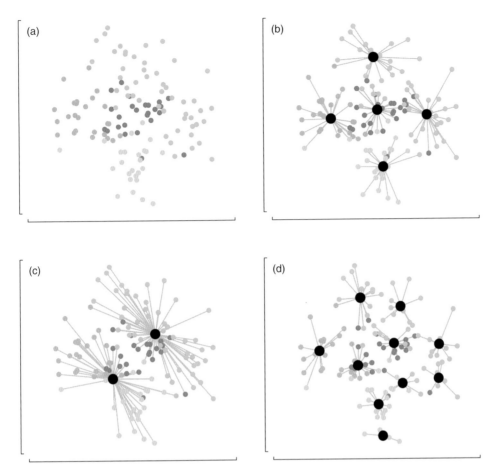

Figure 13.3 Example *k*-means clustering on simulated data. Panel (a) shows data generated from five two-dimensional normal distributions with different means. Panel (b) shows a *k*-means solution with *k* = 5, where black points indicate the centroids, and lines show the residuals for each point. Panels (c) and (d) are for *k* = 2 and *k* = 10 respectively.

such as the residual sums of squares (i.e. the sum of the squared lengths of the residual lines in Figure 13.3(b)–(d)),[1] and add a penalty term. For the AIC, the penalty is *2mk*, where *m* is the number of dimensions (i.e. dependent variables) and *k* is the number of clusters. For the BIC, the penalty is 0.5*log*(*N*)*mk*, where *N* is the number of data points (observations).

For both the AIC and BIC statistics, the best model is the one that produces the lowest score. Generally both statistics behave similarly, meaning that whichever one you use is likely to produce the same outcome, so the choice will not matter for most applications. For the simulated example here, both statistics actually tell us (see Figure 13.4) that *k* = 4 clusters gives the most parsimonious description of the data (despite us actually using five generating distributions).

---

[1] There are variants of AIC and BIC for several different error terms, including the residual sums of squares and the log-likelihood. The key point though is that a penalty is added that is dependent on the number of free parameters in the model, which here are the data dimensions and the number of clusters.

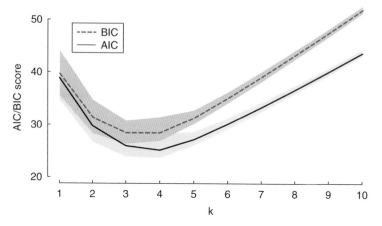

Figure 13.4 Figures of merit as a function of the number of clusters (*k*). Shaded regions indicate 95% confidence intervals for 1000 independent data sets generated from the same underlying distributions.

Example: *k*-means clustering of dinosaur species

To produce an example data set, I compiled a table of dinosaur statistics from the excellent website ZoomDinosaurs.com (my inner 6-year-old had a great time). The data consist of height, length, and weight measurements for 61 dinosaur species, plus whether they were a carnivore or a herbivore (omnivores were classified as carnivores). My very vague childhood dinosaur knowledge characterized herbivores as being generally long and heavy, whereas carnivores were taller and lighter. The data bear this out to some extent, with more blue circles (herbivores) in Figure 13.5 appearing in the top left (heavy and long), and black squares (carnivores) being more prevalent in the lower right (tall and light).

Since we have two types of dinosaur, the first thing we can try is setting *k* = 2 (see Figure 13.5(b)). This doesn't do an amazing job, as there are quite a lot of misclassifications. In particular, there are lots of carnivores included in the upper cluster, which should be mostly herbivores. An alternative might be *k* = 3 (see Figure 13.5(c)), where we could define an

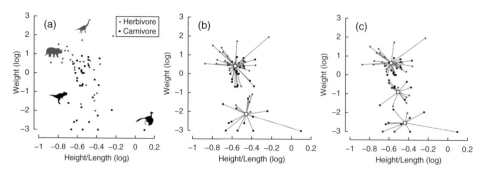

Figure 13.5 *k*-means clustering of dinosaur species. Blue circles represent herbivores, and black squares represent carnivores and omnivores. Panel (a) shows the raw data, with silhouettes indicating some well-known species, and panels (b) and (c) show clusterings for *k* = 2 and *k* = 3.

intermediate cluster. Given the way the data appear, this looks like we now have a 'heavier carnivore' and a 'lighter carnivore' category, as well as a 'herbivore' category. Of course, there are still some errors, but real data are unlikely to cluster perfectly.

If we had the length, height, and weight of a newly discovered species, or one that doesn't appear in our original data set, we might use the cluster arrangement to hazard a decent guess about whether they were a carnivore or a herbivore. For example, my 4-year-old daughter (who knows much more about dinosaurs than I do) really likes the protoceratops, which doesn't feature in the data set. Apparently these weighed 85 kg and were about 0.6 m tall and 1.8 m long. That places them firmly in the top left corner of the plot (at $x = log_{10}(0.6 / 1.8) = -0.48$, $y = log_{10}(85) = 1.93$), closest to the herbivore cluster (they were indeed herbivores).

Variants of *k*-means clustering

There are several variations on the *k*-means clustering algorithm. If our data are skewed or have lots of outliers, we don't necessarily need to use the mean as the measure of central tendency. Instead we could use the median (*k*-medians clustering), which is more robust to outliers and non-normal distributions. Alternatively, for ordinal data the mode (*k*-modes clustering) might be a more appropriate method. This is because the mode is the most common value in a data set, and estimating this does not require metric data, or any assumptions about distributions.

Another variant, called *k*-medioids clustering, has the constraint that the centre of each cluster must be one of the data points, whereas in *k*-means clustering this is only the case for the initial guess. This method is also more robust to outliers than the *k*-means algorithm because the medioid is a plausible (i.e. already observed) data point. Finally, the spherical *k*-means clustering method tries to constrain both the distance and the angle of each point relative to the cluster centroid, so that points are evenly spaced radially. All of these variants work in a broadly similar way, and may be more or less well suited to a particular situation or data type.

There are also several different algorithms for estimating the clusters. In the standard method described at the start of the chapter, the centres of the clusters begin as random samples from the data set, and are iteratively recalculated using the mean of the points allocated to each cluster. This is sometimes referred to as Lloyd's algorithm or the Forgy method (after Lloyd (1982) and Forgy (1965)). One modification to this algorithm, called the *random partition* method, is to assign each data point to a random cluster at the start, instead of choosing *k* data points to form the initial cluster centres. An alternative algorithm proposed by Hartigan and Wong (1979) uses a function minimization approach (see Chapter 9) to determine cluster membership.

Finally, *k*-means clustering is not restricted to working with two dependent variables (i.e. in two dimensions). However, data sets with a large number of variables are very hard to visualize. To help with this, we will next discuss a technique called *multidimensional scaling*, which allows us to collapse multivariate data into a two-dimensional representation (as we saw in our insect example). We can combine this with *k*-means clustering to allow us to visualize clusters even in complex multidimensional data sets.

The multidimensional scaling algorithm

Multidimensional scaling is a technique that is used to reduce the dimensionality of multi-variate data sets to make them easier to visualize. It works by taking a matrix of pairwise dissimilarities (or distances) between data points and mapping it into an abstract space with a defined number of dimensions (usually two). Figure 13.6(a) illustrates five points in a two-dimensional space that are joined by the shortest possible straight lines between each pair of points. Figure 13.6(b) summarizes the distances between each pair of points, with darker blues indicating longer distances. This representation is known as a *dissimilarity matrix*, because it summarizes how far apart the points are in the multidimensional space. If two points are close together, like points 1 and 2, they are quite similar according to the data we have. If two points are far apart, like points 2 and 4, they are quite dissimilar. Note that the dissimilarity matrix is always calculated between pairs of points, regardless of the dimensionality of the data set. This means that the dissimilarity matrix will always have two dimensions, even if the underlying data set has many more dimensions.

The aim of multidimensional scaling is to represent the dissimilarities in a new space with fewer dimensions than the original data set. The idea is that the distances between the points in the lower-dimensional space should correspond as closely as possible to the distances in the original multidimensional space. In other words, the dissimilarity matrices in the original and lower-dimensional spaces should be as similar as possible to each other. The eventual solution will usually be a list of two-dimensional (x,y) coordinates mapping the points in the original data set into the new space. This is achieved by minimizing a statistic called the *strain* (or in some variants the *stress*). The strain is a loss function based on the Euclidean distances between points.

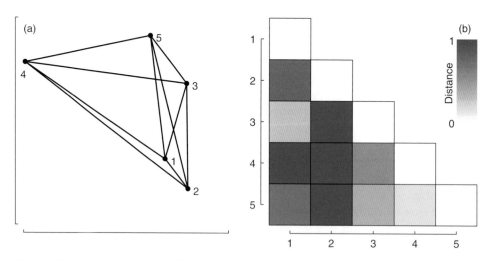

Figure 13.6 Illustration of pairwise Euclidean distances between a set of five points in two dimensions. The lengths of the lines in panel (a) show the Euclidean distances, which are expressed as blue level intensity in panel (b).

RGB   RGBα

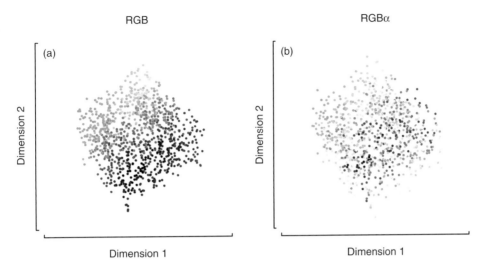

Figure 13.7 Example multidimensional scaling with random colour vectors. Panel (a) shows a two-dimensional solution for three-dimensional RGB vectors, and panel (b) is the solution for four-dimensional RGB-alpha vectors. The x- and y-positions of each point are determined by multidimensional scaling, and the colour is the original colour for each data point. In both cases, similar colours group together.

A good way to illustrate the results of multidimensional scaling is to use random colour vectors. Colour is defined using mixtures of the red, green, and blue pixels on a display. We can therefore create random RGB vectors, and use MDS to reduce from three to two dimensions for plotting. This is shown in Figure 13.7(a), where colours of a similar hue end up being grouped together. A variant in Figure 13.7(b) includes a fourth dimension, the *alpha* (transparency) setting. In this plot the different hues still group together, but the transparency information is clearly being factored in too, for example by placing more transparent points nearer the lower right edge of the cloud.

The starting data for MDS will be an $N \times m$ matrix, where m is the number of dependent variables (dimensions). For example:

```
##                [,1]        [,2]        [,3]         [,4]
## [1,] 0.692666520 0.98132464 0.08029358 0.6964390816
## [2,] 0.802897572 0.13823851 0.93906565 0.6016717958
## [3,] 0.797127023 0.88163599 0.66954995 0.6361913709
## [4,] 0.007445487 0.06651651 0.37043385 0.1724689843
## [5,] 0.621347463 0.68464959 0.11980545 0.0002071382
## [6,] 0.318114384 0.57204209 0.33323977 0.8083065613
```

The output will be an $N \times 2$ matrix, where the two dimensions are x and y coordinates:

```
##              [,1]        [,2]
## [1,] -0.15160650  0.6517739
## [2,]  0.41646691 -0.2108153
## [3,] -0.03000302  0.3792663
## [4,] -0.05064061 -0.6095281
## [5,]  0.09436012  0.1585279
## [6,] -0.22697391  0.1237203
```

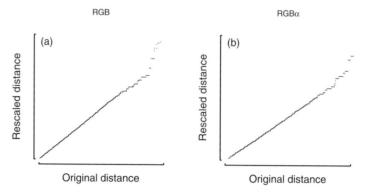

Figure 13.8 Shepard plots for rescaling the three- and four-dimensional colour vectors. Each point represents a pairwise distance between two points. Both panels show strong ordinal relationships between the original and rescaled values, with the largest discrepancies in the upper right corner of each plot, representing pairs of points that were very far apart in both spaces.

We can check the mapping between the original dissimilarities and the dissimilarities between positions in the lower-dimensional space created by the MDS algorithm (the rescaled data) using a *Shepard diagram*. This plots the pairwise distances between points from the original data along the x-axis, and the pairwise distances for the rescaled data along the y-axis. If there is no loss of information due to the rescaling, these values should be perfectly correlated. The amount of scatter around the diagonal is therefore an indication of how faithfully the data have been mapped by the MDS algorithm. One can also calculate statistics, such as Spearman's rank correlation, between the distances in the two spaces. Examples for the colour data are shown in Figure 13.8.

Metric vs non-metric MDS

There are two main variants of multidimensional scaling, known as metric and non-metric MDS. The metric algorithm, also referred to as principal coordinates analysis, tries to maintain the relative distances between data points in the rescaled space. Non-metric algorithms relax this assumption, and try to maintain only the rank orderings of distances between points. In essence, they replace the actual values in the distance matrix with ranks, and so work in a similar way to other non-parametric statistical tests (e.g. the Wilcoxon signed-rank test). This also means that non-metric algorithms can be used with ordinal variables (variables involving categories with a clear ordering), and data that do not meet parametric assumptions (i.e. that are not normally distributed). For many parametric data sets, both methods will produce a similar solution and the choice of which to use is somewhat arbitrary.

Example: multidimensional scaling of viruses

A study by Lopes, Andrade, and Tenreiro Machado (2016) used multidimensional scaling to represent the similarity of 22 viruses. They used quantitative data on several measures including fatality rate, speed of spread, and incubation period, and produced two- and

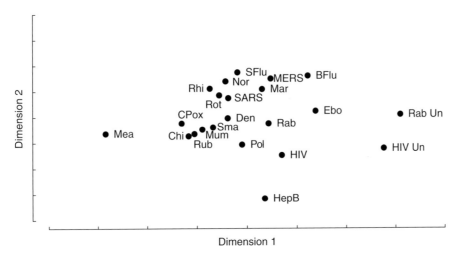

Figure 13.9 Example data on virus characteristics, visualized with metric multidimensional scaling.

The underlying data used for the analysis were obtained from table 1 of Lopes, Andrade, and Tenreiro Machado (2016).

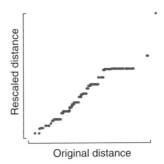

Figure 13.10 Shepard plot for the virus data set.

The underlying data used for the analysis were obtained from table 1 of Lopes, Andrade, and Tenreiro Machado (2016).

three-dimensional MDS plots of virus similarity. Because their data are provided (in table 1 of their paper), we can attempt a simpler version of their analysis ourselves. Figure 13.9 shows a two-dimensional scaling of the virus data. You can see that various types of flu (SFlu = swine flu, BFlu = bird flu) and respiratory disease (SARS, MERS) are grouped together near the top of the plot, and some high fatality rate diseases (HIV, Ebola, and untreated rabies) are clustered towards the lower right. A Shepard plot is shown in Figure 13.10 and indicates a good mapping between the original and rescaled spaces. The original paper reports a more involved analysis with multiple distance measures, and is a good example of how the technique can be used.

Combining *k*-means clustering and MDS

The two techniques described in this chapter can be used together, much as we saw in the example with insect species. For example, a multivariate data set can be clustered using *k*-means clustering, and the results visualized with multidimensional scaling. This is usually the

correct order to apply the techniques, because applying MDS before clustering will mean
that some information has been lost in the rescaling, and the clustering will be less accurate.

An example comes from a study by Coggan et al. (2019), who wanted to create sets of
stimuli with similar image properties to study how the human brain represents objects.
To do this, they took a database of 2761 object images (the *Bank of Standardized Stimuli*,
see Brodeur et al. (2010)). To simplify the representation of these images, they converted
them to greyscale, and then applied the GIST descriptor function (Oliva and Torralba 2001),
which summarizes images using the orientations and sizes of features in each image (see
also Chapter 10 for details of how such image features can be represented). Each image was
thereby reduced to a vector of 4096 numbers. However this is too many dimensions for
k-means clustering to operate on reliably, given the number of examples in the data set. So
principal components analysis (a form of factor analysis) was applied to the GIST vectors to
further reduce them to 20 values per image (the first 20 principal components).

The *k*-means clustering algorithm, with $k = 10$, was then applied to the matrix of 2761 ×
20 numbers. The choice of $k = 10$ clusters was intended to produce a suitable number of
stimulus categories for use in a neuroimaging experiment. The 24 image examples closest to
each of the cluster centroids were chosen for use in the experiment. The distinctness of each
cluster was confirmed using multidimensional scaling to reduce the dimensionality of the
image data set from 20 dimensions to two. It was also clear that images from individual clusters had various properties in common—for example all being roughly circular, or oriented in
a particular direction. A summary of the image selection process is provided in Figure 13.11
(based on figure 2 of Coggan et al. (2019)).

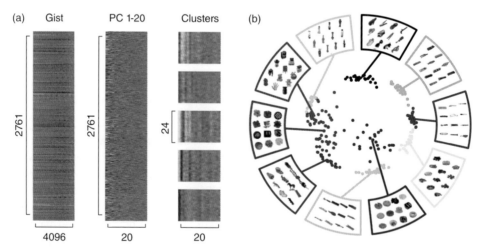

Figure 13.11 Summary of combined *k*-means clustering and multidimensional scaling, following
Coggan et al. (2019). Panel (a) shows a graphical representation (shown in red/green) of the GIST vector
for 2761 images, represented as 4096 values based on their low-level image properties. This was reduced
using principal components analysis to the middle matrix (blue/yellow). Then *k*-means clustering was used
to generate 10 clusters of 24 images. Five example clusters are illustrated to the right of the panel, where
the similarity of components within a cluster is clear. Panel (b) shows a two-dimensional multidimensional
scaling of the principal components, where the clusters are illustrated by coloured points. Image examples
from each cluster are shown around the circumference, and within a cluster have consistent image
properties, such as similar shape, orientation, and aspect ratio.

The final set of 240 images (10 categories × 24 examples) were then presented to participants in a block design fMRI experiment. The study found that neural responses in the ventral visual cortex (a region of the brain believed to be specialized for detecting objects) produced distinct patterns of activity for each cluster. This is important, because distinct patterns are usually associated with specific categories of real-world objects (such as faces, buildings etc.), and this in turn is interpreted as evidence that there are areas of the brain specialized for different semantic object categories, such as faces, bodies, or buildings. By using object clusters defined entirely by their image properties (and not their semantic properties), this study demonstrates that low-level image features (such as orientation, curvature, and so on) are also important in understanding stimulus representations in this part of the brain.

Normalizing multivariate data

Both of the techniques we have discussed in this chapter work best when the dependent variables have consistent units and similar variances. This is because if one variable has much larger units than the others, or a much larger variance, it will contribute much more to estimates of the Euclidean distance between data points and cluster centres (for *k*-means clustering) or between pairs of points (for MDS) than will the other variables. In such cases, it is good practice to *rescale* all of the variables, so that they have similar variances (see 'Normalizing and rescaling data' in Chapter 3). This is achieved by dividing each variable by its standard deviation. It is also possible to subtract the mean from each variable, so that the values are converted to z-scores. Normalization will generally speed up convergence of the algorithm, as well as potentially improving the results.

Examples from the literature of *k*-means clustering and MDS

As we have seen in several examples throughout the chapter, both *k*-means clustering and MDS are very useful when classifying different species of organisms. A good example from the literature is a recent study by Scott Chialvo et al. (2018). These authors investigated varieties of tiger moth that feed on lichens, to understand defensive chemical pathways that rely on substances ingested from the lichens. They studied the *metabolome* (i.e. the chemicals inside the organism) of different species of adult tiger moth, and measured the quantities of chemicals called *phenolics* that were derived from lichen. They used *k*-means clustering to identify eight clusters of chemical profiles, and visualized the results using non-metric multidimensional scaling. As part of a wider analysis that also included genetic measures, this helped them to more accurately reconstruct the evolutionary history of lichen moths.

Multivariate methods are also useful in the commercial sector, in particular for identifying groups of customers that have similar characteristics or preferences, and understanding how those preferences relate to different products. In the food industry, *preference mapping* studies involve asking a sample of consumers about their likes and dislikes, usually in the form of rating scale data for particular products or product categories. Wajrock et al. (2008) evaluated several different clustering methods, including *k*-means and hierarchical clustering algorithms, for 15 preference mapping data sets. They conclude that algorithms such as *k*-means clustering, which involve partitioning individuals into different categories, outperform other approaches. In a different study investigating online shopping behaviour, Jain and Ahuja (2014) used *k*-means clustering to identify four distinct types of online shopper. These were referred to as 'cognizant techno strivers', 'conversant appraisers', 'moderate digital ambivalents', and 'techno savvy impulsive consumers'. The rationale behind such classification is to better target advertising at consumers more likely to be receptive to specific product types. Multidimensional scaling is also useful in marketing and brand analysis—for example Bijmolt, Wedel, and DeSarbo (2021) used it to model consumer perception of different car brands, based on ratings of their perceived similarity. In a highly competitive industry, this type of information is useful for designing new products, as well as for marketing existing ones.

Doing *k*-means clustering in *R*

The *kmeans* function is built into the core *stats* package, and can conduct *k*-means clustering with several different algorithms. It takes an *N* (rows, containing observations) by *m* (columns, containing different variables) matrix as its input, as well as a number to specify the value of *k* (the desired number of clusters). For the example data set from Figure 13.3, we could request a clustering with *k* = 5 and inspect the output as follows:

```
clusters <- kmeans(dataset[,1:2],5)
clusters
## K-means clustering with 5 clusters of sizes 31, 33, 13, 27, 21
##
## Cluster means:
##            X1          X2
## 1 -0.39729954 -0.03181516
## 2  0.39216792  0.06878313
## 3  0.01485053 -0.66225151
## 4 -0.03320486  0.51792763
## 5  0.18295912 -0.30327400
##
## Clustering vector:
##   [1] 5 2 5 1 2 4 2 5 2 2 2 2 2 2 5 2 1 5 4 1 4 1 2 1 2 4 4 4 4 4 4 4
4 4 4 4 4
## [38] 4 4 4 4 4 1 4 4 4 2 4 4 4 2 5 5 5 2 2 2 2 2 2 2 2 2 2 2 2 2 2 2
2 5 5 2 2
## [75] 2 1 1 1 1 1 1 1 1 1 1 1 1 1 1 1 1 1 1 4 1 1 1 1 5 1 1 1 5 5 5 5 1
3 5 3 3 5
## [112] 5 3 5 3 5 3 3 3 5 3 3 3 3 3
##
## Within cluster sum of squares by cluster:
## [1] 2.0466979 2.2336343 0.3633141 2.1496942 1.4766339
##  (between_SS / total_SS = 75.6 %)
##
## Available components:
##
## [1] "cluster"    "centers"    "totss"      "withinss"    "tot.withinss"
## [6] "betweenss"  "size"       "iter"       "ifault"
```

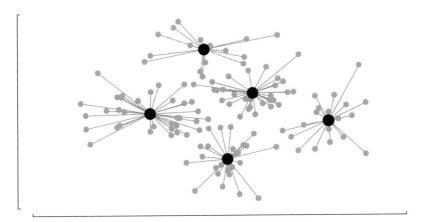

Figure 13.12 Output of *R* code demonstrating *k*-means clustering.

The first line of the output tells us how many clusters we have generated, and their sizes (i.e. how many observations are assigned to each cluster). Then it gives the cluster means as the *x* and *y* coordinates of the cluster centres. Note that for data sets with more than two dependent variables, the means will contain a value for each dependent variable. The clustering vector gives cluster assignments to each of the individual observations from the data set. The summed squared error for each cluster is also provided, and gives an estimate of the residual variance within each cluster. Finally, the ratio of between and total sums of squares is given—this is the same as the R^2 value from ANOVA or regression, and tells us the proportion of the total variance that is explained by cluster assignment.

The output data object allows us to access all of these values, as well as incidental information about things like the number of iterations required for the clustering algorithm to converge. We can use this information to plot the lines between each data point and its assigned cluster centroid as follows (see Figure 13.12 for the output):

```
# set up an empty plot axis
plot(x=NULL,y=NULL,axes=FALSE, ann=FALSE, xlim=c(-1,1), ylim=c(-1,1))
axis(1, at=c(-1,1), tck=0.01, lab=F, lwd=2)
axis(2, at=c(-1,1), tck=0.01, lab=F, lwd=2)

# draw lines between each cluster centre and the assigned data point
for (n in 1:(nrow(dataset))){
 lines(c(clusters$centers[clusters$cluster[n],1],dataset[n,1]),
   c(clusters$centers[clusters$cluster[n],2],dataset[n,2]),
   col='grey')}

# draw individual data points
points(dataset[,1],dataset[,2],pch=16,col='blue')

# draw the cluster centroids
points(clusters$centers[,1],clusters$centers[,2],pch=16,cex=2)
```

Additional colours for each cluster, or for true group membership if this is known, can also be used (as in Figure 13.3).

Hartigan–Wong algorithm   **Lloyd algorithm**   **MacQueen algorithm**

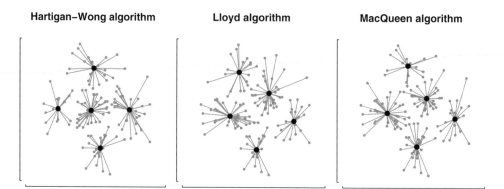

Figure 13.13 Demonstration of different *k*-means clustering algorithms.

There are several optional inputs to the *kmeans* function. One option is to specify the algorithm we use. The default is the Hartigan–Wong algorithm (Hartigan and Wong 1979), which uses a function minimization method to assign cluster membership. Alternatives are the Lloyd (1982) and Forgy (1965) methods described earlier in the chapter (these are identical, but both names are accepted by the function), and the MacQueen (1967) algorithm, which starts with the centroids as far apart as possible. There may be particular data sets that are better suited to one or other algorithm, so it is worth experimenting with different options. Figure 13.13 shows some examples for comparison (the differences are quite subtle), generated by the following code:

```
par(mfrow=c(1,3), las=1)

algorithmlist <- c('Hartigan-Wong','Lloyd','MacQueen')
for (plt in 1:3){
 clusters <- kmeans(dataset[,1:2],5,algorithm = algorithmlist[plt])
 plot(x=NULL,y=NULL,axes=FALSE, ann=FALSE, xlim=c(-1,1),
ylim=c(-1,1))
 axis(1, at=c(-1,1), tck=0.01, lab=F, lwd=2)
 axis(2, at=c(-1,1), tck=0.01, lab=F, lwd=2)
 title(paste(algorithmlist[plt],'algorithm'))
 for (n in 1:(nrow(dataset))){
   lines(c(clusters$centers[clusters$cluster[n],1],dataset[n,1]),
     c(clusters$centers[clusters$cluster[n],2],dataset[n,2]),
     col='grey')}
 points(dataset[,1],dataset[,2],pch=16,col='blue')
 points(clusters$centers[,1],clusters$centers[,2],pch=16,cex=2)
}
```

It is also possible to constrain the number of iterations permitted, and to start from multiple random centroids using the *nstart* option, or even to specify the starting centroids manually. These options allow substantial control over how the clustering proceeds.

If we need to rescale the data before conducting *k*-means clustering, this can be achieved using the *scale* function as described in 'Normalizing and rescaling data' in Chapter 3. Scaled data will typically produce better clustering when the dependent variables are in different units.

Doing multidimensional scaling in *R*

There are several implementations of MDS in *R*, as well as in other software packages and programming languages. Here we will demonstrate the *cmdscale* function in the core *stats* package, which implements classical (metric) scaling, as well as the *isoMDS* function for non-metric multidimensional scaling in the *MASS* package (Venables, Ripley, and Venables 2002).

Multidimensional scaling operates on a matrix of distances between each pair of points in the data set. We need to compute this first using the *dist* function (also part of the core *stats* package). This will take an *N* (observations) × *m* (dependent variables) matrix and calculate the Euclidean (straight line) distances between every possible pairing of data points, producing an *N* × *N* matrix. Note that the output therefore does not depend on *m*, the number of dependent variables. This is because the distance between points is calculated as a single vector in the multidimensional space, regardless of its dimensionality. The data object *colourdata* contains the example data from Figure 13.7(b), arranged in a 1000 (observations) × 4 (RGBα) matrix. We can calculate the distances as follows:

```
coldist <- dist(colourdata)
```

The *coldist* data object is a special class of matrix that contains the pairwise distances in the correct format to use for multidimensional scaling. We can pass this matrix (but *not* the raw data) to the *cmdscale* function as follows:

```
scaledxy <- cmdscale(coldist,2)
head(scaledxy)
##                 [,1]        [,2]
## [1,] -0.15160650   0.6517739
## [2,]  0.41646691  -0.2108153
## [3,] -0.03000302   0.3792663
## [4,] -0.05064061  -0.6095281
## [5,]  0.09436012   0.1585279
## [6,] -0.22697391   0.1237203
```

The second argument for the *cmdscale* function (the number 2) defines the dimensionality of the output. So, if we wanted to produce a 3D plot of the data points, we could change this to 3. It is not clear how one might represent dimensions higher than three graphically, but in principle any number of output dimensions is possible. Each row of the output (*scaledxy*) shows the *x,y* position of a data point (with row number consistent with the original data matrix).

We can then plot the rescaled data. Often it is helpful to colour-code the individual points by some meaningful category. For this example, each data point already has an RGBα colour vector associated with it, which we can use to produce an attractive diagram (see Figure 13.7(b)) as follows:

```
plot(x=NULL,y=NULL,axes=FALSE, ann=FALSE, xlim=c(-1,1), ylim=c(-1,1))
axis(1, at=c(-1,1), tck=0.01, lab=F, lwd=2)
axis(2, at=c(-1,1), tck=0.01, lab=F, lwd=2)
title(xlab="Dimension 1", col.lab=rgb(0,0,0), line=1.2, cex.lab=1.5)
title(ylab="Dimension 2", col.lab=rgb(0,0,0), line=1.5, cex.lab=1.5)
title(expression(paste('RGB',alpha,sep='')))
```

```
points(scaledxy[,1],scaledxy[,2],pch=16,cex=0.5,col=
rgb(colourdata[,1],colourdata[,2],colourdata[,3],alpha=colourd
ata[,4]))
```

An alternative is to use non-metric multidimensional scaling, which has less stringent requirements about the solution. The following code uses the *isoMDS* function from the *MASS* package:

```
library(MASS)

mdsout <- isoMDS(coldist,k=2)
scaledxy <- mdsout$points


plot(x=NULL,y=NULL,axes=FALSE, ann=FALSE, xlim=c(-1,1), ylim=c(-1,1))
axis(1, at=c(-1,1), tck=0.01, lab=F, lwd=2)
axis(2, at=c(-1,1), tck=0.01, lab=F, lwd=2)
title(xlab="Dimension 1", col.lab=rgb(0,0,0), line=1.2, cex.lab=1.5)
title(ylab="Dimension 2", col.lab=rgb(0,0,0), line=1.5, cex.lab=1.5)
title(expression(paste('RGB',alpha,sep='')))


points(scaledxy[,1],scaledxy[,2],pch=16,cex=0.5,col=
rgb(colourdata[,1],colourdata[,2],colourdata[,3],alpha=colourd
ata[,4]))
```

If you run this code, you will notice that the solution is similar to the metric version, but the non-metric diagram has more outliers at the extremes. In both cases, the units of the scaled solution are arbitrary for both dimensions.

Practice questions

1. Standard *k*-means clustering works on what kind of distance between points?
   A) Rank-order
   B) Manhattan
   C) Cartesian
   D) Euclidean

2. The accuracy of a cluster solution is based on:
   A) The distances between data points and cluster centroids
   B) The distances between pairs of data points
   C) The distances between pairs of centroids
   D) The total variance in the data set

3. Which statistic can be used to compare different numbers of clusters, taking into account the degrees of freedom?
   A) Summed squared error
   B) Akaike information criterion
   C) Root mean squared error
   D) R^2

4. Why is it important to normalize (scale) variables that have very different units?

    A) To remove outliers

    B) Distance estimates cannot be calculated when the units are different

    C) Variables with larger units will contribute too much to the distance estimates

    D) Variables with smaller units will contribute too much to the distance estimates

5. Which of the following is not a variant of k-means clustering?

    A) k-medians clustering

    B) k-medioids clustering

    C) inverse k-means clustering

    D) spherical k-means clustering

6. Multidimensional scaling is used to:

    A) Represent multivariate data in a lower-dimensional space

    B) Represent bivariate data in a multidimensional space

    C) Normalize all dimensions in a data set so they have equal variance

    D) Represent multivariate data in a higher-dimensional space

7. We can check how well the distances between points are preserved after multidimensional scaling using:

    A) The summed Euclidean distance between data points

    B) A Shepard diagram and correlation coefficient

    C) The change in the number of dimensions

    D) A Q–Q plot and the summed squared differences

8. Metric multidimensional scaling algorithms:

    A) Use ranks rather than absolute values of the dependent variables

    B) Are typically used for categorical dependent variables

    C) Attempt to preserve the rank ordering of distances between data points

    D) Attempt to preserve the relative distances between data points

9. For a data set with N observations and m dimensions, the distance matrix will have size:

    A) $N \times m$

    B) $m \times m$

    C) $N \times 2$

    D) $N \times N$

10. For a data set with five dimensions, which of the following would not be a valid number of dimensions for the output of multidimensional scaling?

    A) 1

    B) 2

    C) 3

    D) 6

Answers to all questions are provided in the answers to practice questions at the end of the book.

Multivariate pattern analysis

The final multivariate technique we will discuss in this book is called multivariate pattern analysis (MVPA). In recent years it has been widely used to analyse MRI recordings of brain activity, where it is sometimes referred to as multi**voxel** pattern analysis, though the techniques (and acronyms) are much the same. These methods are a subset of machine learning—a family of artificial intelligence (AI) methods that aim to train computer algorithms to perform classification tasks on some sort of complex data. Prominent examples of machine learning include object identification algorithms (i.e. for labelling the contents of photographs) and dictation software that converts speech to text (and in some cases can act on verbal instructions). Machine learning methods also have substantial promise in the area of personalized medicine and automated diagnosis, one prominent example being the diagnosis of eye disease (De Fauw et al. 2018). These techniques will become more widespread and accurate in the future, and at the time of writing (2021) are attracting substantial media attention and commercial investment. In such a fast-moving field, it is always worth keeping up with new developments. However, for a more detailed discussion of the core aspects of pattern analysis and other machine learning methods, the classic text, *Pattern Recognition and Machine Learning* by Bishop (2006), is an excellent resource.

Why use machines?

In some instances (e.g. object identification), the task we are interested in is one that can be easily achieved by humans with a high degree of accuracy. In these cases, the main benefit of automation is the speed and scale offered by computers. For example, the internet contains millions of unlabelled images, and having human operators manually label each one would be prohibitively expensive (not to mention tedious). An algorithm that can automatically identify their contents makes the images searchable using text keywords, without requiring extensive human labour. In other situations, algorithms can be used to identify patterns in data that would be hard for humans to spot, perhaps owing to the complexity of the data. The great promise of this aspect of AI is that it could help improve critical real-world problems such as disease diagnosis and risk prediction in the insurance industry.

Predicting group membership

To introduce the basic concept of classification, we will use a minimal example. Consider some data sampled from two groups with different means, shown in Figure 14.1(a). Most students of statistics know what to do with such data: a t-test (as demonstrated in Chapter 4) can tell us whether the group means differ significantly or not. But what if we wanted to ask a different question? What if we wanted to predict which group a participant is a member of, knowing only their score? This is the basic idea of *classification*, and is the central problem that MVPA algorithms are used to solve.

To think about classification, we could replot the data as a single cloud, as shown in Figure 14.1(b) (the x-position of each point is arbitrary here). A good way to try to classify group membership is to plot a *category boundary* that best separates the two groups. This is shown by the dashed line in Figure 14.1(b), and a sensible decision rule would be to say that data points below the line (most of the grey circles) are more likely to be in group A, and those above the line (most of the blue squares) will be in group B. Of course, this classification is not totally accurate for the current example—there are several blue squares below the line and grey circles above it, and these will be misclassified.

The category boundary is a basic classification algorithm, and it prompts some observations that will generalize to more complex cases. First, we can work out the accuracy of the classification by calculating the percentage of data points that are correctly identified. In the example in Figure 14.1(b), this is something like 90%. It is also clear that if the group means were more similar, accuracy would reduce, and if they were more different, accuracy would increase. Additionally, the variance (spread) of the data points is important. If the variance were greater, classification accuracy would decrease, and if the variance were smaller, classification accuracy would increase.

A convenient way of summarizing the mean difference and variance is to use the Cohen's *d* metric introduced in Chapter 5. The *d* statistic is the difference in means, divided by the

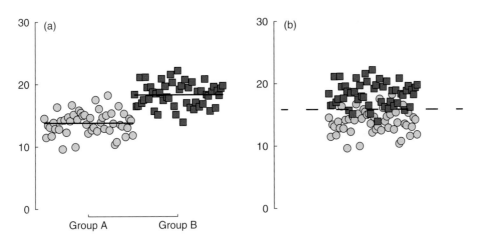

Figure 14.1 Minimal illustration of classification. Panel (a) shows example simulated data from two groups with differing means (black lines). Panel (b) shows the same data replotted with a category boundary (dashed line).

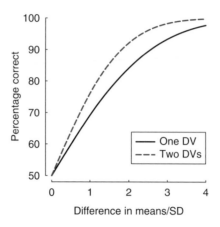

Figure 14.2 Increase in accuracy with Cohen's *d* for one (black) and two (blue) dependent variables (DVs).

standard deviation. We can calculate how classification accuracy (for a two-category data set) changes as a function of *d*, as shown by the black curve in Figure 14.2. As we might expect, increasing the separation between the group means increases the accuracy of our classifications. So far so good, but up until this point our examples have had only a single dependent variable—isn't MVPA supposed to be a multivariate technique?

The same logic as described for a single variable can easily be extended to the case of multiple variables. If we have two variables, we can try to classify data points using both pieces of information by placing a category boundary to separate the two-dimensional space created by plotting the variables against each other. This is shown in Figure 14.3(a) for a linear classifier, where the line separating the white and grey areas indicates the category boundary. Adding extra informative (and uncorrelated) variables increases accuracy (see dashed blue curve in Figure 14.2). In principle this same trick can be applied for any number of variables. In a real data set, some variables will be informative whereas others will not, and there will

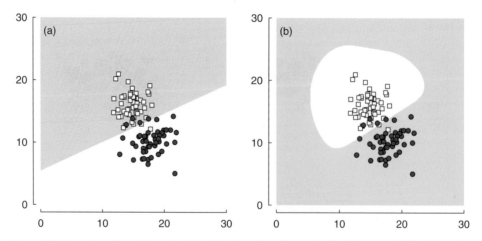

Figure 14.3 Examples of two-dimensional classification using a linear classifier (a) and a radial basis function classifier (b).

usually be some level of covariance between different measures. However this is not generally a problem—classifier algorithms will tend to ignore uninformative variables, and assign more weight to measures that improve accuracy.

Different types of classifier algorithm and pattern analysis

There are many varieties of classifier algorithm available. The straight lines in the examples we have seen so far were determined using a popular algorithm called a linear *support vector machine*. This works by maximizing the distances between the boundary line and the closest example data points from each category (termed the support vectors). Fundamentally this is an optimization problem, and works in a similar way to the non-linear curve fitting methods described in Chapter 9. Support vector machines can also be created with non-linear (radial) *basis functions* (the basis function is the mathematical equation that is used to construct the boundary line). These work by enclosing an 'island' of values from one category (see Figure 14.3(b)). These can be more efficient for some types of data, but also sometimes suffer from problems with generalization to new data sets (Schwarzkopf and Rees 2011).

Other types of algorithm can involve neural networks, which are based on interconnected multilayer processing of the type that happens in biological neural systems. The input layer consists of a set of *detectors* that respond to particular inputs or combinations of inputs. Each successive layer of the network then applies a mathematical operation to the output of the previous layer. These operations are usually fairly basic ones, such as weighted averaging (described in 'Weighted averaging' in Chapter 6 in a different context), choosing the largest input, or a non-linear transform such as squaring. The end result of the network is an *output layer*, from which classification decisions can be read. Neural networks can produce very sophisticated operations (much like the brain), though it can sometimes be difficult to understand fully what a trained neural network is actually doing.

A particularly useful variety is the *deep convolutional neural network*. These are based on the early stages of sensory processing in the brain, and involve taking images (or other natural inputs) and passing them through a bank of *filters* that pick out specific low-level features from the input. An example filter bank is shown in Figure 14.4, involving filters of different orientations and spatial frequencies (see also Chapter 10 for details on how filters are applied to images).

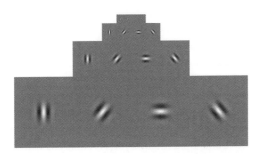

Figure 14.4 Filter bank showing filters of different orientations and spatial frequencies.

Deep neural networks are now advanced enough to classify images into different categories, though it can sometimes be unclear precisely which features of an image set are being used to do this. It is also important to avoid any confounds in the input images that might produce false levels of precision, such as the background of an image. For example, if you wanted to train a network to classify criminals vs non-criminals from their photographs, it would be important to make sure that all the photographs were taken under the same conditions. Otherwise it could be the case that all criminal photographs were taken against the same background (e.g. the height gauge traditionally shown in police mugshots) and these extraneous features would provide the network with a spurious cue.

A rather different approach to MVPA, which has been very influential in the fMRI literature, is to forgo classification algorithms and instead use a correlation-based approach (Haxby et al. 2001). In correlational MVPA, the pattern of brain activity across multiple voxels (a voxel is the volumetric version of a pixel) is correlated between two data sets derived from the same condition, or data sets derived from two separate conditions. The logic is that if there is a distinct and robust pattern of activity in a region of the brain, the correlation scores will be higher when the data come from the same condition than when it comes from separate conditions. There are several variants of this method depending on whether the data sets are derived within a single individual, or averaged across multiple participants. Although this approach sounds very different from the classification-based methods discussed earlier in the chapter, in direct comparisons (e.g. Coggan, Baker, and Andrews 2016; Isik et al. 2014; Grootswagers, Wardle, and Carlson 2017) they behave quite similarly. A related approach is to directly calculate the multivariate effect size (the Mahalanobis distance, see the section 'The Mahalanobis distance for multivariate data' in Chapter 3) between conditions, and use this as a measure of pattern distinctness (Allefeld and Haynes 2014).

Situations with more than two categories

Classifiers are not limited to discriminating between two possible categories. The same principles can be applied to problems with an arbitrary number of categories or conditions. If this is done, it is important to be aware that the baseline guess rate will also change. For two categories, the guess rate will be 0.5 (or 50%), because an algorithm that assigns data randomly to one of two categories will tend to get things right half of the time (assuming a balanced data set). For 10 categories, the guess rate will be 0.1 (or 10%). The guess rate is therefore given by $1/m$, where m is the number of categories (see also 'Removing bias using forced choice paradigms' in Chapter 16 for a related situation).

However, it is worth being aware that in situations with more than two categories, good performance may not be equally distributed between the categories. For example, a situation with three categories (A, B, and C) could involve good discrimination between A and B, but poor discrimination between B and C. This is an analogous problem to interpreting a main effect in ANOVA—the main effect itself does not tell us which pairs of conditions differ. Just as ANOVAs are often followed up by pairwise post hoc comparisons between different conditions, it is also possible to follow up a multiway MVPA analysis with pairwise discrimination between different categories.

Preparing data for analysis

Before performing classification, it is good practice to normalize one's data. Usually this is done by subtracting the group mean from all data points that will be involved in classification, independently for each variable. It is also common to scale by the variance. The idea here is to remove substantial differences in the scale of the different variables, much like converting to z-scores. Some MVPA software does this automatically, but it can also be done manually (see 'Normalizing and rescaling data' in Chapter 3).

Next, the data will usually be split into a training set and a test set. The classifier is trained using the training set, and then tested on the unseen observations in the test set to determine its accuracy. This is an important step to avoid a 'double dipping' confound, where a model is trained and tested on the same data. Such confounds will tend to inflate accuracy because of overfitting. Classification will generally be repeated many times with different allocations of observations to training and test sets, to work out an average accuracy that should be more robust than for a single partitioning. Some MVPA software will implement this automatically using a technique called *k-fold cross validation*. The value of the k parameter determines the number of subsets the data are split into. The model is trained on $k - 1$ subsets, and tested on the remaining subset. This is repeated for all permutations (i.e. each subset is the test set once). We can replace the k with its value when referring to this type of analysis, e.g. 5-fold cross validation, or 10-fold cross validation.

Assessing statistical significance

Once we have determined the classifier accuracy, it is sometimes useful to test whether this exceeds chance level, or perhaps to compare accuracy statistically between different classifications. In situations where we have a single measure of classifier accuracy, the binomial test is an appropriate way to do this. The binomial test can compare the proportion of correct classifications with the guess rate, taking into account the number of observations. This statistic will have greater power (see Chapter 5) the more observations are included in the test set.

An alternative is to generate 95% confidence intervals on the classifier accuracy estimate using a resampling method (see Chapter 8). One way to do this would be to repeat the classification using many random partitionings of the original data into training and test sets, to build up a distribution of classifier accuracy scores. The empirical 95% quantiles (i.e. the points at 2.5% and 97.5% on the distribution) act a bit like a t-test—if these do not overlap the guess rate, the accuracy can be said to differ significantly from chance.

Finally, in experimental designs involving multiple participants, and where the classification is performed separately on each individual participant, traditional statistical tests can be used to assess significance of the mean accuracy level across participants. These might include parametric or non-parametric t-tests, or analysis of variance (ANOVA). Tests can be performed on either the classifier accuracy scores, or measures derived from them, such as d' (see Chapter 16). One would take the score(s) for each participant, and compare these values to chance using a one-sample t-test, or across groups with a two-sample t-test or ANOVA.

In situations where many separate classifications are conducted (e.g. across time, or across different locations in the brain), cluster correction algorithms can be used to correct for multiple comparisons (see Chapter 15).

Ethical issues with machine learning and classification

This is not intended to be a sociology text. However, machine learning techniques carry with them several potential dangers, because they are often presented as being fair and unbiased. After all, how could a computer algorithm be sexist or racist? The reality is that any algorithm can be very easily biased by the prejudices of those who construct them, can perpetuate systemic inequalities, and have the capacity to be used for nefarious purposes. Two very high-profile real-world examples are gender discrimination in automated recruitment software, and algorithms designed to classify based on sexual orientation.

In the first example, in around 2018 a large technology company scrapped its automated recruitment system because it was shown to rate women's CVs less highly than men's for software and technology-related jobs. It did this in some surprisingly blatant ways, such as penalizing graduates of female-only universities, and down-weighting CVs that included the word 'women's'—male candidates would be unlikely to mention being captain of the women's basketball team, for example. Why did this happen? It turns out that the algorithm was trained on historical data from two groups of candidates—those who had been hired, and those who had not. All of those hiring decisions were made in the traditional way, by humans with their own prejudices about what makes a good software engineer. Far from being unbiased, the algorithm inherited and perpetuated the prejudices of the industry it was created to serve.

The second controversial example was a study claiming that deep neural networks could classify sexual orientation from photographs more accurately than humans (Wang and Kosinski 2018). The authors proposed that subtle differences in facial morphology might reveal exposure to various sex hormones in the womb, which also influence sexual orientation in adulthood. This work was criticized on several grounds, including that most of the photographs used to train the algorithm were of Caucasian models, and that the algorithm appeared to be classifying photographs based on cues that were unrelated to what the authors claimed, including make-up and the presence or absence of glasses. But the key point is that the use of machine learning algorithms in this way is extremely unethical. There are many societies where homosexuality is illegal, and tools that can be used to classify sexual orientation (no matter how accurate, or using what cues) could be used to oppress innocent people. Machine learning is a powerful tool, but it is crucial that it is used responsibly and ethically, and that the apparent objectivity of computer algorithms is not used to mask human prejudice and bias.

Doing MVPA in *R* using the *Caret* package

The following examples will use a package called *caret* (pronounced like 'carrot'), which is short for **C**lassification **A**nd **RE**gression **T**raining (Kuhn 2008). *Caret* is a general purpose package that provides access to a wide range of algorithms (over 200) through a consistent interface. It is certainly not the only machine learning package available in *R*, but it is one of the most flexible. There are MVPA packages available in most other contemporary programming languages, including built-in tools in Matlab, the *PyMVPA* toolbox in Python, and the cross-platform *LibSVM* package.

We will use two functions from the *caret* package: *train* and *predict*. These do much as you would expect from the names. The *train* function is used to train a pattern classifier algorithm, and outputs a data object containing the model specification. We can then pass this model specification into the *predict* function, along with some unseen data, to get predictions out. If we know the ground truth categories (i.e. the actual categories) for the unseen data, we can also calculate the classifier's accuracy. Of course *caret* is capable of far more than what we are doing here, but this is a good starting point to demonstrate the basics. There is extensive documentation available on the package's web pages at **http://caret.r-forge.r-project.org**.

Categorizing cell body segmentation

The first scenario we will consider to demonstrate MVPA uses an example data set from the *caret* package, originally from a paper by Hill et al. (2007). The *segmentationData* data set contains data for cells that are classed as either 'well segmented' or 'poorly segmented'. There are over 2000 example cells and 58 predictor variables. The data have been pre-allocated to training and test sets, using the Factor column *segmentationData$Case*. We can therefore load in the *caret* package and the data set, and take a look at what we have:

```
library(caret) # load in the caret package
data(segmentationData)   # load the data set into the Environment
dim(segmentationData)    # request the size of the data frame
## [1] 2019  61
head(segmentationData[,1:6])
##          Cell  Case Class    AngleCh1 AreaCh1 AvgIntenCh1
## 1 207827637  Test    PS 143.247705     185    15.71186
## 2 207932307 Train    PS 133.752037     819    31.92327
## 3 207932463 Train    WS 106.646387     431    28.03883
## 4 207932470 Train    PS  69.150325     298    19.45614
## 5 207932455  Test    PS   2.887837     285    24.27574
## 6 207827656  Test    WS  40.748298     172   325.93333
```

We see that the data set has 2019 rows and 61 columns. The first three columns are the cell ID number, whether it is to be used for test or training, and whether it is well segmented or poorly segmented. The remaining 58 columns contain different measurements like the areas, widths, and lengths of different parts of the cell, obtained by microscope imaging. We can extract the training and test data into separate matrices as follows:

```
# separate the data for the training set and the test set and convert
to matrices
# only include the columns (4 to 61) that contain actual measurements
trainingdata <- as.matrix(segmentationData[which(segmentationData$Ca
se=='Train'),4:61])
testdata <- as.matrix(segmentationData[which(segmentationData$Case==
'Test'),4:61])
```

Each data object is now an $N \times 58$ matrix. In the training set, $N = 1009$, and in the test set $N = 1010$, where N corresponds to the number of cells.

Next, we can store the true categories (well segmented (WS), or poorly segmented (PS)) in separate data objects for the training and test sets. These are found in the third column of the *segmentationData* data frame, which is headed *Class*:

```
traininglabels <- segmentationData[which(segmentationData$Case=='Tr
ain'),3]
testlabels <- segmentationData[which(segmentationData$Case=='T
est'),3]


traininglabels[1:20]    # output some example labels
## [1] PS WS PS WS PS PS PS WS WS WS WS PS WS PS PS PS PS PS PS PS
## Levels: PS WS
```

The labels are a factor variable (as described in 'Recoding categorical data and assigning factor labels' in Chapter 3), just as one would use to specify group membership in an ANOVA design. Now that we have our data prepared for classification, we can train a classifier on the training data set using the *train* function as follows:

```
svmFit <- caret::train(trainingdata, traininglabels, method =
"svmLinear")
svmFit
## Support Vector Machines with Linear Kernel
##
## 1009 samples
##   58 predictor
##    2 classes: 'PS', 'WS'
##
## No pre-processing
## Resampling: Bootstrapped (25 reps)
## Summary of sample sizes: 1009, 1009, 1009, 1009, 1009, 1009, ...
## Resampling results:
##
##   Accuracy   Kappa
##   0.7968955  0.5614672
##
## Tuning parameter 'C' was held constant at a value of 1
```

Note that there are multiple functions called *train* in different *R* packages, so here we specify the one from the *caret* package with the syntax *caret::train*. The model object (*svmFit*) contains some information about what has been done, as shown above, though we rarely need to look at this directly. We can then take the trained model, and ask *caret* to predict categories for the unseen (test) data using the *predict* function. The outcome (stored in a new data object called *p*) is a set of category predictions (PS or WS) for each cell in the test set.

```
p <- predict(svmFit,newdata = testdata)
p[1:20]
## [1] PS PS WS WS PS PS PS WS PS PS WS PS PS WS PS WS PS PS WS PS
## Levels: PS WS
```

We can compare this to the true categories by counting up how many match the true categories (stored in *testlabels*) and converting to a percentage:

```
numbercorrect <- sum(testlabels==p)
totalexamples <- length(testlabels)
100*(numbercorrect/totalexamples)
## [1] 79.80198
```

The classifier has done very well, getting about 80% of the cells in the test set correct. We can see if this is statistically significant using a binomial test to compare this to chance performance (0.5, or 50% correct):

```
binom.test(numbercorrect,totalexamples,0.5)
##
## Exact binomial test
##
## data: numbercorrect and totalexamples
## number of successes = 806, number of trials = 1010, p-value
< 2.2e-16
## alternative hypothesis: true probability of success is not equal
to 0.5
## 95 percent confidence interval:
## 0.7719120 0.8223772
## sample estimates:
## probability of success
##        0.7980198
```

The test is highly significant, which is telling us that a linear support vector machine with the full training set does pretty well. It classifies about 80% of the test set correctly, which is significantly above chance performance of 50% correct.

What if we used a different kernel? Switching to a radial basis function improves things by around 1%:

```
svmFit <- caret::train(trainingdata, traininglabels, method = "svmRa-
dial")
p <- predict(svmFit,newdata = testdata)
100*(sum(testlabels==p)/length(testlabels))
## [1] 80.49505
```

Alternatively, we can try a neural network model called a *Multilayer Perceptron*. For this particular data set, the perceptron does much worse than the support vector machines (as well as taking longer):

```
svmFit <- caret::train(trainingdata, traininglabels, method = "mlp")
p <- predict(svmFit,newdata = testdata)
100*(sum(testlabels==p)/length(testlabels))
## [1] 66.53465
```

Presumably there will be other types of data where the perceptron would be a better choice.

Finally, we can explore how accuracy increases as we include more of the dependent variables in the classification (see Figure 14.5).

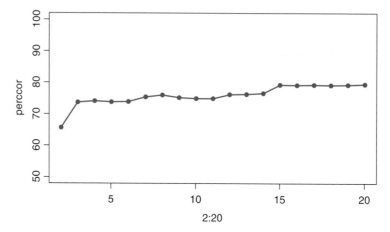

Figure 14.5 Classification accuracy as a function of number of dependent variables, for the cell body segmentation data set.

Data source: Hill et al. (2007).

```
perccor <- NULL
for (n in 1:19){
 trainingdata <- as.matrix(segmentationData[which(segmentationData$C
ase=='Train'),4:(n+4)])
 testdata <- as.matrix(segmentationData[which(segmentationData$Case=
='Test'),4:(n+4)])
 svmFit <- caret::train(trainingdata, traininglabels, method =
"svmLinear")
 p <- predict(svmFit,newdata = testdata)
 perccor[n] <- 100*(sum(testlabels==p)/length(testlabels))
 }
plot(2:20,perccor,type='l',ylim=c(50,100),lwd=3)
points(2:20,perccor,pch=16)
```

As more dependent variables are added, classifier accuracy improves—rapidly at first (e.g. from two to three variables), and then more gradually, before plateauing around 15 dependent variables.

Decoding fMRI data

The second example data set we will use to demonstrate MVPA originates from a classic study by Haxby et al. (2001). In this study, participants viewed images from different image categories, while their brain activity was recorded using an MRI scanner. We will use a subset of conditions from a single participant, consisting of 12 trials in which the stimuli were either faces, houses, or scrambled images. We will compare responses from 30 locations (voxels) in a region of the brain called the *fusiform face area* (FFA), which is thought to be specialized for processing faces. The full data set is available on the PyMVPA website (**http://www.pymvpa. org/datadb/haxby2001.html**) (and note that if you read the original paper, you will see that Haxby et al. actually used an alternative form of MVPA that is based on correlation).

First of all, let's see if the FFA can tell the difference between face images and scrambled images. This is a standard test for 'selectivity' of a particular category. We can set up the *training-data* matrix to contain half of the data from each condition to train the classifier on, as follows:

```
# create an empty matrix
trainingdata <- matrix(0,nrow=12,ncol=30)
# copy half of the face data into the matrix
trainingdata[1:6,] <- facedata[1:6,]
# copy half of the scrambled data into the matrix
trainingdata[7:12,] <- scrambdata[1:6,]
# convert to a data frame
trainingdata <- data.frame(trainingdata)
```

We will also need to create numerical labels to tell the classifier which condition each observation corresponds to. We can just use the numbers 1 and 2 for this as we have two conditions (face and scrambled):

```
# create a factor with two levels and six repetitions
traininglabels <- gl(2,6)
traininglabels
## [1] 1 1 1 1 1 1 2 2 2 2 2 2
## Levels: 1 2
```

Then we can do the same thing with the other half of the data to create a testing set (notice here we choose trials 7 to 12 instead of 1 to 6), for assessing the accuracy of the trained model:

```
testdata <- matrix(0,nrow=12,ncol=30)
testdata[1:6,] <- facedata[7:12,]
testdata[7:12,] <- scrambdata[7:12,]
testdata <- data.frame(testdata)
testlabels <- gl(2,6)
```

As for the previous example, we next train the model using the *train* function and test it on an unseen data set with the *predict* function. We pass in the training data, the labels identifying the conditions, and specify the algorithm we want to use (in this case a linear support vector machine). Then we plug the trained model into the *predict* function along with the test data:

```
svmFit <- caret::train(trainingdata, traininglabels, method =
"svmLinear")
p <- predict(svmFit,newdata = testdata)
p
## [1] 2 1 1 1 1 1 2 2 2 2 2 2
## Levels: 1 2
```

The predictions are stored in the data object *p*, and are condition labels with values of 1 or 2. You can see that five of the 12 examples have been classified as condition 1, and seven have been classified as condition 2. How did this correspond to the true values? As before, we can work out the accuracy by adding up the number of correctly classified examples, and then converting to a percentage:

```
numbercorrect <- sum(testlabels==p)
totalexamples <- length(testlabels)

100*(numbercorrect/totalexamples)
## [1] 91.66667
```

So for this example, we can see that the algorithm has over 90% accuracy. This strongly suggests that the brain region we are looking at responds differently to faces than it does to scrambled images. Next, let's see if it produces a distinct response to pictures of houses. The following code duplicates the example above, except that I have replaced instances of *facedata* with *housedata*.

```
trainingdata <- matrix(0,nrow=12,ncol=30)
trainingdata[1:6,] <- housedata[1:6,]
trainingdata[7:12,] <- scrambdata[1:6,]
trainingdata <- data.frame(trainingdata)
traininglabels <- gl(2,6)


testdata <- matrix(0,nrow=12,ncol=30)
testdata[1:6,] <- housedata[7:12,]
testdata[7:12,] <- scrambdata[7:12,]
testdata <- data.frame(testdata)
testlabels <- gl(2,6)


svmFit <- caret::train(trainingdata, traininglabels, method =
"svmLinear")
p <- predict(svmFit,newdata = testdata)
100*(sum(testlabels==p)/length(testlabels))
## [1] 50
```

For houses, accuracy is at 50% correct, so the classifier has not exceeded chance levels. This indicates that the FFA is not selective for images of houses. Finally, we could do a three-way classification between all stimulus types:

```
trainingdata <- matrix(0,nrow=18,ncol=30)
trainingdata[1:6,] <- facedata[1:6,]
trainingdata[7:12,] <- housedata[1:6,]
trainingdata[13:18,] <- scrambdata[1:6,]
trainingdata <- data.frame(trainingdata)
traininglabels <- gl(3,6)

testdata <- matrix(0,nrow=18,ncol=30)
testdata[1:6,] <- facedata[7:12,]
testdata[7:12,] <- housedata[7:12,]
testdata[13:18,] <- scrambdata[7:12,]
testdata <- data.frame(testdata)
testlabels <- gl(3,6)

svmFit <- caret::train(trainingdata, traininglabels, method =
"svmLinear")

p <- predict(svmFit,newdata = testdata)
100*(sum(testlabels==p)/length(testlabels))
## [1] 44.44444
```

This time around, we have above chance decoding at 44% correct (remember that because there are three categories, the guess rate is 1/3, or 33% correct). This basic MVPA analysis of MRI data has gone pretty well as it has given us quite a clear answer. As mentioned above, in a real MRI study, we would repeat the classifications many times in a loop, randomly reshuffling the examples we use to train and test the model, and then averaging the accuracies that are produced (see Chapter 8 for details of resampling methods). Accuracy scores across multiple participants can then be compared using traditional statistics such as t-tests. For some data sets, perhaps using EEG or MEG methods, we can repeat classification at different moments in time to see how brain signals evolve (see Chapter 15 for further discussion of this).

Practice questions

1. For classification between four categories, what is the baseline (guess) rate?
   A) 4% correct
   B) 25% correct
   C) 50% correct
   D) 70% correct

2. As the Cohen's *d* effect size between two conditions increases, classifier accuracy should:
   A) Increase
   B) Decrease
   C) Stay the same
   D) It will depend on how many dependent variables there are

3. What will a linear classifier use to partition data into categories?
   A) Any arbitrary curve
   B) A straight line, plane, or hyperplane
   C) A radial curve
   D) A sine wave

4. Instead of using a classifier algorithm, MVPA can also be conducted based on:
   A) An extremely fast supercomputer
   B) Scores rounded to the nearest integer
   C) Reduced data from a factor analysis
   D) Correlation

5. An important step in data pre-processing before running MVPA is:
   A) Subtracting the mean differences between conditions
   B) Conducting univariate analyses
   C) Normalization
   D) Squaring all measurements

6. Neural network classifier algorithms involve at least one:
   A) Spatial scale of filter
   B) Simulated calcium channel
   C) Hidden network layer
   D) Real human neuron

7. If a classifier is trained and tested on the same data, what is the most likely outcome?
   A) Accuracy will be perfect
   B) Accuracy will be inflated because of overfitting
   C) Accuracy will be reduced because of overfitting
   D) Accuracy will be at chance

8. A significant three-category classification can be interpreted by:
   A) Running post-hoc pairwise classifications
   B) Running an analysis of variance (ANOVA)
   C) Removing the least informative category
   D) Adjusting the guess rate for a two-category classification

9. In a support vector machine, the support vectors refer to:
   A) The dependent variables
   B) The distance from the category boundary to the nearest points
   C) The distance from the category boundary to each data point
   D) The weights that each dependent variable is multiplied by

10. Deep convolutional neural networks are inspired by the structure of:
    A) The convoluted (folded) structure of the human brain
    B) Complex databases of natural images
    C) A bank of filters with different orientations and spatial frequencies
    D) Biological sensory systems

Answers to all questions are provided in the answers to practice questions at the end of the book.

15 Correcting for multiple comparisons

Most introductory statistics courses introduce the concept of the *familywise error rate*, and correction for multiple comparisons. The idea is that the more statistical tests you run to investigate a given hypothesis, the higher the probability that one of them will be significant, even if there is no true effect. Traditionally, the solution to this problem has been to correct the criterion for significance (i.e. the α level) to account the number of comparisons. We will first discuss several such methods (and their shortcomings), before introducing two newer ideas: the *false discovery rate*, and *cluster correction*. Both approaches deal with multiple comparisons in a principled way, while maintaining statistical power at higher levels than older methods. Controlling the false discovery rate is generally appropriate when the tests are independent, whereas cluster correction should be used in situations where correlations are expected between adjacent levels of an independent variable (e.g. across space or time).

The problem of multiple comparisons

All frequentist statistical tests have a built-in *false positive* rate (or Type I error rate), determined by the criterion for significance (the α level). In many disciplines the widely accepted criterion is $\alpha = 0.05$. This means that the false positive rate for a single test is 5%—1 in 20 statistical tests will return a significant result even when there is no true effect. The overall false positive rate for a family of tests—the chance that at least one test is erroneously significant—increases the more tests we conduct. For tests on independent data sets, this familywise error rate (FWER) is calculated as:

$$FWER = 1 - (1 - \alpha)^m \tag{15.1}$$

where α is the criterion for significance, and m is the number of tests. This function is plotted in Figure 15.1 for $\alpha = 0.05$, and shows that the false positive rate rises rapidly, such that with 14 tests there is a 50% chance of at least one test being significant. With >60 tests, a false positive is virtually guaranteed.

The multiple comparisons problem is frequently presented as justification for avoiding using multiple t-tests, and instead using an omnibus test such as analysis of variance (ANOVA). However, ANOVA itself suffers from issues with familywise error rates that are not widely appreciated (see Cramer et al. 2016; Luck and Gaspelin 2017). Although ANOVA takes into account the number of comparisons *within* an independent variable (i.e. the

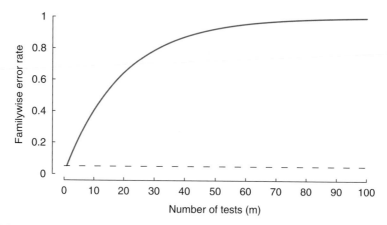

Figure 15.1 Familywise error rate as a function of the number of tests (m), assuming no true effect. The horizontal dashed line indicates the alpha level of 0.05.

number of levels of that variable), it does **not** control for familywise error across the number of main effects and interactions. This means that the chance of a large factorial ANOVA producing a false positive is worryingly high. For example, with four factorially combined independent variables, there will be 6 (2-way) + 4 (3-way) + 1 (4-way) = 11 interaction terms, so 15 effects in total including the main effects. The interdependencies inherent in factorial designs mean that the curve in Figure 15.1 doesn't directly apply to ANOVA, but it's clear that marginally significant effects in large factorial designs should be treated with caution.

The multiple comparisons problem is also an issue for some modern high-throughput and data mining techniques, where large amounts of data are readily available. In genetics, genome-wide association studies involve many thousands of simultaneous comparisons between different SNPs (single-nucleotide polymorphisms) and some phenotype such as disease susceptibility. In brain imaging studies, many thousands of voxels (volumetric pixels) are recorded simultaneously, and often compared across different experimental conditions to determine which brain regions respond to a particular stimulus. Finally, the age of *big data* and widespread internet availability means that researchers have access to an arbitrarily huge number of variables, some of which will be spuriously associated by chance. There is an excellent website (**https://www.tylervigen.com/spurious-correlations**) and book (Vigen 2015) with many absurd examples, such as the correlation (from 1999 to 2009) between people drowning in swimming pools, and films released starring Nicholas Cage ($r = 0.67$).

The issues with Type I errors are well known in their respective fields, and are potentially the cause of numerous published results of dubious provenance. Some especially questionable research practices that take advantage of an inflated false positive rate include *p-hacking* (measuring many variables, but reporting only those that are significant), and *optional stopping* (continuing to collect data until a significant result is found). This may partly explain why large-scale attempts to replicate published findings have extremely low success rates (Open Science Collaboration 2015).

The traditional solution: Bonferroni correction

The simplest solution to the multiple comparisons problem is to modify the α level by dividing it by the number of tests (m):

$$\bar{\alpha} = \frac{\alpha}{m} \tag{15.2}$$

where $\bar{\alpha}$ is the corrected significance criterion. This is known as *Bonferroni correction*, and has a reputation for being a conservative solution. It reduces the familywise error rate to be less than or equal to α, which can be confirmed by calculating:

$$FWER = 1 - \left(1 - \frac{\alpha}{m}\right)^m \tag{15.3}$$

For large values of m, the error rate will always be slightly below α with this formula. For example, with α = 0.05 and m = 100, the error rate is $1 - \left(1 - \frac{0.05}{100}\right)^{100} = 0.0488$. The corrected α level ($\bar{\alpha}$) is then used to threshold the p-values of each statistical test to determine significance.

An alternative implementation of Bonferroni correction, which is the default in some statistical software packages such as SPSS, is to adjust the p-values instead of the α level. This is achieved by multiplying the p-value by m, capping the upper limit at 1. For this approach, the adjusted p-value is then compared to the standard (i.e. uncorrected) threshold value of α.

To illustrate the distinction between these two implementations, imagine that we run eight tests, and one of the p-values comes out at p = 0.0043. We could correct our α level to 0.05/8 = 0.00625, and compare the p-value to this threshold. Since 0.0043 < 0.00625, we would conclude the test is significant. Alternatively we can correct the p-value itself by calculating that 0.0043*8 = 0.034. Since 0.034 < 0.05, we would again conclude the test is significant. These two methods are equivalent, but it is important to report which one you are using so that someone reading your results understands what you have done.

A more exact solution: Sidak correction

A very similar correction that is slightly more precise, and slightly less conservative, than Bonferroni correction is the *Sidak* correction. This sets the α level as the inverse of the equation for calculating familywise error. So the adjusted α value, $\bar{\alpha}$, is given by:

$$\bar{\alpha} = 1 - (1 - \alpha)^{1/m} \tag{15.4}$$

Notice that this is very similar to equation 15.1, except that the exponent is $1/m$ instead of m. It has the effect of reversing the familywise error rate, so that when it is plugged into the familywise error calculation, the false positive rate remains fixed at α (i.e. 0.05), regardless of the number of tests. For our example scenario of eight tests, the Sidak-corrected criterion will be:

$$\bar{\alpha} = 1 - (1 - \alpha)^{1/m} = 1 - (1 - 0.05)^{1/8} = 0.0064 \tag{15.5}$$

A higher power solution: the Holm–Bonferroni correction

Both Bonferroni and Sidak correction mean that each individual test must reach a higher level of significance (i.e. a smaller p-value) before it can be considered significant, compared with if it were run in isolation (and not part of a family of tests). The *Holm–Bonferroni* method is less conservative than this (Holm 1979). It works by taking an ordered list of p-values from all tests, and adjusting the criterion for significance sequentially for each item in the list. The test with the smallest p-value will be assessed against a criterion of $\frac{\alpha}{m}$, the next smallest p-value against a criterion of $\frac{\alpha}{m-1}$ and so on. If we define j as the ranked position of a p-value, the criterion is $\frac{\alpha}{m-(j-1)}$. The largest p-value is assessed against the original uncorrected value of α, because when $j = m$, it follows that $\frac{\alpha}{m-(m-1)} = \frac{\alpha}{1}$. This has the effect of keeping the familywise error rate at or below α, but reducing the Type II error rate (i.e. it is less likely to miss true effects). There is an equivalent method known as the *Holm–Sidak* method, where the Sidak correction is progressively applied instead of the Bonferroni correction.

Adjusting α reduces statistical power

A big problem with corrections that adjust the α level is that they reduce statistical power. As discussed in Chapter 5, power is the ability of a statistical test to detect a true effect. Because power depends directly on the α level, a consequence of adjusting α is that either effect sizes or sample sizes must become much larger to detect significant effects, compared with situations where no correction is applied. For example, Figure 15.2 illustrates how power for an individual test is reduced as the number of tests in the family increases, for Bonferroni-corrected α levels. Figure 15.2(a) shows that for a medium effect size ($d = 0.5$), correcting

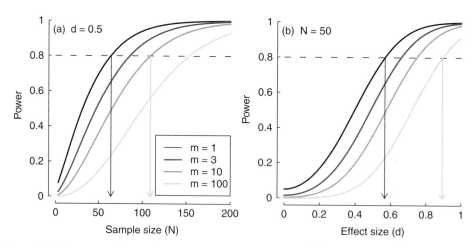

Figure 15.2 Power curves for a single corrected test, for different sizes of test family (m), following Bonferroni correction. Panel (a) shows how power increases as a function of sample size, for an effect size of $d = 0.5$. Panel (b) shows how power increases as a function of effect size, for a sample size of $N = 50$. The dashed horizontal line in both panels shows the target of 80% power.

across 10 comparisons requires an increase in sample size from around N = 64 to N = 109 to maintain power at 0.8 (see the arrows). Figure 15.2(b) shows that for a fixed sample size (N = 50), the minimum effect size that can be detected with a power of 0.8 increases from around $d = 0.57$ with $m = 1$ to $d = 0.89$ with $m = 100$. This means that designing a study that will involve conducting many tests will tend to require a much larger sample size (or be limited to detecting only large effects), compared with a study involving fewer tests.

In the context of making adjustments to the α level, it is worth some consideration of where the criterion of $\alpha = 0.05$ comes from, and whether it remains appropriate. This threshold is often treated as an absolute limit on determining statistical significance, however it is really only a historical convention, and differs across research fields. The value of 0.05 was proposed by Ronald Fisher, the inventor of ANOVA, almost 100 years ago (Fisher 1926). Fisher felt that a 1-in-20 chance of a false positive gave a reasonable balance between identifying a real effect and being misled by noise in the data. However he was not as dogmatic about this value as many more recent scientists (and journals) have become—he also suggested considering α values of 0.02 and 0.01 if the level of 0.05 'does not seem high enough odds'. In recent years, there have been calls to redefine significance at a lower value of $\alpha = 0.005$ (Benjamin et al. 2017), or to provide an explicit rationale for choosing a given value of α (Lakens et al. 2018), or to dispense with the concept of significance entirely (Amrhein, Greenland, and McShane 2019). Ultimately it is the job of a researcher or analyst to choose a criterion that is convincing to both themselves and those reading their work, and this also applies to correction for multiple comparisons. Alternatively, see Chapter 17 for a different approach to statistical hypothesis testing.

Whatever baseline level of α we choose, one potential solution to avoiding overly stringent multiple comparison correction is to reduce the number of tests. In recent years, preregistering an analysis plan in which the anticipated comparisons are specified in advance has been suggested as one way to reduce *researcher degrees of freedom*. But in many situations, such as novel and exploratory work, it is not possible to specify which conditions are expected to produce significant results until after the data have been collected. In such cases, there are two relatively recent developments designed to control the rate of Type I errors but still maintain statistical power as far as possible. These are *false discovery rate* correction, and *cluster correction*. The former method generally makes the assumption that the various tests are independent. The latter method is specifically used when there are correlations between successive tests.

Controlling the false discovery rate

The familywise error rate described above is the probability that one out of a family of tests will be a false positive. The *false discovery rate* (FDR) is subtly different—it is the proportion of all **significant** results that are false positives. A seminal paper by Benjamini and Hochberg (1995) introduced methods for controlling the false discovery rate at a specific level. From a philosophical perspective, this approach accepts that some of our significant results will be false positives (Type I errors), and tries to keep this at a known level, so as to minimize the number of missed true effects (Type II errors). It is particularly appropriate for research designs where significant results are followed up by further experiments, for example in drug

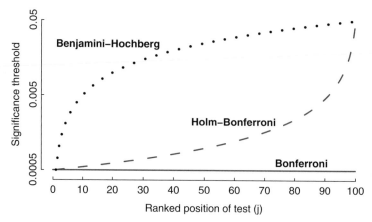

Figure 15.3 Effect of three types of multiple comparison correction across 100 tests on the threshold for significance (y-axis) at each position in a rank-ordered list of p-values (x-axis). Note the logarithmic scaling of the y-axis.

development where a series of candidates might be screened, with those producing significant results subjected to additional tests.

The desired false discovery rate is represented by a value called q^*. We will assume, as Benjamini and Hochberg (1995) did, that $q^* = 0.05$. However for large data sets it is possible to estimate q^* empirically (see Storey and Tibshirani 2003). Much like the Holm corrections, the Benjamini–Hochberg method also takes a rank-ordered list of p-values, but this time compares each value to the criterion $\bar{\alpha} = \dfrac{q^* j}{m}$ (where, again, j is the position in the list, from 1 to m). The threshold for significance across three methods is compared in Figure 15.3, for a family of $m = 100$ tests at each position (j) in the rank-ordered list of p-values. Because the Benjamini–Hochberg curve sits above the other two lines, it is by far the least conservative of the three methods.

Comparison of FDR vs FWER correction

To see how false discovery rate (FDR) and familywise error rate (FWER) correction compare, let's imagine that we run a series of t-tests that produce the following set of 10 p-values:

```
## [1] 0.006 0.754 0.012 0.003 0.005 0.049 0.197 0.022 0.088 0.002
```

We can compare Bonferroni correction, Holm–Bonferroni correction, and FDR correction for these p-values. We do this by sorting the numbers into ascending order, as shown in Table 15.1. The final three columns of the table give the critical threshold values for significance for Bonferroni correction (0.05/10 = 0.005 for all tests), Holm–Bonferroni correction, and FDR correction. A test is significant for each type of correction if the p-value is *lower* than the threshold value in these columns.

For Bonferroni correction, the first two p-values in the list are below the threshold, and so only these two tests are significant—the other eight are not. Holm–Bonferroni correction is a

Table 15.1 Example of multiple comparisons correction for a set of 10 tests. j is the position in the rank-ordered list of p-values, and the column headed p gives the p-values. The columns headed Bonf, Holm, and FDR (thresh) give the critical (threshold) values for Bonferroni correction, Holm–Bonferroni correction, and false discovery rate correction.

j	p	Bonf (thresh)	Holm (thresh)	FDR (thresh)
1	0.002	0.005	0.005	0.005
2	0.003	0.005	0.006	0.010
3	0.005	0.005	0.006	0.015
4	0.006	0.005	0.007	0.020
5	0.012	0.005	0.008	0.025
6	0.022	0.005	0.010	0.030
7	0.049	0.005	0.012	0.035
8	0.088	0.005	0.017	0.040
9	0.197	0.005	0.025	0.045
10	0.754	0.005	0.050	0.050

Table 15.2 Example of multiple comparisons correction for a set of 10 tests. As for the previous table, but the columns headed Bonf, Holm, and FDR (corr) give the corrected p-values for Bonferroni correction, Holm–Bonferroni correction, and false discovery rate correction.

j	p	Bonf (corr)	Holm (corr)	FDR (corr)
1	0.002	0.02	0.020	0.015
2	0.003	0.03	0.027	0.015
3	0.005	0.05	0.040	0.015
4	0.006	0.06	0.042	0.015
5	0.012	0.12	0.072	0.024
6	0.022	0.22	0.110	0.037
7	0.049	0.49	0.196	0.070
8	0.088	0.88	0.264	0.110
9	0.197	1.00	0.394	0.219
10	0.754	1.00	0.754	0.754

little more forgiving, and finds the first four tests significant. False discovery rate correction is the most liberal, with tests 1–6 reaching significance. This direct comparison shows the potential for these three methods to lead to different answers for a given family of tests.

Alternatively, we could correct the p-values themselves, as shown in Table 15.2. Now we can compare each adjusted p-value to the uncorrected α level of 0.05. The same tests are significant as before. Note that the adjustment for the Benjamini–Hochberg method is slightly different from the others. It is a sequential process, where we start with the largest p-value, and work our way through the list. Each adjusted p-value is the lesser of the previous adjusted value, and the current p-value multiplied by an adjustment factor of $\frac{m}{j}$ (where m is the number of tests, and j is the rank-ordered position). The reason we take the lesser of the two numbers is to prevent the adjustments potentially corrupting the rank ordering of p-values.

Correcting the false discovery rate has become a standard approach in many fields, particularly genetics where data sets are large enough to estimate the false discovery rate directly from the data (e.g. Storey and Tibshirani 2003). There also are variants that can cope with dependencies (i.e. correlations) between the different tests (Benjamini and Yekutieli 2001). Overall, it is a suitable choice for dealing with the multiple comparisons problem without having as dramatic an impact on statistical power as other methods.

Cluster correction for contiguous measurements

For many types of data acquisition, sampling occurs in either space or time, or sometimes both. For example, cameras and microscopes create a two-dimensional image sampled at some particular resolution (number of pixels per unit area). Physiological signals like muscle activity, skin conductance, pupil diameter, or electromagnetic brain activity are sampled rapidly on the order of milliseconds. Finally, movie cameras and MRI scanners produce two- or three-dimensional images at multiple points in time.

For all of these types of data, the sample rate is largely arbitrary, and is often determined by hardware restrictions. Imagine you used an infrared video camera to record the movements of fireflies at night. Perhaps your camera returns images that are 256×256 pixels 10 times per second, and that you perform some pixelwise statistics across time to determine the number of fireflies in a sequence of images. At that resolution you would make $256 \times 256 = 65536$ comparisons. But now imagine we upgraded our camera to one with a higher resolution—1024×1024 pixels equates to over a million comparisons. Even a relatively liberal correction for multiple comparisons that adjusts the α level will struggle to reach significance for very many image locations because statistical power will be massively reduced. But there is clearly something wrong here, as a better camera should give us better results! Cluster statistics are a solution to this issue because they group significant observations together into *clusters*, which become the meaningful unit for significance testing regardless of the sampling resolution.

It is reasonable to assume in many situations that adjacent sample points will be correlated to some extent. For example, two adjacent pixels in an image are likely to be more similar in luminance than two randomly selected pixels (Field 1987), and samples from successive moments in time are also likely to be highly correlated (the extent of which can be assessed by calculating the *autocorrelation function*). Sometimes, data preprocessing methods such as low pass filtering (see Chapter 10) or smoothing will deliberately blur adjacent samples together to reduce noise. This is a problem for traditional multiple comparison corrections, which generally assume that tests are independent. Again, cluster correction avoids these issues because it takes into account correlation between adjacent observations.

The basic idea of cluster correction is that we identify clusters—contiguous regions of space and/or time where a test statistic is significant at some threshold level. We then aggregate (add up) the test statistic within each cluster, and compare these values to an empirically derived null distribution. Those clusters that fall outside some quantile (i.e. the 95% region) of the null distribution are retained and considered significant. Various algorithms have been developed along these same lines, but some of them have recently been shown to suffer inflated Type I error rates (see Eklund, Nichols, and Knutsson 2016). The method we will discuss in detail here is a non-parametric cluster correction technique described by Maris and Oostenveld (2007) that does not suffer these issues. We will demonstrate its use on an example data set.

Example of cluster correction for EEG data

Our example data set is an unpublished EEG data set for a cueing experiment, collected in my lab during 2017. The reaction times from the same experiment have been reported by Pirrone et al. (2018). In the experiment, 38 participants were first shown a direction cue (a face with its eyes looking either left or right). Then a visual stimulus (a sine-wave grating) appeared on either the left or the right side of the screen. There were 800 trials per participant, with half of the stimuli appearing randomly on each side of the screen. Brain activity was recorded from 64 locations on the scalp using electroencephalography (EEG), and epoched relative to the time of stimulus onset (0 ms in all plots). Each trial was normalized by subtracting the average of the voltage during the 200 ms before stimulus onset. The waveforms in Figure 15.4 show group averages from electrode P8, which is on the right side of the head near the back (see grey point in upper left insert), adjacent to brain regions that respond to visual stimuli. The black trace is the averaged response to stimuli shown on the right side of the screen, and the blue trace is the response to stimuli shown on the left side of the screen.

There are some substantial differences between the two waveforms, which we can summarize by subtracting the two conditions to calculate a difference waveform. This is shown by the trace in Figure 15.5. There are differences from shortly after stimulus onset, which persist throughout the epoch. But are these differences statistically significant? If we conduct a series of paired t-tests to compare the conditions, these reveal many time points where the waveforms differ, as summarized by the dark blue lines at $y = -2$ in Figure 15.5. However, this involves running 1200 t-tests (because the data are sampled at 1000 Hz), and with $\alpha = 0.05$

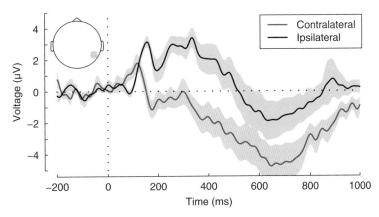

Figure 15.4 Example EEG data from a cueing experiment. Data are averaged across 400 trials per condition for each participant, and 38 participants. Data were bandpass filtered from 0.01 to 30 Hz, and taken from electrode P8 (grey point in upper left insert). Shaded regions show ±1SE across participants.

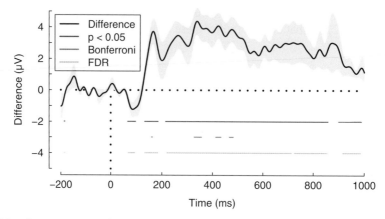

Figure 15.5 Difference waveform (ipsilateral–contralateral). Shaded regions indicate ±1SE across participants. The horizontal lines at y = –2 and below indicate time windows where the difference waveform differed significantly from 0 (one sample t-test), following different corrections (legend).

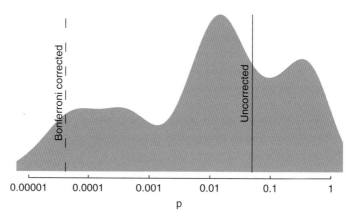

Figure 15.6 Kernel-density histogram of uncorrected p-values (blue shading). The solid vertical line indicates the uncorrected threshold of 0.05, and the dashed line indicates the Bonferroni-corrected threshold of 0.00004. Note the log-scaling of the x-axis.

we would expect at least 60 of these to be significant even if there were no true effects. A good example of a definite false positive is at the very start of the time window (–200 ms). This is before the stimulus was presented, so it must necessarily be a statistical artefact.

Implementing traditional familywise error rate correction will substantially affect our results. The Bonferroni-corrected α level is 0.05/1200 = 0.00004. If we plot a histogram of uncorrected p-values, it shows that very few of the tests remain significant with such severe correction (see Figure 15.6). The blue lines at y = –3 in Figure 15.5 show very few significant time points. Even with the more liberal false discovery rate correction (see light blue lines at y = –4 in Figure 15.5), we still lose quite a lot of our significant time points. Indeed, our largest cluster splits into two, and our other clusters are reduced in size.

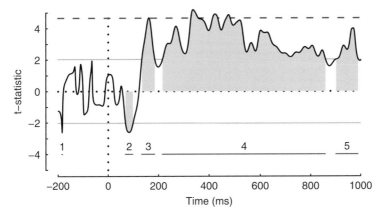

Figure 15.7 Trace of t-statistics as a function of time. The thin blue lines bound the critical t-values, and the blue dashed line shows the critical Bonferroni-corrected t-value. Grey shaded regions indicate periods of significant t-values, which also correspond to the clusters at y = −4.

To perform cluster correction, we need more than just our *p*-values. We also need to think about the test statistic, which in this case is a t-value, though it could equally well be an F-ratio, correlation coefficient, or any other test statistic. Figure 15.7 shows the trace of the t-statistics as a function of time. This looks broadly similar to the difference waveform, because the t-statistic is the mean scaled by the standard error at each time point. The thin blue horizontal lines indicate the critical t-value for a test with 37 degrees of freedom at $\alpha = 0.05$ (which is around t = ±2). The significant clusters correspond to the time periods when the t-statistics are outside of these bounds, as shown by the blue lines at y = −4, and the grey shaded regions.

For the current example, we have five clusters, which have been numbered consecutively in Figure 15.7. These range from very brief durations (i.e. cluster 1) to a very long one (cluster 4). We next calculate the *summed t-value* across all time points within each cluster. This just involves adding up all the t-values in a single cluster. The summed t-statistics for our five clusters are as follows:

```
## [1]  -7.22676 -73.03767 182.08464 2176.61865 254.14413
```

We select the largest of our summed t-values (ignoring the sign), which in this case is cluster 4, with a summed t-statistic of 2177. The raw data from this cluster will be used to generate a *null distribution* using a resampling technique (see also 'Using resampling for hypothesis testing' in Chapter 8). The idea here is that we randomly reassign the condition labels for each participant, and then at each time point within the cluster we repeat the t-test and recalculate the summed t-statistic. This is done at least a thousand times (with different reshufflings of condition labels), to build a distribution of resampled summed t-statistics. Because of the random condition assignments, the distribution should have a mean around 0, and some spread determined by the variability of the data. What we are doing here by randomizing the condition assignments is building up a picture of how we should expect the data to look if there were no true difference between the groups. We will then compare this to our observed data to see if there is evidence for real differences.

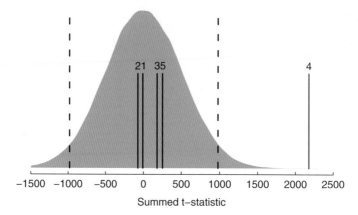

Figure 15.8 Null distribution of resampled summed t-values. The dashed black lines give the upper and lower 95% confidence intervals. The numbered vertical lines correspond to the clusters. Only cluster 4 exceeds the 95% confidence interval, so only this cluster is considered significant.

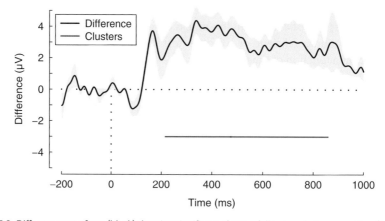

Figure 15.9 Difference waveform (black) showing significant clusters following cluster correction (blue line).

The null distribution is shown by the blue shaded curve in Figure 15.8. The vertical dashed lines indicate the 95% confidence intervals of the null distribution. These are used as thresholds to compare with each individual cluster's summed t-value. The clusters are indicated by the shorter numbered lines. Clusters 1, 2, 3, and 5 are inside the 95% confidence intervals of the null distribution, so these are rejected as not reaching statistical significance. Cluster 4 falls outside of the confidence intervals, and is retained as being significant.

The final difference waveform with the surviving cluster is shown in Figure 15.9, indicated by the horizontal blue line. Note that the smaller clusters have been removed, including the spurious one that occurred before the stimulus was presented. Also, the remaining cluster has not shrunk—its start and end points are preserved. So we would conclude that the significant difference in activity between conditions begins around 200 ms, and extends to around 850 ms.

Cluster correction takes a little getting used to, as it involves very different processes to other multiple comparison corrections. Nevertheless, it has been robustly validated, and achieves a good balance between maintaining the desired familywise error rate, and avoiding Type II errors and cluster shrinkage.

Correcting for multiple comparisons in *R*

As described previously, the two classical approaches to multiple comparison correction are to either correct the α level, or adjust the *p*-values. To make the procedures clear, we will demonstrate both approaches manually, and also describe the use of a built-in *R* function called *p.adjust*, which can be used to adjust *p*-values using various methods. First, let's look at manual Bonferroni correction. For a vector of *p*-values (from Tables 15.1 and 15.2, but these could be collated across all the tests you have run for a data set), we can identify the significant ones at $\alpha = 0.05$ as follows:

```
pvals
## [1] 0.006 0.754 0.012 0.003 0.005 0.049 0.197 0.022 0.088 0.002
alphalevel <- 0.05
pvals < alphalevel
## [1] TRUE FALSE TRUE TRUE TRUE TRUE FALSE TRUE FALSE TRUE
```

With no correction, seven values reach significance. We can Bonferroni-correct our α level by dividing it by the number of tests (which we store in a new data object called *m*, defined as the length of the *pvals* vector):

```
m <- length(pvals)
alphahat <- alphalevel/m
pvals < alphahat
## [1] FALSE FALSE FALSE TRUE FALSE FALSE FALSE FALSE FALSE TRUE
```

Only two of our *p*-values survive this correction. Alternatively, we can apply the Sidak correction, which finds one additional test to be positive:

```
alphahat <- 1 - (1 - alphalevel)^(1/m)
pvals < alphahat
## [1] FALSE FALSE FALSE TRUE TRUE FALSE FALSE FALSE FALSE TRUE
```

The Holm–Bonferroni method is a sequential technique, so we need to rank order our list of *p*-values first using the *sort* function:

```
pvals <- sort(pvals)
pvals
## [1] 0.002 0.003 0.005 0.006 0.012 0.022 0.049 0.088 0.197 0.754
```

The thresholds for significance depend on the position in the list, and can be calculated as follows:

```
holm <- alphalevel/(m - ((1:m)-1))
print(holm, digits=1)
## [1] 0.005 0.006 0.006 0.007 0.008 0.010 0.013 0.017 0.025 0.050
```

Finally, we can compare our *p*-values to the thresholds as follows:

```
pvals < holm
## [1] TRUE TRUE TRUE TRUE FALSE FALSE FALSE FALSE FALSE FALSE
```

The thresholds for false discovery rates are calculated as follows:

```
fdr <- (alphalevel*(1:m))/m
fdr
## [1] 0.005 0.010 0.015 0.020 0.025 0.030 0.035 0.040 0.045 0.050
```

And again can be compared with the rank-ordered *p*-values:

```
pvals < fdr
## [1] TRUE TRUE TRUE TRUE TRUE TRUE FALSE FALSE FALSE FALSE
```

Next, let's look at how to correct the *p*-values themselves, instead of adjusting the α level (recall that these methods produce the same outcome). If we want to apply the false discovery rate algorithm manually, we can do so using a loop (see 'Loops' in Chapter 2) that starts with the second to largest *p*-value, and works its way down the list:

```
fdr <- pvals
for (i in ((m-1):1)){
    fdr[i] <- min(pvals[i]*m/i, fdr[i+1])}
print(fdr, digits=2)
## [1] 0.015 0.015 0.015 0.015 0.024 0.037 0.070 0.110 0.219 0.754
```

However, in R, the *p.adjust* function will do all of our adjustments for us automatically. We just need to give it the vector of *p*-values, and tell it which correction to apply. For false discovery rate correction, we get the same results as above:

```
p.adjust(pvals,method='fdr')
## [1] 0.01500000 0.01500000 0.01500000 0.01500000 0.02400000 0.03666667
## [7] 0.07000000 0.11000000 0.21888889 0.75400000
```

We can also apply Bonferroni correction to adjust our *p*-values as follows:

```
p.adjust(pvals,method='bonferroni')
## [1] 0.02 0.03 0.05 0.06 0.12 0.22 0.49 0.88 1.00 1.00
```

Or we could use Holm–Bonferroni correction:

```
p.adjust(pvals,method='holm')
## [1] 0.020 0.027 0.040 0.042 0.072 0.110 0.196 0.264 0.394 0.754
```

There are several additional correction methods available that you can read about in the *help* file for the *p.adjust* function.

Implementing cluster correction in *R*

Cluster correction is built into most neuroimaging analysis packages, including *FSL*, *SPM*, *AFNI*, *FieldTrip*, *Brainstorm*, and *EEGlab*. Several R packages also implement some form of cluster correction algorithm, including the *eyetrackingR* and *permuco* packages. However, such packages typically require data to be in a specific format, and it is instructive to demonstrate how we can perform the calculations manually so that they can be adapted to different situations and data types.

Let's repeat the cluster correction on the ERP data from earlier in the chapter (see Figure 15.4), which is contained in the file *ERPdata.RData* available from the book's GitHub repository. The file contains a matrix called *data*, which has 38 rows (participants) and 1200 columns (time points):

```
N <- nrow(data)
m <- ncol(data)
N
## [1] 38
m
## [1] 1200
```

Each data point is the voltage difference for a particular participant and time point. We can therefore perform one-sample t-tests (comparing to 0) at each time point, and store the results in vectors of *p*-values and t-statistics as follows:

```
allp <- NULL
allt <- NULL
for (n in 1:m){
    output <- t.test(data[,n])    # run a t-test
    allp[n] <- output$p.value  # store the p-value
    allt[n] <- output$statistic # store the t-value
}

allp[1:5]
## [1] 0.2240673 0.2230249 0.2207536 0.2171818 0.2121924
```

Next, we need to find all the significant time points that are contiguous—i.e. those that occur next to each other. There are doubtless some fancy sophisticated ways of doing this, but we will use a loop, to be as transparent as possible. The loop checks each *p*-value in turn. If it is significant and we are not currently in a cluster, we start a new one, and note the start time for this cluster. If it is non-significant and we are in the middle of a cluster, we note the end time for the cluster. After the loop, if we are still in a cluster, we record the end of the cluster as the end of the data set. At the end of this section of code, we see that five clusters have been identified, along with the start and end indices:

```
clusterstarts <- NULL
clusterends <- NULL
nclusters <- 0    # at the start, we have no clusters
incluster = 0 # at the start, we are not in a cluster
for (n in 1:m){    # loop through all time points
    if (allp[n]<0.05){ # if this test is significant
      if (incluster==0){ # and if we are not in a cluster
        nclusters <- nclusters + 1 # create a new cluster
        clusterstarts[nclusters] <- n # note the start time
        incluster <- 1 # record that we are now in a cluster
      }
    }
    if (allp[n]>=0.05){ # if this test is not significant
      if (incluster==1){ # and if we are in a cluster
```

```
        clusterends[nclusters] <- n-1 # note the end of cluster
        incluster <- 0 # record that we are not in a cluster
    }
  }
}
# if we end the loop in a cluster, record the end of the cluster
if (incluster>0 & nclusters>0){clusterends[nclusters] <- m}

nclusters
## [1] 5
clusterstarts
## [1]   15 269 334 416 1101
clusterends
## [1]   17 298 385 1058 1189
```

Now that we have identified our clusters, we can add up the t-values for each cluster. This uses the vector of t-values that we created earlier, and the start and end indices of the clusters:

```
summedt <- NULL
for (n in 1:nclusters){
    summedt[n] <- sum(allt[clusterstarts[n]:clusterends[n]])}

summedt
## [1]  -7.22676 -73.03767 182.08464 2176.61865 254.14413
```

To generate the null distribution, we need to choose the largest cluster and note its start and end points:

```
biggestcluster <- which(summedt==max(summedt)) # find largest cluster
cstart <- clusterstarts[biggestcluster]      # cluster start index
cend <- clusterends[biggestcluster]      # cluster end index
```

Because we are using a one-sample t-test comparing to zero, we can permute the group labels by randomly inverting the sign of the voltage difference for half of our participants. This is a bit of a shortcut, and if we were doing a between-groups comparison we would explicitly switch the group assignments instead. The following code calculates a t-statistic manually, because this is much faster than calling the *t.test* function.

```
# vector of positive and negative weights to be permuted
signlabels <- sign(c(-(1:(N/2)),(1:(N/2))))
nullT <- NULL
for (i in 1:1000){      # repeat lots of times
    tsum <- 0 # create a data object to store the summed t-value
    randsigns <- sample(signlabels,replace=TRUE) # permute weights
    for (n in cstart:cend){   # loop across all cluster indices
      tempdata <- data[,n]*randsigns      # weight the data
      # calculate the t-value and add to the sum
      tsum <- tsum + (mean(tempdata)/(sd(tempdata)/sqrt(N)))
    }
    nullT[i] <- tsum   # store summed t-value in a list
}
```

The null distribution is now stored in the data object *nullT*, and contains 1000 summed t-statistics. We need to compare the summed t-value for each cluster to this distribution (see Figure 15.8). We can do this using the *quantile* function to identify the lower and upper 95% quantiles (i.e. the points between which 95% of the values in the distribution lie). We can then compare each of our summed t-values to see if they fall outside of these limits, and retain any that meet this criterion:

```
distlims <- quantile(nullT,c(0.05/2,1-(0.05/2)))
distlims
##    2.5%    97.5%
## -988.5799 952.6620
sigclusts <- c(which(summedt<distlims[1]),which(summedt>distl
ims[2]))
sigclusts
## [1] 4
```

Finally, we can retain the start and end indices of any significant clusters (in this case, just cluster 4):

```
if (length(sigclusts)>0){
clustout <- matrix(0,nrow=length(sigclusts),ncol=2)
clustout[,1] <- clusterstarts[sigclusts]
clustout[,2] <- clusterends[sigclusts]}

clustout
##    [,1] [,2]
## [1,] 416 1058
```

So our significant cluster starts at index 416, and ends at 1058. We can use these to index the vector of sample times (from −199 to 1000 ms), to find the true start and end times for the cluster (in ms):

```
timevals <- -199:1000
timevals[clustout]
## [1] 216 858
```

Finally, we can plot the results, along with an indication of the significant cluster (see e.g. Figure 15.9):

```
plot(timevals,colMeans(data),type='l')
lines(timevals[clustout],c(0,0),col='blue')
```

We can package up all of the stages for cluster correction into a single function (a version of this function is included in the *FourierStats* package that we encountered in Chapter 11). The function expects four inputs to define the data, the number of resamples for building the null distribution, and the α levels for forming clusters and determining significance:

```
doclustcorr <- function(data,nresamples,clustformthresh,clustthresh){

clustout <- NULL
N <- nrow(data)
m <- ncol(data)
```

```r
allp <- NULL
allt <- NULL
for (n in 1:m){
    output <- t.test(data[,n])
    allp[n] <- output$p.value
    allt[n] <- output$statistic
}

clusterstarts <- NULL
clusterends <- NULL
nclusters <- 0
incluster = 0
for (n in 1:m){
    if (allp[n]<clustformthresh){
      if (incluster==0){
        nclusters <- nclusters + 1
        clusterstarts[nclusters] <- n
        incluster <- 1
      }
    }
    if (allp[n]>=clustformthresh){
      if (incluster==1){
        clusterends[nclusters] <- n-1
        incluster <- 0
      }
    }
}
if (incluster>0 & nclusters>0){clusterends[nclusters] <- m}

if (nclusters>0){
summedt <- NULL
for (n in 1:nclusters){
    summedt[n] <- sum(allt[clusterstarts[n]:clusterends[n]])}

biggestcluster <- which(summedt==max(summedt))
cstart <- clusterstarts[biggestcluster]
cend <- clusterends[biggestcluster]
signlabels <- sign(c(-(1:(N/2)),(1:(N/2))))
nullT <- NULL
for (i in 1:nresamples){
    tsum <- 0
    randsigns <- sample(signlabels,replace=TRUE)
    for (n in cstart:cend){
      tempdata <- data[,n]*randsigns
      tsum <- tsum + (mean(tempdata)/(sd(tempdata)/sqrt(N)))
    }
    nullT[i] <- tsum
}
distlims <- quantile(nullT,c(clustthresh/2,1-(clustthresh/2)))
sigclusts <- c(which(summedt<distlims[1]),which(summedt>distlims[2]))
```

```
if (length(sigclusts)>0){
clustout <- matrix(0,nrow=length(sigclusts),ncol=2)
clustout[,1] <- clusterstarts[sigclusts]
clustout[,2] <- clusterends[sigclusts]}
}

return(clustout)}
```

We can then perform cluster correction with a single line of code:

```
doclustcorr(data,1000,0.05,0.05)
##    [,1] [,2]
## [1,] 416 1058
```

The first column contains the cluster start indices, and the second contains the cluster end indices. If there were multiple significant clusters, there would be additional rows in the output matrix.

Cluster correction is a very flexible technique, and I hope that by including a basic implementation, readers will see how to apply the method to their own data analysis. Although the example here involves a signal that varies in time, the same principle applies to other dimensions, such as spatial position. For a two-dimensional image (or three-dimensional volume), determining the adjacency of points in a cluster is somewhat more challenging.

Practice questions

1. What is the familywise error rate for six tests, with $\alpha = 0.01$?
    A) 0.059
    B) 0.265
    C) 0.002
    D) 0.008

2. What is the Bonferroni-corrected α value for 20 tests and an uncorrected value of $\alpha = 0.05$?
    A) 0.0642
    B) 0.0025
    C) 0.0049
    D) 1.0000

3. For Holm–Bonferroni correction of a family of 12 tests, the adjusted value of $\alpha = 0.05$ for the ninth smallest p-value is:
    A) 0.025
    B) 0.005
    C) 0.013
    D) 0.004

4. Two consequences of adjusting the α level using Sidak correction are:
    A) Power is increased and the Type I error rate is increased
    B) Power is reduced and the Type I error rate is reduced

C) Power is increased but the Type I error rate remains constant

D) Power is reduced but the Type I error rate remains constant

5. The false discovery rate is defined as:
   A) The proportion of significant tests that are Type I errors
   B) The proportion of all tests that are Type I errors
   C) The proportion of significant tests that are Type II errors
   D) The proportion of all tests that are Type II errors

6. The FDR-adjusted α level for the 50th p-value in a family of 88 tests, assuming $q^* = 0.05$, is:
   A) 0.05/(88-(50-1))
   B) 0.05*50/88
   C) 0.05*88/50
   D) 0.05/88

7. Which of the following would not be a plausible situation for using cluster analysis to compare between two experimental conditions?
   A) Continuous measures of pupil diameter from a video eye-tracker
   B) Optogenetic neural activity across a 1 cm^2 section of mouse cortex
   C) A clinical trial in which each participant contributes one outcome measure
   D) Differences in brain activity measured with an fMRI scan

8. Non-parametric cluster statistics can be calculated using:
   A) The largest p-value within each cluster
   B) The largest test statistic within each cluster
   C) The summed p-value within each cluster
   D) The summed test statistic within each cluster

9. An advantage of cluster correction over other methods of correcting for multiple comparisons is:
   A) Significant clusters retain their original extent
   B) The largest cluster is always significant
   C) The smallest clusters can always be rejected
   D) Clusters can be broken into smaller sub-clusters

10. In cluster correction, a null distribution is generated by:
    A) Resampling data from a random subset of observations within the largest cluster
    B) Resampling data from the largest cluster, with random condition assignments
    C) Summing the test statistics across all clusters
    D) Summing the test statistics across a random selection of clusters

Answers to all questions are provided in the answers to practice questions at the end of the book.

Signal detection theory

Signal detection theory (Green and Swets 1966; Macmillan and Creelman 1991) is a method for formalizing how humans (and other systems) make decisions under uncertainty. By uncertainty I mean that either the information being used to make the decision, or the decision process itself, is *noisy*—that is, it fluctuates over time. Everyday examples might include trying to understand what someone is saying on a loud train, spot a faint star in the sky at night, or work out whether you have added salt to your dinner. The theory was developed in the 1950s, primarily to characterize the performance of radar operators, but it has much wider applicability. It is the foundation of modern psychophysical studies of perception and memory, and in recent years the same concepts have been applied in artificial intelligence (machine learning) research. This chapter will cover the basic concepts of signal detection theory, and also discuss common experimental designs. However, we will begin with an example where we consider how best to determine the sex of a chicken.

The curious world of chicken sexing

The global poultry industry is unimaginably huge. Current estimates at time of writing are that there are in excess of 24 billion chickens on the planet, meaning they outnumber humans 3:1. But there is an inherent problem in chicken farming—half of all fertilized eggs hatch into male chicks, and these are useless for egg and meat production. Chicken farmers are therefore highly motivated to identify the female hatchlings as soon as possible. Although recent methods have been developed to determine sex before the eggs hatch (based on tiny emissions of gas from the shell), these are expensive and not in widespread use. Instead, the standard technique for many years has been to employ highly trained, and highly paid, expert chicken sexers.

The job of a chicken sexer is to take trays of day-old hatchlings (see Figure 16.1) and sort them by sex. They do this by inspecting the nether regions (the *vent*) of the chick under a bright light for specific features that indicate sex. The precise features are to do with the shape of a small protruberance called the *bead*, which is pointier in female chicks, and rounder in male chicks (see Biederman and Shiffrar 1987 for diagrams). This is an expert task, and it typically requires many years of training to reach the levels of performance (>99% accuracy) expected by the industry. Furthermore, each chick is inspected for only a few seconds, so professional sexers will process around 1000 chicks per hour (and over a million each year).

Figure 16.1 Recently hatched chicks. Most of them turned out to be male.

Hits, misses, false alarms, and correct rejections

Chicken sexing can be considered a signal detection problem. Suppose that our signal is whatever visual information indicates that a chick is female (i.e. the pointiness of the bead). There are then four possible outcomes for every chick that is inspected.

The first possibility is that we decide that a chick appears to be female, and it turns out (when it grows up) to actually be female. In signal detection terms, we consider this a *hit*. We correctly identified the signal we were looking for. The second possibility is that we decide that a chick is not female, and it turns out (when it grows up) to indeed be male as we predicted. This is known as a *correct rejection*, because we are saying the 'female' signal is not there, and it isn't. It is very common to calculate an overall (percentage) accuracy in detection tasks, and this is taken as the sum of hits and correct rejections, divided by the total number of trials (chicks). However, it is also important to think about instances when we are incorrect, as there are also two different types of these.

One way we can be wrong is by deciding that a chick is female, when it actually turns out to be male. This is referred to as a *false alarm*, because we have incorrectly decided that the 'female' signal is present. The converse situation is where we decide that a chick is male, but it turns out to be female. This is known as a *miss* (because we have missed the 'female' signal).

Thinking about these four possible outcomes is important, because it gives us information about how an individual is going about the task. For example, imagine three student chicken sexers: Aya, Bia, and Che. Each student is given a tray of 100 chicks to sex (half of which are female), and all three have the same overall accuracy: 80/100 chicks are correctly sexed, so their percentage accuracy is 80%. Yet as we will see, the pattern of responses is very different across the three students.

Aya is a good all-rounder. She correctly identifies 40 of the female chicks (hits) and 40 of the male chicks (correct rejections). She incorrectly identifies 10 of the female chicks as being male (misses), and 10 of the male chicks as being female (false alarms). We can summarize her performance with the grid shown on the left of Figure 16.2.

Bia has a somewhat different strategy. She is overall more likely to assign a chick as being female. Out of all her chicks, she assigns 68 as being female, and the remaining 32 as being male. In other words she has a *bias* towards assigning chicks as being female. This strategy works well for spotting the female chicks, as it means she correctly identifies nearly all of them (49 hits, 1 miss). However, because of her bias she incorrectly assigns 19 of the male chicks

Assigned sex

Aya	F	M	Total
True sex F	40 (Hits)	10 (Misses)	50
True sex M	10 (FAs)	40 (CRs)	50
Total	50	50	80% correct

Assigned sex

Bia	F	M	Total
True sex F	49 (Hits)	1 (Misses)	50
True sex M	19 (FAs)	31 (CRs)	50
Total	68	32	80% correct

Assigned sex

Che	F	M	Total
True sex F	35 (Hits)	15 (Misses)	50
True sex M	5 (FAs)	45 (CRs)	50
Total	40	60	80% correct

Figure 16.2 Grids showing the accuracy of three student chicken sexers (see text for details). FAs: false alarms; CRs: correct rejections.

to the female category (19 false alarms) and correctly identifies the rest as male (31 correct rejections). Bia's performance is summarized in the middle grid of Figure 16.2.

Che has a bias in the opposite direction to Bia. He is more likely to assign a chick as being male (perhaps to avoid getting fined for letting male chicks through as female). This strategy means that he misses 15 of the female chicks, incorrectly assigning them as male. However he correctly identifies most of the male chicks. His results are shown in the final grid of Figure 16.2.

What is striking is that all three students have the same percentage accuracy, even though both Bia and Che have substantial biases. It would be very useful if we could quantify these biases, and if we could separate them from the true sensitivity at doing the task. This is precisely what signal detection theory aims to do, as we will describe in the following section.

Internal responses and decision criteria

Signal detection theory makes some assumptions about what is happening inside the *receiver*, in other words the mind of the person making the decision. The first assumption is that the sensory information used to make the decision is mapped onto a monotonic scale termed the *internal response*. If we are comparing male and female chicks, once we have learned to do the task, we expect that some group of neurons in our brain might respond more to female chicks and less to male chicks (on average). We could then use the magnitude of this response to make a decision. The same idea applies if we were trying to detect a weak signal— for example a faint light, or a quiet sound. When the signal is present, the internal response should be larger on average than when it is absent. It is not necessary to know precisely which population of neurons are doing this task, or where they are in the brain.

The second assumption is that incoming information is processed imperfectly, and is subject to *noise*. This noise could come from a number of sources, including external noise associated with the stimulus itself (perhaps the chick moves while it is being inspected), and internal noise from the inherent variability of biological nervous systems. A consequence of noise is that repeated presentations of even the same stimulus will produce different amounts of response.

We can characterize the internal response to the two stimulus categories (here male and female) with two *probability distributions* (see examples in Figure 16.3). One distribution tells

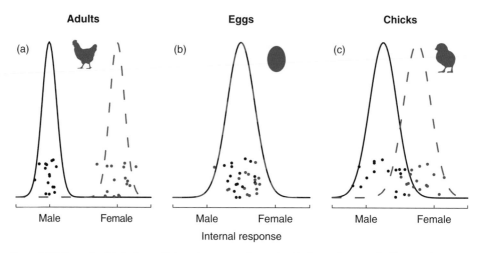

Figure 16.3 Example distributions of internal responses (curves), with data points sampled from each distribution (points). In panel (a), the distributions are far apart and narrow, consistent with an easy task. In panel (b), the distributions overlap entirely, and distinguishing male from female animals is impossible. In panel (c), the distributions overlap somewhat, indicating that there will be some animals that are misclassified.

us how likely it is that a male chick will produce a given response, and the other tells us how likely it is that a female chick will produce a given response. The midpoint of the distribution is the average internal response for that category, and the spread is determined by the noise. Different example stimuli (i.e. chicks, or trials in some experiment) can be thought of as drawing samples from these underlying distributions.

Let's start off by thinking about sexing adult birds. This should be very easy, as the males have very prominent wattles and crow loudly, and the females lay eggs. The distributions in Figure 16.3(a) are very far apart, and random samples (individual birds) from the two distributions (points) are very easy to categorize correctly. In this situation, we would expect perfect accuracy—all animals would be either hits (correctly identified females) or correct rejections (correctly identified males).

Next let's think about the opposite extreme. Imagine we have to categorize eggs as male or female before they have hatched, just by looking at them. Obviously this is impossible, and so the two distributions will completely overlap (see Figure 16.3(b)), with any given egg being equally likely to be male or female. In this situation, hits, misses, false alarms, and correct rejections are all equally likely (each consisting of 25% of trials), and accuracy will be at chance (50% correct).

A more interesting situation occurs for day-old chicks, where sexing is possible but it is challenging. The probability distributions in Figure 16.3(c) overlap somewhat. This means that some individual chicks will be difficult to categorize, and may produce misses or false alarms, but accuracy should be somewhere above chance.

So what is the best strategy for categorizing male and female chicks? The example student chicken sexers described above illustrate that there are different criteria we can adopt when doing this task. Recall that Bia is more likely to assign chicks as being female, Che more likely to assign them as male, and Aya somewhere in between. Figure 16.4 shows us the overlapping distributions in a bit more detail, with an extra feature—the vertical dotted line indicates the

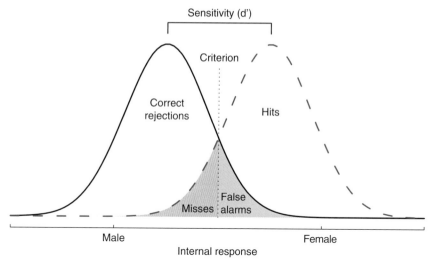

Figure 16.4 Internal response distributions with additional features. The vertical dotted line indicates the criterion—samples to the left of this line will be classified as male, and those to the right as female.

criterion. Any individual observations that fall to the left of this line will be classified as male, and any falling to the right as female. Of the chicks classified as female, some will be hits (chicks that really are female) and some will be false alarms (chicks that are actually male). Of the chicks classified as male, some will be correct rejections (chicks that really are male) and some will be misses (chicks that are actually female).

A key idea in signal detection theory is that we can move our criterion around if we want to. So we could instruct our students that misses are a very bad thing (because potentially productive female chicks would be missed). This would make them shift their criterion to the left to avoid misses (and increase hits). However the number of false alarms would also increase (and correct rejections would decrease)—in other words, they would be like Bia, who has a bias towards saying chicks are female. Alternatively we could instruct the students that false alarms are a bigger problem (because it wastes money raising male chicks), so that they shift their criterion to the right. Then false alarms (and hits) would decrease, but misses (and correct rejections) would increase. On the other hand, the sensitivity at the task is determined by the overlap of the two internal response distributions, and is not affected by changing the criterion. So this is generally assumed to be fixed for a given observer and situation (though learning and training are possible).

Calculating sensitivity and bias

We can fully characterize the pattern of performance of our student chicken sexers (and indeed any rater in a similar situation) by calculating two properties: *sensitivity* and *bias*. The sensitivity is the amount of overlap between the two distributions in Figure 16.4. This depends on both the offset of the means, and the spread of the distributions. We can standardize this by taking the difference between the means, and dividing by the standard deviation. This results in a statistic called d' (pronounced d-prime). Note that d' is conceptually very similar to measures of effect size such as Cohen's *d* (which is calculated as the difference between group means divided by the pooled standard deviation—see Chapter 5). If the distributions are narrow and far apart (e.g. Figure 16.3(a)), this will result in a large d' value. If they are wide and close together (and therefore overlap a lot), we will get a small d' value. Because d' is based on the underlying internal response, it is referred to as the *sensitivity index*.

The other component is the *bias* (C). This depends on the placement of the criterion, and can be altered by instruction. As described in the previous section, we can often choose a particularly strict or lax criterion to achieve a particular outcome. Sometimes rewards or penalties can be used to shift the criterion in a specific direction.

How can we calculate d' and C if we just have information about hits, misses, false alarms, and correct rejections? The first thing we need to do is convert our trial counts from each category into *proportions* of the total number of trials where the signal was either present or absent (we can also think of these values as probabilities). Let's do this for Aya's ratings in the first grid in Figure 16.2—these are shown in Figure 16.5.

Something we notice in the proportions grid is that the proportions within each category of true sex (i.e. the rows) must sum to one. If we have a fixed number of chicks that are truly female, each of them must either be a hit or a miss. Similarly, for a fixed number of truly male chicks, each of them must be classified as either a correct rejection or a false alarm. This means that $P(Miss) = 1 - P(Hit)$ and also $P(CR) = 1 - P(FA)$. Because of these interdependencies, once the data are converted to proportions, we actually only need to know the hit rate and the false alarm rate to fully characterize performance.

In order to calculate d', we next need to convert the proportions to z-scores using an inverse cumulative normal distribution (we will demonstrate how to do this in *R* later). Recall that a z-score is a standardized score that tells you how far an observation is from the group mean, expressed in multiples of the standard deviation. These are unitless measures, which can easily be compared across data sets. There is a straightforward monotonic transform between probabilities and z-score (standard deviation) units as shown in Figure 16.6. For Aya, the probability for hits (0.8) converts to $z(Hit) = 0.84$ and the probability for false alarms (0.2) converts to $z(FA) = -0.84$. The equation for d' is the difference between these values:

$$d' = z(Hit) - z(FA) = 0.84 - (-0.84) = 1.68 \quad (16.1)$$

True sex / **Aya**	Assigned sex: F	Assigned sex: M	Total
F	0.8 P(Hit)	0.2 P(Miss)	1
M	0.2 P(FA)	0.8 P(CR)	1

Figure 16.5 Detection performance for chick sexing, converted to proportions.

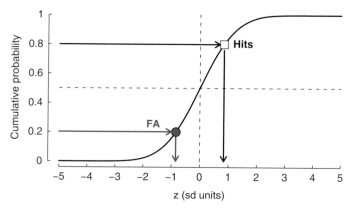

Figure 16.6 Cumulative normal distribution, mapping standardized units (z-scores) to probabilities. This function (and its inverse) are used to convert between proportions and z-scores.

We can also convert the performance of Bia and Che (our biased chicken sexers) to z-scores and d' values in the same way. For Bia, her hit rate is 49/50 = 0.98, and her false alarm rate is 19/50 = 0.38. This works out as:

$$d' = z(Hit) - z(FA) = z(0.98) - z(0.38) = 2.05 - (-0.31) = 2.36 \qquad (16.2)$$

So despite being heavily biased, Bia is actually more sensitive (and therefore potentially better at the task) than Aya. Che's hit rate is 35/50 = 0.7 and his false alarm rate is 5/50 = 0.1. His sensitivity can be calculated as:

$$d' = z(Hit) - z(FA) = z(0.7) - z(0.1) = 0.52 - (-1.28) = 1.81 \qquad (16.3)$$

This means that Che is slightly more sensitive than Aya, but less sensitive than Bia.
We can also quantify the bias of each rater by calculating the criterion. The equation for the criterion is:

$$C = -(z(Hit) + z(FA))/2 \qquad (16.4)$$

For Aya (who has no bias), this reassuringly works out to 0:

$$C = -(z(Hit) + z(FA))/2 = -(0.84 + (-0.84))/2 = 0 \qquad (16.5)$$

For Bia, who has a substantial bias, this works out as:

$$C = -(2.05 + (-0.31))/2 = -0.87 \qquad (16.6)$$

And for Che, whose bias is a bit smaller (and in the opposite direction), this works out as:

$$C = -(0.52 + (-1.28))/2 = 0.38 \qquad (16.7)$$

In many contexts, signal detection theory is used to estimate sensitivity (d') for a task, by removing the confounding effect of bias (C). For our chicken sexing example, we might want to decide which student to offer a job to. Although they all have the same accuracy, Bia was the most sensitive, and so she might be our best choice—after all, we could ask her to adjust her bias if necessary, but her underlying ability was the strongest. Without dividing accuracy data into sensitivity and bias, we would have wrongly judged all three students to be equally good at the task. Having a good understanding of signal detection theory might help to avoid analogous errors in your own research.

Estimating sensitivity and bias is important in many lab-based studies of human perceptual and memory abilities, and more recently has been applied to assessing the ability of artificial intelligence systems to judge the content of images. Sometimes an unbiased estimate of sensitivity can be of substantial practical importance: examples include a football (soccer) referee's ability to detect a handball, an airport security officer's ability to detect a weapon in a luggage X-ray, a passport control officer's ability to detect a fake passport, or disease diagnosis in the context of medical screening. The following section describes an example from the literature on cancer diagnosis.

Radiology example: rating scales and ROC curves

Radiologists must interpret diagnostic scans (usually X-ray or CT scans) to determine whether patients have cancer. In signal detection terms, the images will be in two categories: cancer (signal) present and cancer (signal) absent. The radiologists' decisions will fall into four categories:

- cancer diagnosed when cancer present (hits)
- cancer diagnosed when cancer absent (false alarms)
- cancer not diagnosed when cancer present (misses)
- cancer not diagnosed when cancer absent (correct rejections)

Exactly the same approach can be taken to analyse these data as described above. However in this context, the consequences of bias (or criterion) are worth further consideration. If a radiologist is very liberal, they might diagnose cancer even when the evidence is quite marginal. This will catch most real cancer cases, but also result in a lot of false alarms. These will cause anxiety to patients, and use additional resources running further tests (such as biopsies) to confirm the diagnosis. If a radiologist is very conservative, they might diagnose cancer only when the evidence is overwhelming. This will miss some true cancer cases, which could be fatal if not caught early. On the other hand, there will be few false alarms, so resources will not be wasted running additional tests.

A recent study (Evans et al. 2016) tested 49 radiologists in a series of controlled experiments. The stimuli were mammograms (breast images) where the eventual true diagnosis of the patient was known (i.e. the experimenters knew if the patient really did have cancer or not, based on follow-up tests). Although in clinical practice radiologists can inspect images for as long as they like, in the experiment the images were presented very briefly for 500 ms so that all participants had the same inspection time. Also, to provide more information about each radiologist's ratings, instead of a binary classification (cancer vs no cancer), participants provided ratings on a 101-point scale, from 0 (clearly abnormal) to 100 (clearly normal).

This finer-grained information allows a more detailed analysis because we can place criteria at multiple locations on the rating scale. For example, if we place a criterion at a rating of 10, and count all images with ratings below this value as cancer cases and all those above as clear, this will produce very few hits and very few false alarms. On the other hand, if we place a criterion at a rating of 90, this will produce many hits, but also many false alarms. A useful way of presenting such data is to plot the hit rate against the false alarm rate for a range of criterion levels, as shown in Figure 16.7(a).

This plotting convention is called a *receiver operating characteristic* (ROC) plot. Note that only the criterion changes across the points—the sensitivity (d') remains constant. So the ROC curve that we produce should correspond to a fixed value of d'. Figure 16.7(b) shows ROC curves for three different levels of d'. Notice how the highest sensitivity ($d' = 2$, dashed line) stretches closer to the top left corner, which represents perfect performance (i.e. a hit rate of 1 and a false alarm rate of 0). The lowest sensitivity ($d' = 0.5$, dotted line) is much closer to the diagonal, which represents chance performance (hits = false alarms). Also note that any single point on the graph corresponds to a unique combination of hit and false alarm rates, which maps onto a single estimate of d' using the equations we encountered above.

In the Evans et al. (2016) study, this method was used to estimate sensitivity for cancer diagnosis with various stimuli. When comparing breast images containing tumours with those that did not, trained radiologists had a sensitivity around $d' = 1.2$ (recall that presentation was very brief). Most strikingly, when comparing scans from the *opposite* breast of cancer patients (i.e. the one without a tumour) to breasts from healthy controls, it was still possible to identify the cancer patients with a sensitivity of $d' = 0.6$. This suggests that global changes in breast tissue provide a distributed signal that can potentially indicate patients likely to have (or to develop) cancer, even in the absence of visible tumours.

ROC curves can be constructed in other ways besides using a rating scale (and also by using rating scales with fewer levels), and applied in many contexts besides cancer diagnosis. In some lab experiments, performance can be aggregated across many participants who all naturally have different levels of bias, mapping out an ROC curve for the population. In other studies, participants can be instructed to change their criterion in different blocks

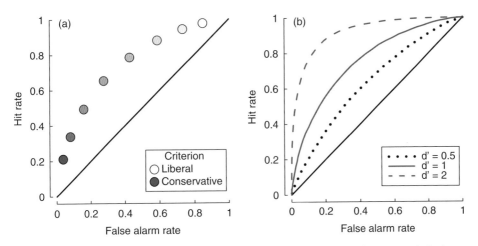

Figure 16.7 Example ROC curves. Panel (a) shows hit rates against false alarm rates for a range of criterion levels, with sensitivity fixed at $d' = 1$. Panel (b) shows ROC curves for three different levels of d'.

of the experiment (Morgan et al. 2012). This can be done explicitly, by asking participants to be more liberal or more conservative (e.g. 'Only respond "yes" if you are really sure'), or implicitly by implementing a reward and penalty structure (this is popular in behavioural economics studies). For example, rewarding hits will encourage a more liberal criterion, whereas penalizing false alarms will encourage a more conservative criterion. The data from blocks with different instructions should map out a single ROC curve.

Removing bias using forced choice paradigms

Signal detection theory gives us a principled way to estimate both sensitivity and bias in a class of experimental methods referred to as yes/no paradigms. In these experiments, a single stimulus is presented on each trial (for the examples above, this was a single chick, or a single breast image). But there is another class of method that aims to avoid the issue of bias completely. This is called a *forced choice* paradigm. The most common version is the two-alternative forced choice (2AFC) paradigm.

In a 2AFC experiment, *two* stimuli will be presented (either one after the other, or simultaneously but in different locations). One will always come from the null distribution (i.e. a male chick, or a healthy breast image), and the other will always come from the target distribution (i.e. a female chick, or a cancerous breast image). The participant's task is to indicate which of the two stimuli contains the target. This seems like a very minor difference to yes/no designs, but it has an important consequence. In terms of the internal response, we now have two discrete values to compare (see Figure 16.8). The optimal strategy is **always** to pick the stimulus that produces the largest internal response, as that is more likely to be the target. There is no need for us to have a criterion, and no opportunity for bias. Forced choice paradigms are therefore often referred to as *bias free*.

Forced choice methods are not limited to having two alternatives, but the number of alternatives will determine the baseline (chance) performance level. For 2AFC, the guess rate is 0.5 (50% correct). For 3AFC, the guess rate is 0.33 (33% correct). In general, the guess

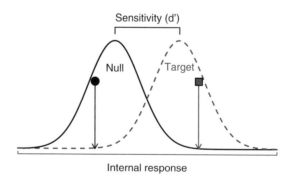

Internal response

Figure 16.8 Example internal response distributions (curves) and samples (points) for a 2AFC paradigm. The black solid curve is for the null stimulus (e.g. target absent), and the blue dashed curve is for the target stimulus (e.g. target present). There is no criterion; the optimal strategy is to select the larger of the two responses, as this is more likely to arise from the target distribution.

rate is $1/m$, where m is the number of alternatives (see also 'Situations with more than two categories' in Chapter 14). Depending on the other constraints of an experiment, there may be good reasons for choosing designs where $m > 2$. For example, 3AFC can often be explained to participants as an *odd one out* paradigm, where they choose the stimulus that appears different from the other two.

Because bias is irrelevant in mAFC designs, there is a monotonic mapping between accuracy and sensitivity. This means we can convert directly between the proportion of correct trials and d'. This is done by rescaling the accuracy, and then mapping it to the cumulative Gaussian distribution illustrated in Figure 16.6. For 2AFC, we calculate d' as:

$$d' = \phi^{-1}(P_c)\sqrt{2} \tag{16.8}$$

where P_c is the proportion of correct responses, and ϕ^{-1} indicates the inverse of the cumulative Gaussian function shown in Figure 16.6.

Note that it is very common to see single interval experiments described as being 2AFC when they are actually yes/no. The confusion stems from the two possible response options that a participant is 'forced' to choose between—for example, being shown a line stimulus and having to indicate if it is tilted leftwards or rightwards. However, this is really a yes/no paradigm in disguise, because there is only one stimulus being presented. For true 2AFC there must always be two distinct stimuli, and the participant selects one of them. If bias is negligibly small then accuracy and d' are closely related even for yes/no paradigms, so this distinction becomes largely semantic, but without estimating bias for a yes/no task this might be difficult to argue convincingly.

Manipulating stimulus intensity

The examples we have encountered so far in this chapter have involved a fixed signal level, and therefore a single corresponding level of sensitivity (d'). But it is quite common in empirical studies to manipulate properties of the signal to measure performance at a range of stimulus intensities. Consider an experiment designed to estimate a detection threshold—that is, the weakest signal strength that can be reliably detected at a given level of performance. We would typically present a range of signal intensity levels on different trials, measuring accuracy at each one. Performance (accuracy or d') can then be plotted as a function of signal strength. Figure 16.9 illustrates the underlying distributions of internal response for each signal level (panel (a)), and the corresponding proportion correct scores (panel (b)) that form a *psychometric function*.

In such an experiment, we are more interested in finding out the signal level that corresponds to some particular threshold, such as 82% correct (the midpoint of a cumulative Weibull function), or a sensitivity of $d' = 1$. This is traditionally estimated by fitting a sigmoidal function to our discrete data points (such as the curve Figure 16.9(b); see Chapter 9 for details of how non-linear models can be fitted to data), and interpolating to find the required performance level (indicated by the dashed lines). Thresholds can then be used as a dependent variable and compared across individuals, or between different experimental conditions (e.g. for different types of stimuli). These paradigms are widely used in psychophysical studies in both humans and animals, and form the basis of our contemporary understanding of sensation and perception (Kingdom and Prins 2010).

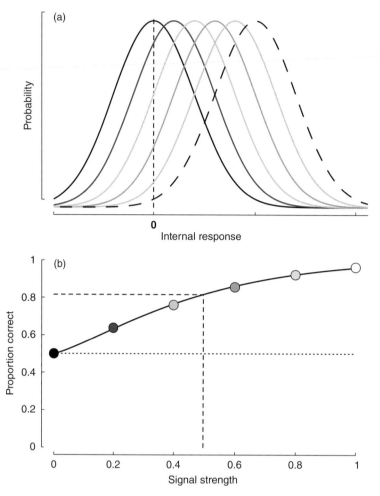

Figure 16.9 Probability distributions of internal response for different signal strengths (top panel) and proportion correct scores for a simulated 2AFC experiment (bottom panel). The black distribution is the null interval in each condition, and d' scores for the six stimulus levels range from 0 to 2.5 in steps of 0.5. The curve in the lower panel is a fitted Weibull function.

Type II signal detection theory (metacognition)

The preceding sections describe what is known as Type I signal detection theory, which is what most people think of when signal detection theory is mentioned—the quantification of performance in detection tasks. However, another aspect to the theory has gained widespread interest in recent years. This is known as *Type II signal detection theory* and involves having participants make confidence judgements about their own performance. This is a form of *metacognition*, or knowledge about our own cognitive processes. Broadly speaking, the same types of calculations can be applied to Type I and Type II measures, though the equivalent sensitivity measure is called *meta-d'*.

The additional information from Type II measures can be used to test theories and models about the source of different effects. The most famous example is the case of *blindsight*, a curious phenomenon experienced by patients with brain damage to specific parts of their visual cortex (Weiskrantz 2007). These patients do not experience conscious visual perception in the affected regions of space. If you ask them what is there, they cannot tell you. However, they are able to respond to visual information in the affected region, and their accuracy in some tasks is above chance (though they report that they are guessing). In terms of signal detection theory, Type I performance is above chance, but Type II performance is not.

Practical problems with ceiling performance

One common practical issue with calculating *d'* occurs when participants get every trial correct in one or other category of trials in a yes/no design (i.e. the hit rate or correct rejection rate is 1), or every trial is correct in an *m*AFC experiment. This can sometimes occur by chance for easy tasks with finite numbers of trials, but it implies that $d' = \infty$. This is implausible in reality, and the true *d'* value is most likely lower. Possible solutions include adding one 'fake' incorrect trial for each participant, or establishing a maximum possible *d'* score, such as $d' = 8$.

A related issue that can interfere with fitting psychometric functions (as in Figure 16.9(b)) is when performance at high stimulus intensity levels asymptotes at a proportion correct slightly below 1. This happens because of *lapse* trials, where the participant gets the trial wrong even though it is very easy, perhaps because of inattention, tiredness, or unpredictable events such as blinks or sneezes. Lapses can skew psychometric function fits by making them much shallower than they would usually be (to capture the reduced performance at high stimulus intensities). This can make the threshold estimate incorrect. Many software tools for fitting psychometric functions now give the option of setting an upper asymptote slightly below 1 to counteract this problem (see section 4.2 of Kingdom and Prins 2010).

Calculating *d'* in *R*

For yes/no tasks, we can calculate z-scores from proportions directly using the *qnorm* function to generate quantiles from the normal distribution. This is the curve shown in Figure 16.6. Here are the z-scores for a range of proportion values:

```
proportions <- seq(0.1,0.9,0.1)
proportions
## [1] 0.1 0.2 0.3 0.4 0.5 0.6 0.7 0.8 0.9
zscores <- qnorm(proportions)
round(zscores,digits=2)
## [1] -1.28 -0.84 -0.52 -0.25 0.00 0.25 0.52 0.84 1.28
```

We can then calculate d' manually by subtracting z-scores for hits and false alarms. For a hit rate of 0.95 and a false alarm rate of 0.1, d' is calculated as follows:

```
hits <- 0.95
FA <- 0.1
dprime <- qnorm(hits) - qnorm(FA)
dprime
## [1] 2.926405
```

Similarly, bias (C) is calculated using the same values:

```
C <- -(qnorm(hits) + qnorm(FA))/2
C
## [1] -0.181651
```

For 2AFC, the appropriate conversion is:

```
dprime2afc <- qnorm(proportions)*sqrt(2)
round(dprime2afc,digits=2) # round to 2 decimal places
## [1] -1.81 -1.19 -0.74 -0.36 0.00 0.36 0.74 1.19 1.81
```

Notice that the proportions below chance performance (0.5) produce negative d' scores. Most participants will tend to perform above chance in 2AFC experiments, so this should occur only rarely. Because there is only one proportion involved in this calculation, we can convert in the opposite direction as well, by scaling the d' scores by $\sqrt{2}$ and using the *pnorm* function:

```
proportionsconverted <- pnorm(dprime2afc/sqrt(2))
proportionsconverted
## [1] 0.1 0.2 0.3 0.4 0.5 0.6 0.7 0.8 0.9
```

For m AFC tasks (where $m > 2$), the calculations become more complex (see section 6.3.2 of Kingdom and Prins (2010) for a detailed explanation). However, there is a function called *dprime.mAFC* available in the *psyphy* package. It will convert proportion correct scores to d', provided m is also specified. However it can only convert one score at a time (it does not work with vectors), so to convert multiple values we can either use a loop, or use the *sapply* function (a built-in function that applies a function to all values in a vector), as follows:

```
library(psyphy)
dprime3afc <- sapply(proportions,dprime.mAFC,3)
dprime4afc <- sapply(proportions,dprime.mAFC,4)
dprime5afc <- sapply(proportions,dprime.mAFC,5)
```

If we plot the d' scores (in Figure 16.10), you can see (from the vertical dotted lines) that chance performance at the baseline (of $1/m$) always corresponds to $d' = 0$.

These tools will allow a signal detection analysis to be applied to data from yes/no and forced choice experiments. Similar tools exist in other programming languages and toolboxes: for example the *Palamedes* toolbox provides signal detection functions, along with tools for fitting psychometric functions, in the Matlab environment. In *R*, the *quickpsy* package (Linares and López-Moliner 2016) can be used to fit psychometric functions. For further reading on signal detection theory, I recommend Kingdom and Prins (2010), Macmillan and Creelman (1991), and (for an excellent historical account) Wixted (2020).

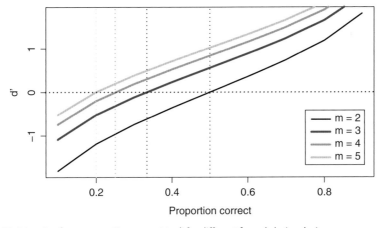

Figure 16.10 Mapping from proportion correct to *d'* for different forced choice designs.

Practice questions

1. In a yes/no task, responding 'no' when the signal is absent would be a:
   A) Hit
   B) Miss
   C) False alarm
   D) Correct rejection

2. In a yes/no task, responding 'no' when the signal is present would be a:
   A) Hit
   B) Miss
   C) False alarm
   D) Correct rejection

3. In a yes/no task with 28 hits and 15 misses, the hit rate (as a proportion) is:
   A) 0.65
   B) 0.35
   C) 0.54
   D) 0.45

4. The sensitivity index in signal detection theory is called:
   A) Cohen's *d*
   B) The criterion
   C) *d'*
   D) *z*

5. For a yes/no task, *d'* is calculated as:
   A) $z(\text{Hit}) - z(\text{Miss})$
   B) $z(\text{Hit}) - z(\text{FA})$
   C) $z(\text{Hit}) - z(\text{CR})$
   D) $z(\text{FA}) - z(\text{Hit})$

6. Using the graph in Figure 16.6, what is the approximate z-score for a probability of 0.9?
   A) 0.5
   B) 1.3
   C) 3.1
   D) 4.5

7. By collecting data on a task at a range of criterion levels, we can plot:
   A) A psychometric function
   B) The change in d' with bias
   C) A normal distribution
   D) An ROC curve

8. An experiment is described as being 2AFC if there are:
   A) Two possible responses
   B) Two different stimuli to choose between
   C) Two trials, and the participant responds to both of them
   D) Additional responses to indicate confidence

9. Use the *qnorm* function in R to calculate d' for a yes/no experiment with a hit rate of 0.95 and a correct rejection rate of 0.8.
   A) $d' = 0.80$
   B) $d' = 2.32$
   C) $d' = 2.49$
   D) $d' = 3.28$

10. Use the *qnorm* function in R to calculate d' for a 2AFC experiment where the participant got 76% of trials correct.
    A) $d' = 1.00$
    B) $d' = 0.95$
    C) $d' = 0.77$
    D) $d' = 0.71$

Answers to all questions are provided in the answers to practice questions at the end of the book.

Bayesian statistics

This chapter will introduce an approach to statistics that is quite different from that traditionally taught. Statistical tests that are evaluated for significance using a *p-value* are known as *frequentist statistics*. This category probably includes most of the tests you are familiar with—things like t-tests, ANOVA, correlation, and regression. Bayesian methods involve quite different assumptions and different procedures, and there are situations where they could be a sensible choice in data analysis. But before we go into detail about how they work, it is worth discussing some of the basic assumptions of, and problems with, frequentist methods. It might seem counterintuitive to start a chapter on Bayesian statistics by talking about frequentist methods, but these are issues that people who routinely use statistics have often not thought about in depth.

The philosophy of frequentist statistics

When we run a frequentist test, we are asking whether the data we have collected provide evidence to support our *experimental hypothesis*. Depending on the study design, this will often be of the form that the group mean differs from some stated value, or that the means of multiple groups or conditions differ from each other. The *null hypothesis* is that there are no differences. If the data provide evidence for a difference, we say that our experimental hypothesis is supported by the data, and we *reject* the null hypothesis. On the other hand, if the data do not provide evidence for a difference we have *failed to reject* the null hypothesis.

Failing to reject the null is **not** the same as accepting the null hypothesis, and we cannot conclude with confidence that there is no true difference. This might be the case, but it could also be that our experiment was *underpowered* to detect the effect we are interested in (termed a Type II error). This might happen if either the true effect size or the sample size were too small (see Chapter 5). With frequentist methods, we generally cannot distinguish whether there really is no difference, or if our study design lacks power.

The *p*-value is usually reported alongside a test statistic as a way of summarizing the outcome. Formally, the *p*-value is a statement about the probability that we would have observed our data if the experimental hypothesis were untrue. A small *p*-value suggests that it is very unlikely that we would have observed the data were the hypothesis untrue, and so this allows us to reject the null hypothesis. A large *p*-value means we have failed to reject the null but, as mentioned above, we do not obtain an estimate of how likely the null hypothesis is to be true.

An unfortunate consequence of the fixed Type I error rate

A rather curious feature of frequentist methods is that the Type I error rate is fixed, no matter the sample size. The Type I error rate is the false positive rate—the probability of reporting an effect as significant when in fact there is no true effect. It is determined by the α level, the threshold value below which a p-value is considered significant. In many disciplines, this is fixed at $\alpha = 0.05$, meaning there is a 1 in 20 (or $1/\alpha$) chance of getting an apparently significant effect when there is no true effect. Other disciplines use a more stringent threshold—for example some research in high energy physics uses the 5σ limit, which corresponds to a threshold of $\alpha = 0.0000003$, and a Type I error rate of about 1 in 3.3 million.

The curious thing about the Type I error rate is that it does not change at all if we collect more data. Whether we have 10 participants or 10000, there is still the same probability that we will get a false positive when there is no true effect. The same is not true of situations where there is a real effect. Here, our statistical power will increase with sample size, and so we will be more likely to detect a (true) effect the more observations we make. We demonstrate this by stochastic simulation (see Chapter 8) in Figure 17.1. Simulated experiments with no true effect (a Cohen's d effect size of 0) have a constant (false) positive rate of 0.05 (black circles), whereas simulated experiments involving real effects (blue squares, $d = 0.25$; white diamonds, $d = 0.5$) are more likely to be accurate with larger sample sizes.

This is a rather counterintuitive feature of frequentist methods. We might reasonably expect that as we gather more evidence, our tests should become more accurate. That is the case in the presence of a true effect, but not in the presence of a null effect. But this seems irrational, since in the real world two things can be the same, and we would ideally like our statistical tests to be able to show this.

So what has caused this curious state of affairs? Primarily it is due to a philosophical position taken by many of the originators of frequentist tests. Jacob Cohen, for example, who did seminal work on effect sizes (and from whom Cohen's d gets its name), took the position that every variable always influences every other variable to some extent, even if the effect size is minuscule. Ronald Fisher (who developed analysis of variance) argued that the null hypothesis could never be proven or established, but that each experiment had the potential to disprove the null.

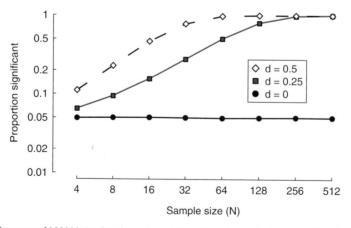

Figure 17.1 Outcome of 100000 simulated experiments for different sample sizes and effect sizes (Cohen's d).

If this position seems unbalanced, that's because it is. Indeed, in recent years it has been realized that combining Type I errors with publication bias (the preference for both journals and researchers to publish results that are statistically significant) has had a distorting effect on the scientific literature. Imagine that 20 laboratories across the world independently investigate an effect that does not truly exist. Purely by chance, one laboratory is likely to find an apparently significant result (assuming $\alpha = 0.05$). If this is the only study that gets published (because of publication bias) then a spurious effect will enter the literature. This will be the case even if all the studies have a large sample size.

Bayesian statistics: an alternative approach

The methods we are about to introduce take a different philosophical position. Instead of assuming that the experimental hypothesis is either true or unproven (as in frequentist methods), Bayesian techniques try to estimate the probabilities of both the experimental and the null hypotheses being true. In other words, in Bayesian analysis we use our observations about the world (data) to assign probabilities to our competing hypotheses. As we will discover in the following sections, this avoids the problems outlined above with frequentist methods, as well as several other shortcomings of the frequentist approach (see also chapter 11 of Kruschke 2014; Dienes and Mclatchie 2018; Wagenmakers 2007).

The legacy of Thomas Bayes: Reverend

Thomas Bayes (probably not pictured in Figure 17.2) was an eighteenth-century English clergyman, who had a particular interest in mathematics. He developed a mathematical theorem that specifies how *conditional probabilities* can be combined. This is the basis of

Figure 17.2 Portrait of a Presbyterian clergyman, claimed (controversially) to be Thomas Bayes.

Bayesian statistics, and the reason why the method bears his name. However, much of the theoretical elaboration of the approach was developed after his death by the French mathematician Pierre-Simon Laplace. Whereas frequentist statistics can often be straightforwardly conducted by hand (being based on the means and variances of the data), Bayesian methods are often more computationally intensive. This means that, although some aspects of Bayesian statistics predate their frequentist cousins, they were not commonly used until computers became widely available.

The dominance of frequentist methods in scientific research for many decades has lent Bayesian statistics a curious outsider status. Scientists who use them often self-identify as 'Bayesians', and there is even a society that meets regularly to discuss methodological development (the International Society for Bayesian Analysis). Visits to Bayes's grave in a London cemetery are common (indeed, I have even been myself). Most interestingly, Bayesians are almost evangelical about the use of their methods, and the literature contains many articles arguing that frequentist methods are fundamentally flawed and should no longer be used. I will leave the reader to come to their own conclusions about this debate.

Bayes' theorem

Bayes' theorem is an equation that governs the combination of conditional probabilities. A conditional probability is written $P(A|B)$, and expresses the probability of A *given* B. Everyday examples might be the probability that it will rain *given* that it is February, or the probability that you have an infectious disease *given* a positive test result. Where possible this book avoids equations, but Bayes' theorem (also called Bayes' rule) is sufficiently important that we will discuss it explicitly. It is defined as:

$$P(A|B) = \frac{P(B|A)P(A)}{P(B)} \tag{17.1}$$

In plain language, this equation tells us how to work out the probability of A given B, if we already know the general probabilities of both A and B, and also the probability of B given A. That probably doesn't sound very plain! A more concrete example might be more useful to introduce how this works.

Imagine that you are a general practitioner, and you want to know how likely it is that a patient has a specific disease. If you know nothing else about that patient, your estimate will be based on the prevalence of that disease in the population. Of course, in real life doctors know lots of other things about a patient, not least whether they have any symptoms of the disease, and this information would also factor into their diagnosis. But for the purposes of this example, let's imagine that the best estimate from the medical literature of the baseline probability is $P(A) = 0.01$ (e.g. that one in a hundred individuals have this disease). So your initial expectation, in the absence of any symptoms, is a 1% chance your patient has the disease, and a 99% chance they do not.

To improve your estimate, you take a blood sample from the patient and send it off to be tested. You know that the test will correctly identify 90% of people who have the disease; this is referred to as the *sensitivity* of the test, or the true positive rate, and is conceptually similar to statistical power (see Chapter 5). This also means that the test will miss 10% of people who have the disease—they will still be infected, but the test result will be negative, meaning they appear to be free of the disease. When real clinical tests are developed, the aim is to get the sensitivity as high as possible, though there will usually be technical limitations that prevent perfect sensitivity.

The other aspect of test performance is what happens with uninfected patients, who do not actually have the disease. This is summarized by the *specificity*, or true negative rate of the test. This is how often the test will correctly identify healthy patients as not having the disease—let's say it's 95% of the time for our example. The remaining 5% of healthy patients will produce a false positive, and might be incorrectly diagnosed as having the disease. Balancing the trade-off between these two values (sensitivity and specificity) is a major challenge in test development, as increasing sensitivity can often decrease specificity. A test that is very sensitive but has low specificity might catch most people with the disease, but also give false positives to lots of people who do not, perhaps resulting in unnecessary treatment. An insensitive test that has high specificity would avoid false positives, but might also miss many people who are truly ill.

A good way to think about the probabilities involved is to construct a table (see Figure 17.3) showing the four possible outcomes for a hypothetical group of 1000 people tested. This table will look very familiar to those who have just read Chapter 16 on signal detection theory, and indeed this can be thought of as a signal detection problem (where the 'signal' is the presence of the disease). Out of our 1000 people, we expect that 10 of them (1%) actually have the disease. So the top row of the table (showing people with a true positive disease status) adds up to 10 individuals, and the lower row (those with a true negative disease status) is the remaining 990 individuals. The columns indicate the test results. In the left column are the positive test results. This is 90% of the 10 people who actually have the disease (so 9 of them), and 5% (the false positive rate) of the people who don't have it (~50 people). In the right column are the negative test results. These comprise the one missed person who actually has the disease, and 940 *correct rejections*—people who don't have the disease and tested negative.

Now let's imagine we get back a positive test result for our patient. Can we update our estimate from the baseline of 0.01 to incorporate the information from the test, as well as the information we have about how accurate the test is? We already know the baseline probability $P(A)$ is 0.01. The other part of the numerator of equation 17.1 is the probability of getting

Test result

		Positive	Negative	Total
True status	Positive	9 Hits	1 Misses	10
	Negative	50 FAs	940 CRs	990
	Total	59	941	

Figure 17.3 Grid showing test results and true disease status for a group of 1000 individuals. FAs: false alarms; CRs: correct rejections.

a positive test result if we actually have the disease. This is the sensitivity value we identified above, and so $P(B|A) = 0.9$.

The denominator of the equation is $P(B)$. This is the probability overall of getting a positive test result. We don't know this yet, but we can work it out, because we have enough information about the sensitivity and specificity, as well as the baseline probability. The overall probability of getting a positive test result will be a combination of the probability of getting a positive result if you have the disease, and the probability of a positive result if you don't have the disease. Mathematically:

$$P(B) = P(B|A)P(A) + P(B|\bar{A})P(\bar{A})$$ (17.2)

The first part of this equation is just the numerator term again ($P(B|A)P(A)$). The second part is the probability of getting a positive test result if you *don't* have the disease. The bars over the A symbols are indicating the probability of **not** having the disease. So the second section of the equation is the probability of getting a positive result if you do not have the disease (the false positive rate), multiplied by the probability of not having the disease in the first place. Overall, the equation we need to calculate is:

$$P(A|B) = \frac{P(B|A)P(A)}{P(B|A)P(A) + P(B|\bar{A})P(\bar{A})}$$ (17.3)

or, more specifically for this example:

$$P(disease \mid positive) = \frac{P(positive \mid disease)P(disease)}{P(positive \mid disease)P(disease) + P(positive \mid healthy)P(healthy)}$$ (17.4)

We can now plug in the probabilities from the above scenario to calculate the *posterior probability* (in this instance, the probability of having the disease if you get a positive test result). $P(positive|disease)$ is the sensitivity, which is 90%, or 0.9. $P(disease)$ is the *prior probability*, or baseline prevalence rate of 0.01 (1 in 100). $P(healthy)$ is the rest of the population who don't have the disease, so it must be $1 - P(disease) = 0.99$. Finally, $P(positive|healthy)$ is $1 -$ specificity $= 1 - 0.95 = 0.05$. Completing the calculation with these numbers gives:

$$P(A|B) = \frac{0.9 \times 0.01}{0.9 \times 0.01 + (1 - 0.95) \times 0.99} = 0.15$$ (17.5)

This outcome is known as the *posterior* probability. In this case, it is telling us that if you get a positive result on this test, you only have a 15% chance of actually having the disease. In some respects, that might not seem like an especially great test, even though the sensitivity and specificity are quite high. But note that the final probability we calculated depended not just on the specificity and sensitivity of the test, but also on the prevalence of the disease in the population. If any one of these factors changes, for example if the test becomes more sensitive or the disease becomes more or less prevalent, it will affect our calculation. Figure 17.4 shows how different values of baseline and sensitivity alter the posterior probability ($P(A|B)$). A test is most likely to correctly identify a disease when the sensitivity and specificity are both high, and the disease has a high prevalence in the population.

As a topical aside—as previously mentioned, this book was mostly written during 2020 when a novel coronavirus caused a deadly global pandemic—crucial to attempts to control the spread of the virus was the development of rapid and accurate testing. Of course, in the

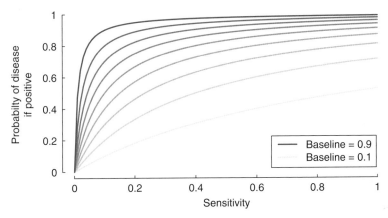

Figure 17.4 Change in posterior probability for different values of test sensitivity and baseline prevalence. The specificity was fixed at 0.9.

early stages of the pandemic, the baseline prevalence rate was changing by the day and in fact was acknowledged to be inaccurate in most countries because only patients with severe symptoms were being tested. For individual tests, sensitivity and specificity were often unclear, and were certainly not communicated to the general public. In such situations, assessing the likelihood of a test result being accurate is very challenging. It was also the case that many scientists and politicians had *prior* expectations about the virus that were ultimately incorrect. It was assumed to be like a severe flu with a relatively low fatality rate, which meant that implementing countermeasures such as lockdowns and mass testing were delayed in many countries. In the next section we will introduce probability distributions, which allow us to quantify our uncertainty about a given situation and consider a range of possibilities.

Probability distributions

The example of Bayes' theorem above applies to point probabilities—exact values of a probability. But typically our knowledge of probabilities is not exact, and we would ideally like to incorporate this uncertainty into our calculations. An important concept in Bayesian statistics that allows us to do this is the *probability distribution*. These distributions specify the mean and spread of relevant probabilities, allowing us to estimate our confidence in the final outcome.

Three critical concepts in Bayesian statistics are the *prior*, *evidence*, and *posterior* distributions. The *prior* distribution quantifies our expectations before we have made any new observations. This is equivalent to the baseline probability of having a disease in the above example ($P(A)$). In an experimental setting, it might be generated using data (such as effect sizes) from previous studies, or we might use a default (neutral) prior if other information is not available. The *evidence* distribution (also called the *likelihood function*) is derived from our new observations, which in an experimental or observational study would be the data we collect. The *posterior* distribution is the outcome, and is calculated by multiplying the prior and evidence distributions together according to Bayes' rule. Example distributions are shown in Figure 17.5.

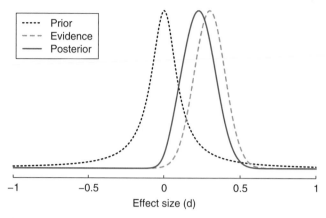

Figure 17.5 Hypothetical probability distributions for the prior (dotted curve), empirical evidence (dashed curve), and posterior (solid curve). Curves have been vertically scaled for visualization.

The prior distribution is an interesting feature of the Bayesian approach. It allows us to incorporate our existing knowledge into the statistic we are calculating. One criticism of Bayesian methods is that this is very subjective, and two different researchers could easily derive different priors. However a counterargument is that many aspects of statistics involve subjective decisions, which are often hidden. Examples might include the choice of which test to conduct, or the decision about when to stop collecting data, particularly if the experimenter is aware of the results they have collected so far (known as optional stopping). The prior is at least explicit, allowing other researchers to see if the outcomes remain constant when the prior is changed. As we will see in the following section, contemporary implementations of Bayesian methods typically offer a default prior to make them more directly comparable to standard frequentist tests. However the option remains for an experienced researcher to modify the prior based on previous evidence. These two approaches are referred to as *subjective Bayes* (when a prior is informed by earlier research) and *objective Bayes* (when the prior is generic).

The Bayesian approach to statistical inference

Now that we have introduced the idea of conditional probabilities, they provide a convenient way to think about frequentist *p*-values. A *p*-value is the probability that we will observe some data *given* the null hypothesis (i.e. assuming that two groups really are equal): $P(data \mid null)$. When the *p*-value is small, this is telling us that it is very unlikely that we would have observed the data if the null hypothesis were correct. Under these circumstances we *reject* the null hypothesis, and conclude that the alternative (experimental) hypothesis is more likely to be correct. This approach treats the data as a random variable, which is compared to a fixed model (the null).

The Bayesian perspective is rather different. In this tradition, we treat the data as a fixed observation about the world, and instead ask which of our hypotheses (experimental or null) is more likely. In other words, we use Bayes' theorem to calculate $P(experimental \mid data)$ and $P(null \mid data)$. Note that here we are no longer calculating the probability of observing the data (because we have definitely observed it already), but instead working out the probability that a model (or hypothesis) is true, given the data we have already observed. This is the crucial distinction between the two traditions (frequentist and Bayesian).

Exactly how these posterior probabilities are calculated depends on the type of data you are working with and your choice of prior. For some relatively simple data such as proportions (e.g. coin tosses, patients infected in a vaccine trial, correct trials, flies killed by a pesticide, etc.), the posterior probabilities can be derived analytically using specific distributions. For a long time, until computers were invented, this was really the only way of doing Bayesian statistics, and explains why the approach was often ignored in favour of frequentist methods. For more complex types of data, like continuous and categorical variables, the mathematics required to derive the posterior becomes intractable. In those situations, the posterior is estimated using a sophisticated stochastic method known as Markov chain Monte Carlo (MCMC). Very briefly, this combines aspects of the methods discussed in Chapters 8 and 9 to randomly sample the posterior distribution using random numbers. A detailed explanation of this would require a whole book, so the interested reader is referred to Kruschke (2014) and Lambert (2018).

Once the probabilities have been calculated, a useful summary statistic is to take their ratio: $\frac{P(experimental \mid data)}{P(null \mid data)}$. This statistic is known as the *Bayes factor* (Jeffreys 1961), and is often reported as the outcome measure for Bayesian tests.[1] If the experimental hypothesis is more likely, the Bayes factor score will be >1. If the null hypothesis is more likely, the Bayes factor score will be <1. Confusingly, some software calculates the ratio the other way up, which inverts the interpretation of the scores. It is possible to indicate which way round the probabilities were considered by using the subscripts BF_{10} and BF_{01} (where 1 is the experimental and 0 is the null hypothesis, and the first value corresponds to the numerator). The Bayes factor tells us which model is best supported by the data.

Heuristics for Bayes factor scores

Unlike the *p*-values ubiquitous in frequentist statistics, Bayes factor scores do not have a hard threshold that determines statistical 'significance'. However, Jeffreys (1961) proposed some heuristics that have been widely adopted, and are given in Table 17.1. Bayes factors around 1 are inconclusive, typically as a result of small sample size or very noisy data. As Bayes factors increase or decrease from 1, this provides evidence in support of one or other hypothesis. Larger values support the experimental hypothesis, with values around 3 considered reasonably convincing evidence. These descriptors will often be referenced when reporting the results of Bayesian statistics. See Dienes and Mclatchie (2018) for an in-depth discussion of the Bayes factor.

---

[1] Actually it's a bit more complex than that, but explaining it this way makes for a clearer comparison with frequentist methods. This value is technically called the *posterior odds*, and is the product of the *prior odds* and the ratio of *marginal likelihoods*. In most implementations, we set the prior odds to equal 1, because we assume that the null and alternative hypotheses are equally likely. The posterior odds therefore depends entirely on the ratio of marginal likelihoods, which is also called the Bayes factor. So in situations where we consider the null and alternative hypotheses equally likely, the posterior odds and the Bayes factor are equivalent.

Table 17.1 Table of Bayes factor scores corresponding to different levels of evidence.

Amount of evidence	Supporting null hypothesis	Supporting experimental hypothesis
Inconclusive	1.00	1
Some evidence	0.33	3
Strong evidence	0.10	10
Very strong evidence	0.03	30

Bayes factors can support the null hypothesis

A useful feature of Bayes factors is that they can provide evidence in support of the null hypothesis (see Table 17.1). To illustrate this, we can repeat our earlier simulations of experiments with different sample and effect sizes, but this time calculating Bayes factor scores instead of frequentist *p*-values. These are shown in Figure 17.6. Bayes factors for the two conditions with positive effect sizes (blue squares and white diamonds) increase as a function of sample size, just as with their frequentist counterparts (Figure 17.1). But in the case where the effect size is *d* = 0, the Bayes factors decrease as a function of sample size (black circles). This means that some data sets will provide evidence for the null hypothesis (Bayes factors <1/3), and this will not be conflated with inconclusive data sets that may have insufficient sample size (Bayes factors around 1).

Bayesian implementations of different statistical tests

The general Bayesian approach has been applied to various statistical tests. In brief, empirical observations (data) are combined with an appropriate prior to estimate a marginal likelihood for both the null and experimental hypotheses. The ratio of these likelihood estimates

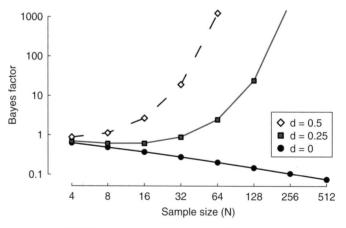

Figure 17.6 Bayes factors for 100000 simulated experiments with different sample sizes and effect sizes.

can then be used to calculate a Bayes factor. Specific implementations vary, and these methods are still under active development, particularly with respect to a suitable choice of prior (see Jeffreys 1961; Kass and Raftery 1995; Rouder et al. 2009; Gronau, Ly, and Wagenmakers 2019). The approach we will consider here uses a *default prior*, which involves as few assumptions as possible (following the objective Bayes tradition). The idea behind a default prior is that it covers the range of plausible non-zero effect sizes (often using a Gaussian-like distribution), and is compared to a model where the prior is zero (i.e. no effect). A detailed breakdown of the mathematics involved in these calculations is beyond the scope of this chapter, and the interested reader is directed to recent work on this topic (Rouder et al. 2009; Gronau, Ly, and Wagenmakers 2019). For more complex experimental designs, estimating the posterior distributions is typically done using the stochastic MCMC method, which means the implementations are computationally intensive, and can take some time to calculate.

So which approach is the best?

We have seen that frequentist tests have two key shortcomings: a fixed false positive rate (regardless of sample size), and an inability to provide evidence to support the null hypothesis. Bayesian methods fix both of these problems, and furthermore allow (in subjective Bayes) the incorporation of prior knowledge, so that research can build iteratively upon previous findings. However, Bayesian methods can be harder to implement than their frequentist counterparts, and are less well understood by the wider research community.

Much has been written about the relative strengths of each approach, and we will not attempt to resolve this question here. However, a practical suggestion that may be of value is to start to report Bayes factors alongside frequentist tests, in much the same way that effect sizes are now routinely reported in many papers. In most cases, these will simply reinforce the frequentist outcome. But in situations where a non-significant result is found, reporting the Bayes factor can help to distinguish between an inconclusive result and one in which the evidence favours the null hypothesis. This adds substantial value to the interpretation of null results, at very little cost.

Software for performing Bayesian analyses

There are several software options for conducting Bayesian statistics. For those more comfortable with spreadsheet-based graphical interfaces, the *JASP* package is freeware, and offers both frequentist and Bayesian analyses. Within *R*, there are many options depending on the level of detail required. Advanced users can programme their own bespoke hierarchical models using advanced stochastic MCMC methods, available in packages such as *rstan*, *rjags*, and *brms*. For those who wish to access Bayesian alternatives to popular frequentist tests, the *BayesFactor* package offers a selection of functions that are designed to mirror the generic frequentist *R* functions as closely as possible. We will demonstrate some examples from this package in the following section.

Calculating Bayes factors in *R* using the *BayesFactor* package

First, we will generate some synthetic data and run a one-sample frequentist t-test. The rationale for using random data (rather than one of the built-in data sets in *R*) is that we know the ground truth population mean that we have drawn our sample from (here a mean of 2, with a standard deviation of 3).

```
data <- rnorm(20, mean=2, sd=3)
t.test(data)
##
##   One Sample t-test
##
## data: data
## t = 4.644, df = 19, p-value = 0.000177
## alternative hypothesis: true mean is not equal to 0
## 95 percent confidence interval:
##   1.800734 4.755699
## sample estimates:
## mean of x
##   3.278217
```

The effect size for the synthetic data is large, so this test gives a highly significant result ($p < 0.05$). We can run a Bayesian t-test on the same data by calling the *ttestBF* function from the *BayesFactor* package:

```
library(BayesFactor)
ttestBF(data)
## Bayes factor analysis
## --------------
## [1] Alt., r=0.707 : 167.5407 ±0%
##
## Against denominator:
##   Null, mu = 0
## ---
## Bayes factor type: BFoneSample, JZS
```

This produces a substantial Bayes factor score of 167.5, which is given on the third line of the output. This is consistent with the significant frequentist result, and indicates very strong evidence for the experimental hypothesis.

The other parts of the output give us information about the prior and the value we are comparing to. The value of *r* (0.707) is the prior scale parameter—changing this as an optional input to the *ttestBF* function affects the behaviour of the test. Larger values of the *rscale* argument might be appropriate if large effect sizes are expected (and vice versa). The value of mu (0 for this example) is the comparison value for the test—in other words it is the prediction of the null hypothesis, which is why it is described as being 'against denominator'—the null appears on the denominator of the Bayes factor equation. The final line of the output tells us the type of test (a one-sample Bayes factor) and the type of prior (JZS). The JZS prior (standing for Jeffreys–Zellner–Siow) is a particular type of default prior constructed using a Cauchy distribution for the effect size and a Jeffreys prior on the variance, as described by Rouder et al. (2009).

What about if we have a null result? To explore this, we can generate a new data set with a true mean of zero, and run both tests again:

```
data <- rnorm(20, mean=0, sd=3)
t.test(data)
##
## One Sample t-test
##
## data: data
## t = -0.18955, df = 19, p-value = 0.8517
## alternative hypothesis: true mean is not equal to 0
## 95 percent confidence interval:
##  -1.442209 1.202684
## sample estimates:
##  mean of x
## -0.1197628
ttestBF(data)
## Bayes factor analysis
## --------------
## [1] Alt., r=0.707 : 0.2361331 ±0.02%
##
## Against denominator:
##   Null, mu = 0
## ---
## Bayes factor type: BFoneSample, JZS
```

This time, there is no significant difference in the frequentist test, and the Bayes factor score is much less than 1 (it is 0.24). This indicates support for the null hypothesis, which of course in this case we know is true (because we drew values from a distribution with zero mean).

Finally, how about a situation where the result is ambiguous? The following simulation has a very small sample size (N = 5):

```
data <- rnorm(5, mean=0.6, sd=1)
t.test(data)
##
##  One Sample t-test
##
## data: data
## t = 1.3714, df = 4, p-value = 0.2422
## alternative hypothesis: true mean is not equal to 0
## 95 percent confidence interval:
##  -0.6562693 1.9373199
## sample estimates:
## mean of x
## 0.6405253
ttestBF(data)
## Bayes factor analysis
## --------------
## [1] Alt., r=0.707 : 0.7510559 ±0%
##
## Against denominator:
```

```
##    Null, mu = 0
## ---
## Bayes factor type: BFoneSample, JZS
```

As you can see from the output, this produces a non-significant *p*-value, and a Bayes factor score around 1 (0.75). Because the sample size is so small, we simply don't have enough data from our observations to conclude that either the null or alternative hypothesis is more convincing.

If we have more than two groups to compare, we can conduct a one-way ANOVA using the frequentist *aov* function, and the Bayesian *anovaBF* function. First, we simulate and plot some data (see Figure 17.7):

```
# generate three levels of a dependent variable
dv1 <- rnorm(60, mean = 3, sd = 3)
dv2 <- rnorm(60, mean = 4, sd = 3)
dv3 <- rnorm(60, mean = 5, sd = 3)
alldata <- c(dv1,dv2,dv3)   # combine together
group <- gl(3,60,labels = c("DV1", "DV2", "DV3")) # make condition
labels
dataset <- data.frame(group,alldata) # combine into a data frame
plot(alldata ~ group)   # plot as box plots
```

A frequentist ANOVA shows a significant main effect of group:

```
summary(aov(alldata ~ group, data = dataset))
##        Df Sum Sq Mean Sq F value  Pr(>F)
## group       2  127.4   63.70   7.19 0.000995 ***
## Residuals 177 1568.0    8.86
## ---
## Signif. codes: 0 '***' 0.001 '**' 0.01 '*' 0.05 '.' 0.1 ' ' 1
```

(For those unfamiliar with *R*'s linear model syntax, the term *alldata ~ group* is interpreted as 'alldata is predicted by group'. In other words, *alldata* is the dependent variable, and *group* is

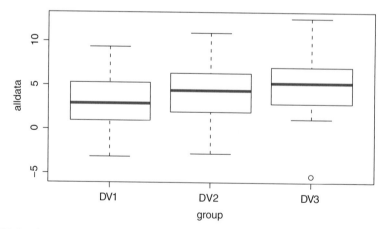

Figure 17.7 Simulated data for a one-way ANOVA.

the independent variable—see Chapter 4). The frequentist effect is complemented by a large Bayes factor score of 26.7:

```
bf <- anovaBF(alldata ~ group, data = dataset)
summary(bf)
## Bayes factor analysis
## --------------
## [1] group : 26.69677 ±0.01%
##
## Against denominator:
##    Intercept only
## ---
## Bayes factor type: BFlinearModel, JZS
```

Finally, we can calculate a Bayes factor score for correlation and linear regression. We first generate two weakly correlated data sets, by making the dependent variable incorporate some proportion of the independent variable:

```
IV <- rnorm(60, mean = 3, sd = 1) # create independent variable
# create dependent variable with some portion of the IV
DV <- rnorm(60, mean = 4, sd = 1) + 0.5*IV
regressiondata <- data.frame(IV,DV) # combine into a data frame
plot(IV,DV,type='p') # plot the data
cor.test(IV,DV) # run a correlation test
##
##    Pearson's product-moment correlation
##
## data: IV and DV
## t = 3.6749, df = 58, p-value = 0.0005212
## alternative hypothesis: true correlation is not equal to 0
## 95 percent confidence interval:
##   0.2030679 0.6200815
## sample estimates:
##    cor
## 0.4345837
```

The data show a significant positive correlation (not plotted here). There is also a Bayesian correlation test (the *correlationBF* function) that can be called as follows:

```
correlationBF(IV,DV)
## Bayes factor analysis
## --------------
## [1] Alt., r=0.333 : 72.60094 ±0%
##
## Against denominator:
##    Null, rho = 0
## ---
## Bayes factor type: BFcorrelation, Jeffreys-beta*
```

Again, the Bayes factor is substantial. A frequentist linear regression can be conducted on the same data using the *lm* function as follows:

```
rsreg <- lm(DV ~ IV, data = regressiondata)
summary(rsreg)
##
## Call:
## lm(formula = DV ~ IV, data = regressiondata)
##
## Residuals:
##      Min       1Q   Median       3Q      Max
## -2.57064 -0.83313  0.06715  0.65323  2.38935
##
## Coefficients:
##             Estimate Std. Error t value Pr(>|t|)
## (Intercept)   4.2038     0.4216   9.971 3.46e-14 ***
## IV            0.4933     0.1342   3.675 0.000521 ***
## ---
## Signif. codes: 0 '***' 0.001 '**' 0.01 '*' 0.05 '.' 0.1 ' ' 1
##
## Residual standard error: 1.041 on 58 degrees of freedom
## Multiple R-squared: 0.1889, Adjusted R-squared: 0.1749
## F-statistic: 13.5 on 1 and 58 DF, p-value: 0.0005212
```

As we would expect based on the correlation results, there is a significant effect of the independent variable on the dependent variable, with a significant *p*-value in the IV row of the coefficients table. The Bayesian equivalent is the *lmBF* function, which uses the same syntax to define the relationships between variables:

```
bflr <- lmBF(DV ~ IV, data = regressiondata)
summary(bflr)
## Bayes factor analysis
## --------------
## [1] IV : 53.64218 ±0%
##
## Against denominator:
##    Intercept only
## ---
## Bayes factor type: BFlinearModel, JZS
```

Again, the Bayes factor score of 53.64 suggests strong evidence for the alternative hypothesis, consistent with the significant frequentist test result and positive correlation coefficient.

Further resources on Bayesian methods

The above examples give a flavour of how straightforward it can be to implement Bayesian versions of standard statistical tests. Of course, more elaborate experimental designs will require the use of other packages, and there are some very good books available that explain how this can be done (Kruschke 2014; Lambert 2018). Bespoke Bayesian models usually require using stochastic methods to estimate the shape of the posterior distribution, and these two books (known as 'the dogs book' and 'the chillis book' because of the pictures on their front covers) describe the current state of the art on how to go about this. Despite their potential complexity, Bayesian methods have some clear advantages over their frequentist counterparts, and are likely to become more heavily used as they become better developed, standardized, and more widely understood.

Practice questions

1. In frequentist methods, the *p*-value is a statement about:
   A) The probability of the experimental hypothesis being true
   B) The probability of the null hypothesis being true
   C) The probability of observing the data if the null hypothesis is true
   D) The probability of observing the data if the experimental hypothesis is true

2. In frequentist methods, the false positive rate:
   A) Increases with sample size
   B) Decreases with sample size
   C) Stays constant regardless of sample size
   D) Depends on statistical power

3. In Bayesian methods, information we already know is called:
   A) The posterior
   B) The prior
   C) The likelihood
   D) The data

4. Bayes's theorem allows us to combine:
   A) Conditional probabilities
   B) *p*-values from t-tests
   C) Bayes factor scores
   D) Data from experiments with different sample sizes

5. A Bayes factor denoted BF_{01} indicates:
   A) The ratio of the experimental over the null hypothesis
   B) The probability of the experimental hypothesis being true
   C) The probability of the null hypothesis being true
   D) The ratio of the null over the experimental hypothesis

6. A Bayes factor of $BF_{10} = 7.5$ indicates:
   A) Some evidence in support of the alternative hypothesis
   B) Some evidence in support of the null hypothesis
   C) Strong evidence in support of the alternative hypothesis
   D) Strong evidence in support of the null hypothesis

7. In a situation with an effect size of $d = 0$, the Bayes factor (BF_{10}) score:
   A) Increases with sample size
   B) Decreases with sample size
   C) Stays constant regardless of sample size
   D) Depends on statistical power

8. In the case of a non-significant frequentist test, running a Bayesian alternative would help to:
   A) Distinguish between the null hypothesis and an underpowered (inconclusive) study
   B) Distinguish between the experimental and null hypotheses
   C) Increase the statistical power
   D) Get a significant result

9. If some data support the experimental hypothesis with a probability of 0.67, and the null with a probability of 0.02, what is the Bayes factor score BF_{10}?
   A) 0.03
   B) 0.01
   C) 33.5
   D) 0.65

10. Suppose that the probability of a man sporting a moustache is 5%. However, during November lots more people grow moustaches for charity, and the probability of any individual man having a moustache increases to 30%. Use Bayes' rule to calculate the conditional probability that it is November given that you see a man with a moustache. The probability is:
   A) 0.08
   B) 0.25
   C) 0.35
   D) 0.49

Answers to all questions are provided in the answers to practice questions at the end of the book.

18 Plotting graphs and data visualization

An essential part of scientific communication is to be able to display your data in a clear and interpretable way. Despite the importance of this skill, it is rarely taught explicitly, and most researchers have to pick up some general heuristics about graph plotting by osmosis. This chapter will attempt to distil some principles of data visualization that I have found useful, though of course such guidelines are necessarily subjective—others may disagree on what looks good.

There are several plotting packages in *R* that one might choose, the most popular being the *ggplot2* package (Wickham 2016). This is considered part of the *tidyverse* suite of packages created by the makers of *RStudio*. The *ggplot2* functions use a plotting grammar, whereby a plot is constructed by adding layers of components. This can produce some excellent results, however in my personal experience I find that there is usually something I need to change from the default way that *ggplot2* behaves. This often proves to be very difficult indeed, and so I have found that the built-in plotting functions from the base *R graphics* package afford more flexibility. We will use these functions throughout the chapter. Remember that the code to produce all figures in the book, including this chapter, is available on *GitHub* at: **https:// github.com/bakerdh/ARMbookOUP**.

We will start by discussing some key principles of data visualization, and considerations regarding the choice of colour palettes. We will then build up a plot with multiple components using the basic *plot* command, demonstrating various options. Finally, we will discuss how to export graphs, and also how to combine several graphs into a multipart figure. These tools are sufficient for assembling publication-quality plots entirely within *R*, without the need for additional software (such as Adobe Illustrator).

Four principles of clear data visualization

The purpose of plotting graphs is to communicate data in a clear and unbiased way, so that it can be easily interpreted by your reader with a minimum of effort. The following principles are things I have picked up informally over a period of many years. Of course others may have a different aesthetic sense and disagree with many of my specific choices, but hopefully this section will at least prompt readers to consider how they construct figures, and why they make the choices they do.

Different conditions should be maximally distinguishable

When I first started writing up the results of experiments, I had in my head the idea that 'proper' science needed to involve black and white graphs with thin lines joining some very small data points (see an example from my own undergraduate work in Figure 18.1). I probably got this idea from reading too many papers from the 1980s and before, when figures were created by hand using a ruler, and colour reproduction was not available in most journals (or was prohibitively expensive). This is no longer the case, and modern tools can make graphs that are more visually appealing, and also clearer to the reader.

The main problems with Figure 18.1 are that the data points are very small, and extremely hard to distinguish. Different symbols were used to represent different conditions (circles, squares, and triangles), but the reader needs to work quite hard to realize this, and to see which symbol is which. We can replot these data using larger symbols, thicker lines, and also giving each condition its own colour and line style. Note that in common with many of the figures in this book, I have been quite sparing with colour. This is partly due to using the two-colour printing system common in publishing, but it is also good practice to avoid overusing colour when it is not strictly necessary. However, the reader should not be dissuaded from using colour when it is appropriate, and when it makes a figure easier to understand.

Figure 18.2 shows the same data replotted with these changes, and it is immediately much clearer which condition is which. Notice that the symbols are now large enough to easily distinguish, and the use of colour makes linking with the figure legend effortless and automatic. Finally, the axis labels are more appropriate in the revised figure, avoiding block capitals, and the tick marks point in towards the data.

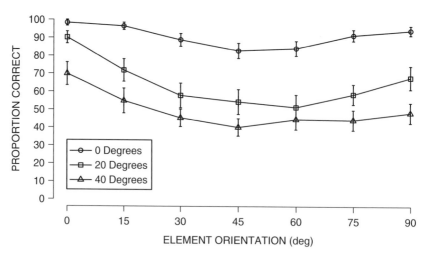

Figure 18.1 Reproduction of a rubbish graph from my own final year undergraduate project. The experiment involved finding contours in a field of randomly oriented elements. Performance is best when all elements in the contour had a similar orientation (circles), but rapidly degrades as each element's orientation is jittered. Error bars give 95% confidence intervals.

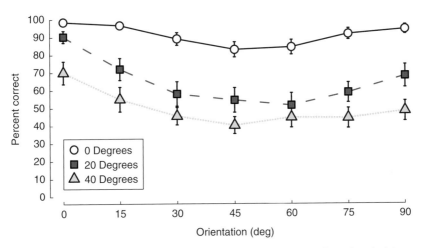

Figure 18.2 Replotted data from my undergraduate project, using colour, line style, and symbol size to maximally distinguish conditions.

Axes should be labelled and span an informative range

It is critical that axis labels explain clearly what is plotted on that axis, including appropriate units. Although it may seem obvious, the axis tick marks should always be plotted at regular intervals in the chosen units (see Figure 18.3(a))—there are some truly awful examples in the media that violate this basic rule. For interval and ratio data, the units of measurement are meaningful, and so these need to be represented appropriately in a plot. Some spreadsheet software that is widely used to produce graphs has a default setting where the x-axis is categorical even for continuous data, which can result in an unequal spacing of values that is very misleading.

For some types of data, a logarithmic scale may be more appropriate than a linear one. Log-scalings can be used when the data have a positive skew in linear units, as is often the case with values involving time (such as reaction times), and also when the data are ratios. There are several ways to plot a log axis. Figure 18.3(b) shows the tick marks in linear increments, but on a log axis, so that successive spaces become smaller (as used on old logarithmic graph paper). An alternative is to log-transform the data, and plot this on a linear axis, where the labels are in log units (Figure 18.3(c)). A disadvantage of this method is that the log units themselves are arbitrary, and hard to relate to the original values. The method I find most transparent is to give tick labels in the original linear units (rather than in log units), but spaced in logarithmic steps, such as factors of 2 or 10 (see Figure 18.3(d)). Conventions differ across disciplines about which approach to use, so it is always best to check some related papers in your field.

There is also some disagreement about whether it is necessary for a linear axis to extend to zero (especially for bar charts). Sometimes starting the axis at an arbitrary point closer to the means can exaggerate an apparent difference between conditions (see Figure 18.4). This is bad practice if the intention is to mislead, and exaggerate the apparent size of an effect.

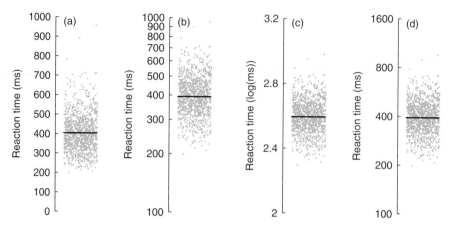

Figure 18.3 Examples of linear and logarithmic axes. Panel (a) shows a linear axis for reaction time data in milliseconds (ms). Panel (b) shows a logarithmic axis for the same data, with tick marks spaced every 100 ms, resulting in uneven tick placement. Panel (c) shows the logarithmic transform of the data—the units are now much harder to interpret. Finally, panel (d) shows logarithmic scaling with tick marks spaced at equal factors, but values given in linear units. Position on the x-axis is arbitrary.

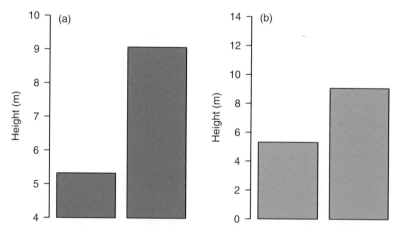

Figure 18.4 Illustration of how apparent differences between groups can be exaggerated by starting the y-axis at a value other than zero (a), compared with at zero (b). The data are identical in both panels.

Yet it is not necessarily the case that zero should always be included. For many variables, zero is not a plausible value for any of our observations—reaction times being a good example again. Stretching an axis to zero arbitrarily can obscure real and meaningful differences in the data. Instead, I favour using axes that span an informative range, and combining this with error bars to give a clear visual indication of which differences are meaningful, as we will discuss in the following section. A good rule of thumb might be to have the axes extend around 2 standard deviations above the highest point, and around 2 standard deviations below the lowest point you are plotting. However, this heuristic also needs a good dose of common sense—round the upper and lower values to a sensible value, such as an integer, or a factor of 10 or 100, depending on the natural scale of your data. You should ideally use the same range for multiple plots of the same type of data.

A measure of variability should always be included

A third principle is that we should include some indication of the variability of our data. This might feel like second nature to anyone with a scientific background, but it is surprising how often graphs are shown in the media, and even in technical reports (such as market research reports), that have no visual indication of how precise our estimates of means or medians might be. For example, the data shown in Figure 18.5(a) might suggest a large difference between the two conditions, yet without an indication of how variable each estimate is, we cannot tell whether the apparent differences are real, or just due to sampling error.

The most widely used measures of variability are the standard deviation, standard error, and confidence intervals. The *standard deviation* (SD, or σ) is the square root of the variance,

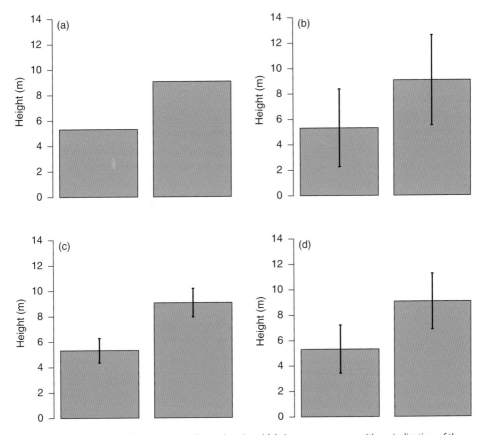

Figure 18.5 Illustration of different types of error bar. Panel (a) shows two means with no indication of the variance. Panel (b) shows the same data with standard deviations plotted. The lower row shows the same data with standard errors (c) and 95% confidence intervals (d) plotted. In this example, the two groups are significantly different ($t = 2.5$, $df = 18$, $p = 0.02$, $d = 1.1$).

so it is the most direct indication of the underlying variability in the data (see Figure 18.5(b)). It does not depend on the number of estimates, meaning it should remain approximately constant as the sample size changes. On the other hand, the *standard error* (SE) scales the standard deviation by the square root of the sample size (see Figure 18.5(c)). This means that as we collect more data the standard error becomes smaller, reflecting the greater precision of our estimate of the mean (or other measure of central tendency).

Alternatively, *confidence intervals* (CIs) explicitly represent our confidence in the location of the mean. Actually, standard errors are a type of confidence interval—they are the 68% confidence intervals of the mean. But it is more typical to plot 95% confidence intervals (see Figure 18.5(d)), because these can act a bit like a t-test—if the confidence intervals do not overlap with some particular point, then the mean is likely to differ significantly from that point (at $p < 0.05$). Confidence intervals can be estimated with the parametric assumption of $1.96 \times SE$, or by bootstrap resampling of the data (see Chapter 8).

Something worth observing in Figure 18.5 is that the different types of error bars can appear to indicate different amounts of variance, if we don't take into account what they are actually showing (remember it is the same data set in each case). For example, standard deviations will always produce larger error bars than standard errors, and without clarity about what is being shown a reader might develop a spurious understanding of the variability of the data. It turns out that even trained researchers often misinterpret what error bars are showing, and treat standard errors and 95% confidence intervals very similarly (Belia et al. 2005). Many people appear to use heuristics, such as whether the error bars 'just touch', which are not a good indicator of statistical significance. All of this confirms that we always need to explicitly state in the figure caption what our error bars represent.

In terms of graphical presentation, error bar whiskers such as those shown in Figure 18.5(b)-(d) are widely used and well understood. However in some types of plot they overlap across conditions, which can make them hard to see properly (see for instance the error bars for the square and triangle at 60 degrees orientation in Figure 18.1). An alternative visualization is to plot a shaded region that encompasses the error bars. This solves the problem of overlap, and also looks visually appealing (see Figure 18.6).

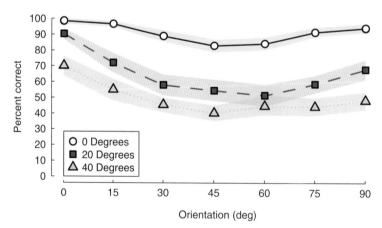

Figure 18.6 Replotted data from my undergraduate project, with shading to represent the error region (95% confidence intervals).

Where possible, include data at the finest meaningful level

Traditional bar plots (e.g. Figure 18.7(a)) can hide a multitude of sins. The main problem is that without seeing the original data points that went into calculating the means, it is impossible to tell whether one or more outliers might be responsible for the results. Or there could be unexpected patterns in the data (such as bimodally distributed results) that are hidden by the averaging process. A classic demonstration is Anscombe's quartet (Anscombe 1973)—a set of four scatterplots that have identical summary statistics, including the correlation coefficient, but appear very differently when the individual data points are plotted.

The most straightforward solution is to plot the individual data points that went into calculating the averages being displayed (see Figure 18.7(b)). Interestingly, it has been shown that the presence of a bar can distort the reader's interpretation of individual data points. Newman and Scholl (2012) found that people judge points falling within the bar to be more likely to be part of the underlying distribution than those falling outside of it. Since the bar itself is not informative, it can be safely removed, and replaced by a line or point to indicate the average. In practical terms, individual data points should not get in the way of the representation of the mean, and so alpha transparency can be useful. Plotting data points semi-transparently also allows us to see when multiple points with similar values are overlapping with each other.

For some data sets, it can also be helpful to plot distributions, or kernel density functions (smoothed histograms). There are several conventions for plotting these. For example, with bivariate scatterplots showing correlations, it is common to plot histograms along the margins of the plot (we will see an example of this later in the chapter). For univariate data, the violin plot replaces bars with mirrored distributions (Figure 18.7(c)). A more recent suggestion is the *raincloud plot* (Allen et al. 2019), which shows a distribution with individual data points either underneath or to one side, and a mean and error bar in between (Figure 18.7(d)).

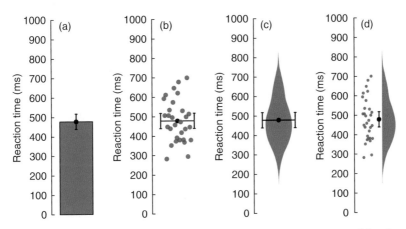

Figure 18.7 Example visualizations that provide more information about the distribution of data. Panel (a) shows a generic bar plot (error bars indicate 95% confidence intervals). Panel (b) shows the raw data points. Panel (c) shows a violin plot, illustrating the distribution of points as a symmetrical kernel density function. Panel (d) shows a raincloud plot, which features the distribution and the raw data.

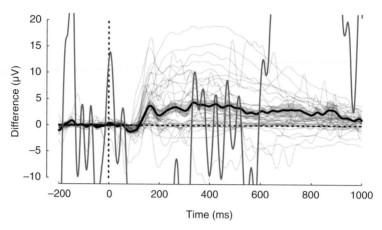

Figure 18.8 Replotted ERP difference data, showing results for individual participants (thin traces) and the group mean (thick line). The shaded region indicates ±1SE across participants. The blue curve highlights an outlier participant that might sensibly be excluded.

For data types that involve continuous functions rather than individual points (e.g. where a time course has been measured), individual functions can still be plotted with a thin line-width, and the grand mean overlaid using a thicker line and more salient colour (Rousselet, Foxe, and Bolam 2016). An example of this is shown in Figure 18.8, replotting the ERP difference data described in 'Example of cluster correction for EEG data' in Chapter 15. For this data set there is one clear outlier (highlighted in blue) that is contributing only noise to the average, and might reasonably be excluded on various criteria. In the absence of outliers, this method of presentation allows the reader to confirm that nothing has been hidden by the averaging process, giving them more confidence in the results of any inferential statistics, and the conclusions drawn from them.

Choosing colour palettes

The choice of colour palette for plotting a figure is more than a purely aesthetic one. Two key factors that should be considered are accessibility to readers with atypical colour vision, and the introduction of spurious features through a colour palette that is not perceptually uniform. In this section, we discuss both issues.

Palettes that are safe for colour-blind individuals

As most readers will be aware, some individuals are 'colour-blind' (indeed, you may be yourself). It is not necessarily the case that these individuals see the world like a black and white photograph. Such severe colour blindness (known as *achromatopsia*) is quite rare (around 1 in 30,000), though it can be acquired through injury or partial asphyxiation. A more common form of genetic colour blindness affects around 2% of men, but far fewer women because the genes involved are linked to the X-chromosome. This is red/green colour blindness, with two variants: *deuteranopia* and *protanopia*. Affected individuals lack a class of cone photoreceptor (a cell that detects light) in their eye. In general terms, these individuals cannot tell the difference between red and green, with both appearing as a yellowish brown colour. A rarer variant of colour blindness (*tritanopia*) involves similar problems with perceiving the colour blue. Finally, many individuals (>5% of men) have anomalous colour vision, where they do not lack classes of photoreceptor, but the photopigments are shifted in their sensitivity, making discrimination between some colours more difficult.

These various types of colour blindness will have important consequences for how graphs are perceived by some readers (see also Crameri, Shephard, and Heron 2020). A very simple heuristic that will aid the largest population of colour-blind readers is to avoid using red and green to distinguish two different conditions. If red and green need to be used together, they should be paired with different symbol types (as in Figure 18.2) and/or use different lightnesses—for example a light red and a dark green. This will allow the brightness and shape of the symbols to differentiate conditions even if the colour information is lost (which also happens when using a black and white printer).

We can simulate how a colour-blind individual will perceive a figure or image through our understanding of the conditions. Websites and applications exist that can do this for individual images, for example *Vischeck* (**http://www.vischeck.com**) and *Color Oracle* (**https://colororacle.org**). For a given colour palette, the likely appearance for colour-blind individuals can also be simulated, as demonstrated in Figure 18.9 for the *rainbow* palette. Notice the presence of peaks and discontinuities of luminance in the second row. For all four simulated varieties of colour blindness there are regions of the palette that are spaced far apart but have a similar appearance. These features will make it hard to interpret images plotted with this palette.

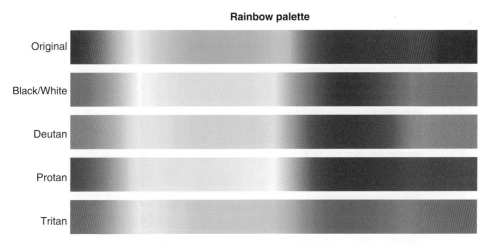

Rainbow palette

Original

Black/White

Deutan

Protan

Tritan

Figure 18.9 Simulation of colour blindness for a rainbow colour palette. The first row shows the original palette, which transitions through the colours of the rainbow. The second row shows a greyscale rendering, as might be perceived by an individual with achromatopsia. The third and fourth rows show simulations for individuals with two types of red/green colour blindness—reds, greens, and yellows all appear similar. The final row simulates tritanopia, where blue and green cannot be distinguished.

Because of the different profiles of various types of colour blindness, it is challenging to produce palettes that are appropriate for everyone. Palettes such as the rainbow palette shown in Figure 18.9 are problematic partly because they involve many different hues. A better alternative would be colour maps that pass smoothly from one hue to another, and simultaneously change in brightness. One example, the *parula* palette, is shown in Figure 18.10. Both ends of the palette are clearly distinguishable for all four simulated varieties of colour blindness. When used to plot real data (e.g. contour maps), the parula palette is still attractive and clear, even when the colour information is removed (see Figure 18.11).

Perceptually uniform palettes

Another important consideration when choosing a colour palette is whether the palette is *perceptually uniform*. Many standard colour maps vary in luminance (brightness) across their range (see e.g. Figure 18.9, and Ware (1988)), and also ignore features of the human visual system that mean we are less sensitive to changes in some regions of colour space (Borland and Taylor 2007). Generating colour maps that avoid these problems is a challenge, and some useful principles are described by Kovesi (2015).

In general, most colour palettes involve linearly changing the values of the red, green, and blue colour channels in various combinations. But this often produces an overall luminance profile that has unwanted peaks in different parts of the palette, as shown by the black curve in the left panel of Figure 18.12. Perceptually uniform palettes reverse these priorities, and aim to create consistent changes in luminance (and perceived hue) by manipulating the red, green, and blue levels non-linearly (see right panel of Figure 18.12). Notice how the standard ramp on the left has bands of higher brightness (especially in the reddest part of the palette), but the perceptually uniform palette has a much smoother transition across the whole range. This is important when representing data using a colour map, because a non-uniform palette will introduce spurious features that are not present in the data.

Figure 18.10 Simulation of colour blindness for the parula colour palette.

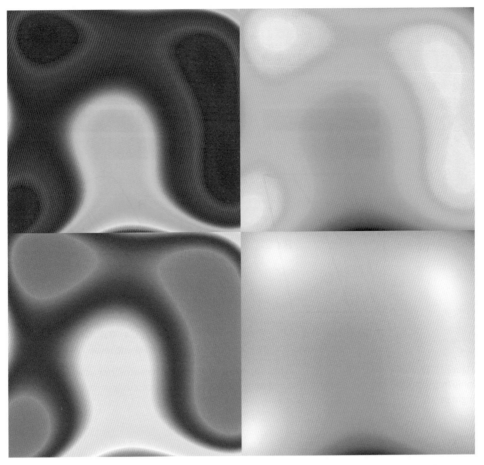

Figure 18.11 The Himmelblau function, plotted with two colour palettes and their greyscale equivalents—the rainbow palette (left) and the parula palette (right). The upper row shows the original palettes, and the lower row shows the same values but with colour information removed (simulating achromatopsia). Notice how the rainbow palette introduces sharp boundaries of brightness between regions that are not truly present in the data, whereas the parula palette has a smoother transition that is more accurate.

To demonstrate this, Figure 18.13 shows a time–frequency plot of neural responses to a visual stimulus, plotted using a linear ramp (left) and a perceptually uniform palette (right). Notice how the linear ramp produces banding and distortions in the data, making the peaks and troughs appear more salient and oversaturated. The perceptually uniform palette produces smoother transitions between features, and the true location of the peak in the lower left corner is clearer. For most types of scientific image, these characteristics are to be preferred for the accurate communication of results.

As well as being used for displaying scientific image data and surfaces, perceptually uniform colour palettes can also be used for selecting colours to plot data points and curves. We will demonstrate this in the practical sections of the chapter.

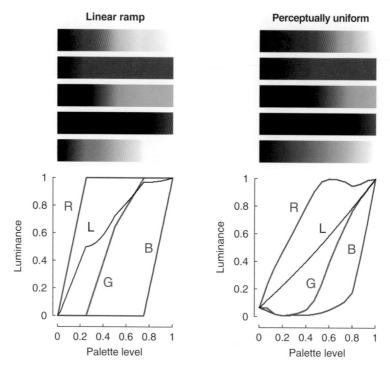

Figure 18.12 Profile of red (R), green (G), and blue (B) values, and overall luminance (L) for a standard linear ramp from black through red, orange, yellow, and white (left panel), and a perceptually uniform equivalent (right panel).

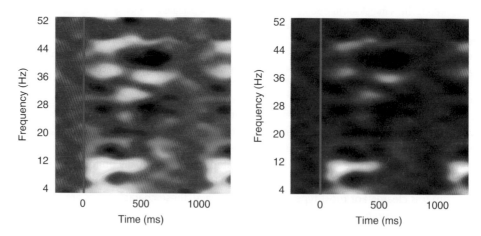

Figure 18.13 Time–frequency plots of neural responses to a visual stimulus presented at 0 ms (indicated by the blue line). Data were recorded using MEG (magnetoencephalography) and are shown for a single sensor. Left panel shows a linear colour palette; right panel shows a perceptually uniform colour palette.

Building plots from scratch in *R*

My preferred approach to creating plots in *R* is to create a blank axis first, and add individual features to it. In the following section we will construct the plots shown in Figure 18.14 one component at a time. This will illustrate how to build and label axes, and how to add widely used features such as symbols, points, lines, error bars, and arbitrary shapes.

For all plots, we first need to create a new empty figure, and add appropriate axes. This requires us to know the dimensions of the plot (i.e. the range of the x- and y-axes), and then to call an empty plot with the axes and annotations suppressed:

```
# create an empty axis of the correct dimensions
plotlims <- c(0,1,0,1)
```

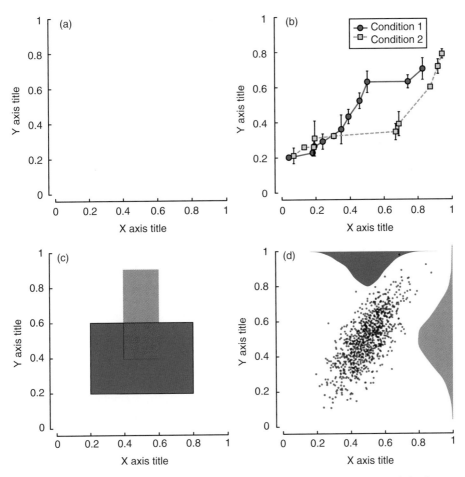

Figure 18.14 Example outputs from the plotting function. Panel (a) shows an empty axis, ready for data to be plotted. Panel (b) shows an example of two conditions plotted with lines, points, error bars, and a figure legend. Panel (c) shows an example of the polygon function with alpha transparency. Panel (d) shows an example of kernel density functions for an x–y scatterplot.

```
plot(x=NULL,y=NULL,axes=FALSE, ann=FALSE, xlim=plotlims[1:2],
ylim=plotlims[3:4])
```

The above code will appear to do nothing at all, but it will define an empty axis in the plot window of *RStudio*. We can then add axes with tick marks at sensible intervals using the *axis* command, tick labels with the *mtext* command, and axis labels with the *title* command:

```
ticklocs <- seq(0,1,0.2)    # locations of tick marks
axis(1, at=ticklocs, tck=0.01, lab=F, lwd=2)
axis(2, at=ticklocs, tck=0.01, lab=F, lwd=2)
mtext(text = ticklocs, side = 1, at=ticklocs) # add the tick labels
# the 'line' command moves away from the axis, the 'las' command
rotates to vertical
mtext(text = ticklocs, side = 2, at=ticklocs, line=0.2, las=1)
title(xlab="X axis title", col.lab=rgb(0,0,0), line=1.2, cex.lab=1.5)
title(ylab="Y axis title", col.lab=rgb(0,0,0), line=1.5, cex.lab=1.5)
```

This code will produce the axes shown in Figure 18.14(a). These axes are quite minimal—it is also possible to add the off-axes (again using the *axis* command, but with the *side* option of 3 or 4). If required, we can also force the axes to be square by adding *par(pty="s")* before the call to the *plot* command.

Additional text can be added anywhere in the plot using the *text* function. This is often useful for adding panel labels (a, b, etc.) for example:

```
text(-0.05,0.95,'(a)',cex=1.8, pos=4)
```

The first two inputs are the x- and y-coordinates, and the third input is the text string we want to plot. The *cex* option specifies the font size, and the *pos* option has several possible values that determine the position of the text relative to the coordinates: 4 means that the string is plotted to the right of the coordinates.

A brief note about font size and style—many readers have visual impairments that make small text particularly difficult to read. Therefore it is a good idea to make text within your figure as large as is practical. Some people also find that fonts without serifs (the little decorative lines added to some letters) are easier to read. Popular *sans serif* fonts include Arial, Calibri, and Helvetica.

Adding data to plots

Once our axes are defined, it is time to add data to the plot. The examples here will illustrate plotting data points, error bars, lines, and semi-transparent polygons. First, let's create some synthetic data to plot using a random number generator, as follows:

```
# create some synthetic data to plot as points and lines
datax <- sort(runif(10,min=0,max=1))
datay <- sort(runif(10,min=0.2,max=0.8))
SEdata <- runif(10,min=0,max=0.1)
```

The generic plotting functions do not require the data to be stored in a specific format such as a data frame. They work fine with vectors of numbers. Lines and error bars should ideally be plotted first so that they appear behind the data points. A line connecting the data points can be defined using the *lines* function:

```
# draw a line connecting the points
lines(datax, datay, col='cornflowerblue', lwd=3, lty=1)
```

We provide the x- and y-values as vectors, define the colour using the *col* option, and the width using the *lwd* option. There are also several line styles that can be selected with the *lty* option, though here we have chosen the default option (1), which plots a continuous straight line. Other options include dashed (option 2) and dotted (option 3) lines.

Error bars are plotted using the *arrows* function. We need to manually add and subtract the standard error to define the upper and lower extents of the bars as follows:

```
# add lower error bars
arrows(datax, datay, x1=datax, y1=datay-SEdata, length=0.015,
angle=90, lwd=2)
# add upper error bars
arrows(datax, datay, x1=datax, y1=datay+SEdata, length=0.015,
angle=90, lwd=2)
```

Notice that we plot all of the lower error bars at once, followed by all of the upper error bars. The *arrows* function allows us to enter a vector of values for each input, just like the *lines* function. As the name suggests, the arrows function is designed to plot actual arrows, but setting the angle of the arrow to 90 degrees (as in the above code) produces error bars that are flat at the ends.

Next, we can add our data points with the *points* function. There are many different point styles available that can be selected through the *pch* option. I tend to favour styles 15–20 (solid shapes of a single colour) and styles 21–25 (filled shapes with an outline). Details on the available shapes are given in the *help* file for the *points* function. For solid shapes of a single colour, the colour is specified by the *col* option. For filled shapes, we specify the outline colour with the *col* option, and the fill colour with the *bg* (background) option, allowing the two parts of the shape to have different colours.

```
# draw the data points themselves
points(datax, datay, pch=21, col='black', bg='cornflowerblue',
cex=1.6, lwd=3)
```

We can also include a legend to indicate what different symbols refer to. The *legend* function requires x and y coordinates corresponding to its upper left corner. We also provide a list of text labels for the conditions, and repeat some of the options from the *points* and *lines* functions to specify the symbol properties.

```
legend(0, 1, c('Condition 1','Condition 2'), col=c('black','black'),
    pt.cex=1.6, pt.bg=c('cornflowerblue','grey'),lty=1, lwd=3,
    pch=21:22, pt.lwd=3, cex=1, box.lwd=2)
```

Putting all of the above lines of code together (with some additional data for a second condition) gives us the plot in Figure 18.14(b).

Setting alpha transparency for points and shapes

Whenever a colour is defined in *R*, we also have the option of setting its transparency using the *alpha* channel. Images on a computer are usually defined in terms of the intensity of the red, green, and blue pixels on the display. But we can also specify a fourth value, *alpha* (or α), which sets transparency between 0 (fully transparent, and therefore invisible) and 1 (fully opaque). Some level of transparency is useful for two reasons. First, it allows us to see plot features that are layered on top of each other, so that we can estimate the density of clouds of many small points. Second, it is a handy way of reducing the saturation of a colour. There is evidence that highly saturated colours can introduce perceptual distortions that affect how data are interpreted, whereas less highly saturated colours are perceived more veridically (Cleveland and McGill 1983).

We can set the *alpha* level using the *rgb* function, which also requires that we specify the red, green, and blue values. The following code specifies semi-transparent blue and black colours:

```
colour1 <- rgb(0,0,1,alpha=0.5)
colour2 <- rgb(0,0,0,alpha=0.5)

colour1
## [1] "#0000FF80"
colour2
## [1] "#00000080"
```

Notice that the colours are converted to a hexadecimal representation. If we don't know the RGB values of the colour we want to use, but instead have the colour name, the following helper function (*addalpha*) will convert to the correct format and add the alpha level:

```
addalpha <- function(col, alpha=1){
apply(sapply(col, col2rgb)/255, 2, function(x) rgb(x[1], x[2],
x[3],alpha=alpha))}

colour1 <- addalpha('cornflowerblue',0.5)
colour2 <- addalpha('black',0.5)
```

We can then use these colour values with lines, points, and polygons. Polygons are two-dimensional shapes of arbitrary specification. For example, we could draw two rectangles using the *polygon* function as follows:

```
polygon(c(0.2,0.8,0.8,0.2),c(0.2,0.2,0.6,0.6),col=colour1)
polygon(c(0.4,0.6,0.6,0.4),c(0.4,0.4,0.9,0.9),col=colour2)
```

In these two lines of code, the first two vectors of numbers are x- and y- values that specify the corners (vertices) of the polygons. Adding the code to an empty plot produces the image in Figure 18.14(c). Of course we are not limited to only four vertices, and a common use of the *polygon* function is to plot histograms as kernel density functions. With a larger set of correlated data, we can add histograms to the margins of a scatterplot as follows:

```
datax <- rnorm(1000,mean=0.5,sd=0.1)
datay <- rnorm(1000,mean=0.5,sd=0.1) + (datax-0.5)
a <- density(datax)
```

```
a$y <- 0.2*(a$y/max(a$y))
polygon(a$x, 1-a$y, col=colour1,border=NA)
a <- density(datay)
a$y <- 0.2*(a$y/max(a$y))
polygon(1-a$y, a$x, col=colour2,border=NA)
```

The *density* function generates the kernel density distribution, and the subsequent line of code rescales the height to be between 0 and 0.2 so that it fits neatly onto our plot. Notice that the x- and y- variables are reversed for the second polygon so that it appears on the right side of the plot. Figure 18.14(d) plots these polygons along with the data points, which are also semi-transparent.

Choosing, checking, and using colour palettes

There are many ways to define colours in *R*. We have already seen that we can define a colour by its RGBα values, using a name (such as 'red'), or using a hexadecimal number. There are various functions for converting between these different formats. For example, the *rgb* function takes values for the individual colours (scaled from 0 to 1) and outputs a hexadecimal code:

```
rgb(1,0.5,0,alpha=0.8)
## [1] "#FF8000CC"
```

The *col2rgb* function converts a colour name or hexadecimal code back to red, green, and blue values (scaled from 0 to 255):

```
col2rgb('orange')
##       [,1]
## red   255
## green 165
## blue    0
```

```
col2rgb("#FF8000CC")

##       [,1]
## red   255
## green 128
## blue    0
```

Note that this function removes the transparency information.

To obtain a custom colour palette, we can use the *colorRamp* function to transition smoothly between two or more colours, sampling the ramp at an arbitrary number of intermediate points. For example, the following code produces the palette shown on the left side of Figure 18.12:

```
cr <- colorRamp(c("black","red","orange","yellow","white"))
thispal <- rgb(cr(seq(0, 1, length = 256)), max = 255)
```

The *colorRamp* function is unusual in that it actually returns a custom function (named *cr* in the above example) rather than a standard data object, which we can then sample at a sequence of values between 0 and 1.

There are also many built-in and default palettes that can be accessed through the *grDevices* package. Built-in palettes include *rainbow* (sometimes called *jet*, and shown in Figure 18.9), *heat.colors*, *terrain.colors*, *topo.colors*, and *cm.colors* (cyan-magenta). Further details are available in the help for the *Palettes* function.

Finally, a great many additional packages provide further colour palettes. The *pals* package is particularly useful, as it contains a series of perceptually uniform colour palettes, as well as tools for assessing the properties of a colour palette. The perceptually uniform colour maps are described in the *kovesi* function (there are around 50), and can be used to generate a palette with an arbitrary number of levels, for example:

```
thispal <- kovesi.linear_bgyw_15_100_c67(5)
thispal
## [1] "#1B0084" "#2441D4" "#4C9A41" "#C0CC23" "#FFFFFF"
```

When plotting data from several conditions, setting the number of colours to equal the number of conditions is a sensible approach. This will give a set of hues that should distinguish the conditions, but also appear to be drawn from a consistent set.

The *pal.test* function produces a series of test images for a given palette, as shown in Figure 18.15, which shows the output of the following expression:

```
pal.test(kovesi.linear_bgyw_15_100_c67)
```

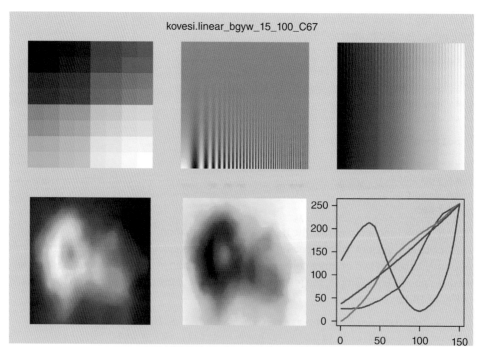

Figure 18.15 Test image for colour palettes, generated by the *pal.test* function for the Kovesi linear blue-green-yellow-white palette.

The graph in the lower right shows the red, green, blue, and luminance ramps in the same format as Figure 18.12. The upper right image shows the full palette with a sine wave modulation along the upper edge. If the oscillations are harder to see in some parts of the palette, this indicates perceptual non-uniformity. The volcano images in the lower row are the same image with the colour palette ranging from 0 to 1 and from 1 to 0. If some features are visible in only one of these images, it again suggests problems with the palette. Since the Kovesi palettes are all perceptually uniform, they pass most of these tests. It is worth also viewing the output of the test function for a non-uniform palette (see Figure 18.16). Notice that the two volcanoes appear to differ in size, and that parts of the sine-wave oscillation are not visible (particularly in the green band).

To check how a palette will appear to individuals with several types of colour blindness, the *pal.safe* function produces images like those shown in Figures 18.9 and 18.10.

Once we have selected a suitable colour palette, we can apply it to an image using the *image* function. A suitable test image is the interesting two-dimensional mathematical function known as Mishra's Bird, defined by the equation:

$$f(x,y) = \sin(y)e^{(1-\cos(x))^2} + \cos(x)e^{(1-\sin(y))^2} + (x-y)^2 \tag{18.1}$$

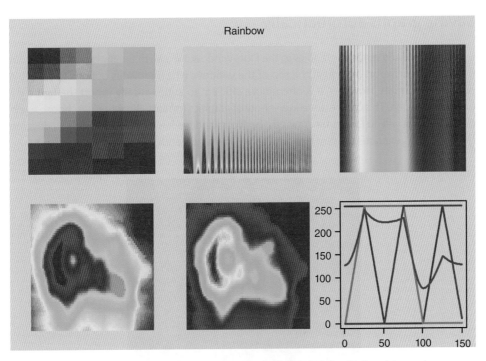

Figure 18.16 Test image for colour palettes, generated by the *pal.test* function for the rainbow palette. Observe in particular that the sinusoidal oscillation in the upper right panel is very hard to see in some regions of the palette (e.g. the green band).

The following code plots this function using the palette stored in the *rpal* object, with the output shown in Figure 18.17:

```
x <- matrix(rep(seq(-10,0,length.out=200),200),nrow=200,ncol=200)
y <- t(matrix(rep(seq(-6.5,0,length.out=200),200),nrow=200,ncol=200))
z <- sin(y)*exp((1 - cos(x))^2) + cos(x)*exp((1 - sin(y))^2) + (x - y)^2

rpal <- kovesi.diverging_bky_60_10_c30(256)
image(x[,1], y[1,], z, col=rpal, useRaster=TRUE, axes=FALSE, ann=FALSE)
```

The *image* function requires vectors of *x* and *y* values, and a matrix of *z* (intensity) values, as well as a palette specification for the *col* option. When it is called, the function generates a new plot, for which axes and labels can be plotted if required. The *useRaster* option tells the function to plot either as a raster image (if TRUE) or a vector image (if FALSE). Raster images are like photographs, where the RGB value of each pixel is stored at a particular resolution. Vector graphics use graphical primitives such as lines, points, and polygons to represent images. For most plots (i.e. lines, points, polygons), vector graphics are preferred because they can easily be enlarged with no loss of quality. However the vector rendering of images such as the one shown in Figure 18.17 often look worse than a raster version, because the individual elements of the texture surface get separated by little white lines. It is worth trying both options to see what looks best.

Saving graphs automatically

By default, *RStudio* sends all graphics to the Plots window. We can manually export a plot from here using the 'Export' button at the top of the window, which provides options to save in a number of formats. However, it is often more efficient to automatically export the figure we have constructed. This can be done in a variety of graphics formats. But when we choose to export a figure, it will no longer be drawn to the Plots window. It is therefore sensible to develop the code that generates the plot first, and add in the commands to export it later.

Figure 18.17 The Mishra Bird function, plotted using the image function and the Kovesi diverging bky palette.

In computer graphics there are two main types of file format: *vector* and *raster* files. As we mentioned in the previous section, raster images are like photographs. They have a set resolution (i.e. number of pixels), and the colour value of each pixel is represented explicitly as a triplet of red, green, and blue values. It is appropriate to use raster formats (such as *.jpeg*, *.tiff*, *.png*, and *.bmp* files) to store photographs, medical images, and similar types of data. However there are two big downsides to raster formats. First, for high-resolution images, the file sizes can become very large. Second, for a fixed resolution, if we zoom into a raster image too far it will start to look blocky and pixelated, because we can see each individual pixel (and sometimes compression artefacts). This can look particularly bad for text, as shown in the example in Figure 18.18(a).

An alternative are vector file formats (including *.pdf*, *.svg*, *.ps*, and *.eps* formats). Vector images are lists of instructions about how to draw an image using a collection of lines, shapes, and text—much like the way we actually construct graphs in R. Vector files do not have a set resolution, and if we zoom in on a particular part the computer can redraw it with a high level of detail (see Figure 18.18(b)). This is the natural format for scientific figures, and most journals prefer you to submit artwork in a vector file format. An added advantage is that vector images typically have much smaller file sizes than their raster equivalents. It is possible to embed raster images as part of a vector file, though these will still have all the limitations of the original image (e.g. fixed resolution and large file size).

We can export a graph as a vector pdf file (pdf stands for portable document format) using the *pdf* function:

```
pdf('filename.pdf', bg='transparent', height = 5.5, width = 5.5)
```

The *height* and *width* parameters are in inches, and other background colours can be set using the *bg* option if required. All plots you create will then be exported to this file up until the following command is called:

```
dev.off()
```

The *svg* function exports in scalable vector graphics format in a similar way. Other functions can export as raster graphics formats (*jpeg*, *tiff*, *png*, and *bmp* functions). For all of the raster functions the file size is set in pixels by default. Finally, we can export in *ps* and *eps* format using the *postscript* function:

```
postscript('filename.ps', horizontal = FALSE, onefile = FALSE,
      paper = 'special', height = 5.5, width = 5.5)
```

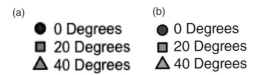

Figure 18.18 Comparison of raster and vector art. Panel (a) shows a figure legend from a graph that was saved in a raster format (jpeg) and then zoomed in. Panel (b) shows the same thing for a vector format (pdf)—the text and objects are sharper, with no visible pixels.

For this function the *paper* option needs to be set to 'special' for the *height* and *width* options to be used. Postscript files are vector images (though they can have raster images embedded inside them). In the next section we will discuss how to reimport individual graphs that have been saved in *ps* format, to combine into a single figure without losing the vector information. There is one downside to this format—postscript files cannot store alpha transparency information. However we will suggest a workaround for this issue.

Combining multiple graphs

In most published journal articles, figures contain multiple subpanels to show different aspects of some data. If these components are all the same size, there are some useful *R* functions and packages that will allow us to define these individual panels within the same figure. For example, many of the multipanel figures in this book were constructed using the *par* function:

```
par(mfrow=c(2,2))
```

This code will generate a 2x2 plot like Figure 18.14, with each subsequent panel being added in sequence. The *layout* function is slightly more sophisticated, and can allow for plots of different sizes. However, sometimes publication-quality figures need more flexibility than is permitted by regular layouts. For this reason, this section will demonstrate an alternative approach to combining plots that is built on functions from the *grImport* and *grid* packages.

The *grImport* package allows us to import a postscript (*.ps*) file by first converting it to a custom *xml* format, as follows:

```
library(grImport)

PostScriptTrace('filename.ps')
e1 <- readPicture('filename.ps.xml')
```

Note that for this to work, you will need the free *Ghostscript* tools installed on your system (see **https://www.ghostscript.com/**). The imported figure is described as a set of lines, text, and filled surfaces stored in a systematic way—in other words it is a vector image. We can draw this figure in its entirety onto a new plot, either in the Plots window or (more typically) a pdf file that we wish to export to. To do this, we use the *grid.picture* function from the *grid* package as follows:

```
grid.picture(e1,x=0.5,y=0.5,width=0.5,height=1)
```

The advantage of this approach is that we have total control over the size and placement of the figure panel in the new plot. By default, the image will be centred on the *x* and *y* coordinates provided, with the *width* value specifying a proportion of the width of the plot window. Somewhat unintuitively, the *height* option specifies a proportion of the width, and so should generally be set to 1 to avoid distorting the aspect ratio of the figure.

Putting everything together, we can create a plot, save it to a postscript file, reimport it, and plot it using the following code, which produces Figure 18.19.

```
# generate the graph and save as postscript
postscript('outputfile.ps', horizontal = FALSE, onefile = FALSE,
paper = 'special', height = 5.5, width = 5.5)
```

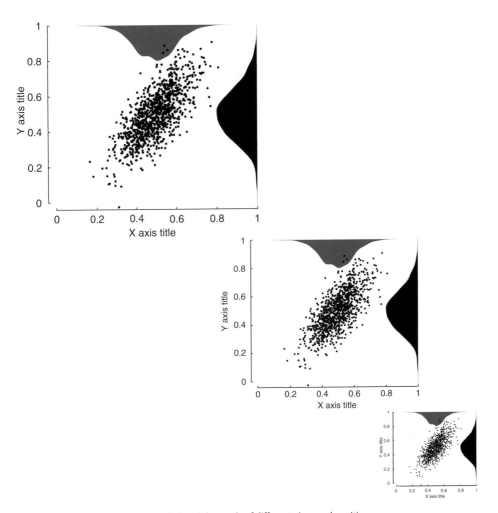

Figure 18.19 Example of multipanel plot with panels of different sizes and positions.

```
plot(x=NULL,y=NULL,axes=FALSE, ann=FALSE, xlim=plotlims[1:2],
ylim=plotlims[3:4])
axis(1, at=ticklocs, tck=0.01, lab=F, lwd=2)
axis(2, at=ticklocs, tck=0.01, lab=F, lwd=2)
mtext(text = ticklocs, side = 1, at=ticklocs)
mtext(text = ticklocs, side = 2, at=ticklocs, line=0.2, las=1)
title(xlab="X axis title", col.lab=rgb(0,0,0), line=1.2, cex.lab=1.5)
title(ylab="Y axis title", col.lab=rgb(0,0,0), line=1.5, cex.lab=1.5)

datax <- rnorm(1000,mean=0.5,sd=0.1)
datay <- rnorm(1000,mean=0.5,sd=0.1) + (datax-0.5)

a <- density(datax)
a$y <- 0.2*(a$y/max(a$y))
polygon(a$x, 1-a$y, col='blue',border=NA)
```

```
a <- density(datay)
a$y <- 0.2*(a$y/max(a$y))
polygon(1-a$y, a$x, col='black',border=NA)

points(datax,datay,col=rgb(0,0,0),pch=16,cex=0.6)

dev.off()

# import the figure
library(grImport)
PostScriptTrace('outputfile.ps')
e1 <- readPicture('outputfile.ps.xml')

# create a new plot and add the graph at several scales
plot(x=NULL,y=NULL,axes=FALSE, ann=FALSE, xlim=plotlims[1:2],
ylim=plotlims[3:4])
grid.picture(e1,x=0.8,y=0.2,width=0.2,height=1)
grid.picture(e1,x=0.6,y=0.4,width=0.4,height=1)
grid.picture(e1,x=0.25,y=0.75,width=0.5,height=1)

# clean up the external files we generated
file.remove(c('outputfile.ps','outputfile.ps.xml'))
```

This approach to collating plots works very well, but there is one problem—the postscript format does not support alpha transparency. This means that when we create our individual subplots, we cannot include any transparent features as they will generate an error. However, we can specify transparency when the plots are loaded back in with the *readPicture* function. The picture object created by this function is a list of all the components that make up a plot. We can loop through this list, find the features we want to be transparent, and change their colour settings (this will produce the image shown in Figure 18.20):

```
for (n in 1:length(e1@paths)){   # loop through all features
temp <- class(e1@paths[n]$path)[1] # look at just one feature
# check if this feature is a filled shape
if (pmatch(temp,"PictureFill",nomatch=0)){
# set any filled shape to be semi-transparent
e1@paths[n]$path@rgb <- addalpha(e1@paths[n]$path@rgb,alpha=0.3)}}
# plot everything again
plot(x=NULL,y=NULL,axes=FALSE, ann=FALSE, xlim=plotlims[1:2],
ylim=plotlims[3:4])
grid.picture(e1,x=0.8,y=0.2,width=0.2,height=1)
grid.picture(e1,x=0.6,y=0.4,width=0.4,height=1)
grid.picture(e1,x=0.25,y=0.75,width=0.5,height=1)
```

Adding raster images to a plot

Sometimes it is necessary to include a raster image as part of a figure, for example to illustrate experimental stimuli, or show something like a microscope image or Western blot. We can load images from a wide range of graphics formats using packages such as *jpeg* (for *.jpg* files) and *tiff* (for *.tiff* files). These images are stored as matrices in R, either with two dimensions (x and y) for a greyscale image, or with three dimensions for a colour image (x, y and the

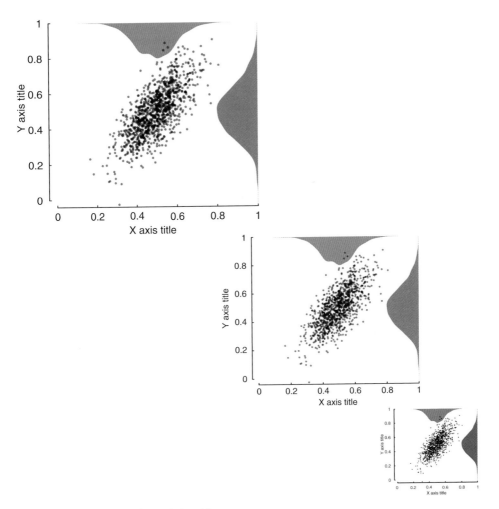

Figure 18.20 Example of multipanel plot with transparency.

colour channels: red, green, and blue). For example, we can load in a greyscale jpeg image of a violin (from the *Bank of Standardized Stimuli*, see Brodeur et al. (2010)) using the *readJPEG* function as follows:

```
library(jpeg)
violin <- readJPEG('images/BOSSr/violin.jpg')
dim(violin)
```

```
## [1] 256 256
```

We see using the *dim* function that the image is 256x256 pixels (I have resized it—the original database is at 2000x2000 pixels). This image can then be inserted into a plot using the *rasterImage* function from the *graphics* package as follows:

```
rasterImage(violin,0.2,0.2,0.5,0.5)
```

This will plot the image at the x- and y-coordinates given by the first two numbers for the bottom left corner, to the x- and y-coordinates given by the last two numbers for the top right corner. Note that graphs in which the axes have different limits or units will require careful thought about how to set these coordinates to preserve the correct aspect ratio of the image. For example, if the x-axis runs from 0 to 1, but the y-axis spans from 0 to 2, the extent of a square image in the y-direction must be twice that in the x-direction.

It is also possible to rotate the image if required using the additional *angle* option, specified in degrees. Rotation is about the lower left corner of the image (not the centre), so this will also shift the centre of the image, meaning that the coordinates often need to be adjusted for appropriate placement. A good way to think about this is that the second pair of coordinates are really specifying the size of the image (rather than its top right corner). For example, consider the code:

```
rasterImage(violin,0.8,0.8,1.1,1.1,angle=180)
```

This actually plots the image between $x = 0.5$ and $x = 0.8$, because the rotation through 180 degrees about the lower left-hand corner means that the top right corner has moved: Figure 18.21 shows the violin image plotted at both rotations. One practical consideration when plotting raster images is that they will cover up anything plotted underneath them. It is often pragmatic to plot the images first, so that any points, lines, and other features of the figure appear on top of the image.

The methods described in this chapter can be combined to produce publication-quality plots entirely within *R*. I have used this approach for a number of papers, and example scripts are available for several of them. For example, scripts to produce a dozen figures from Baker et al. (2021) are available on the *Open Science Foundation website* at: **https://osf.io/ebhnk/**. I hope that the principles for clear data visualization and choosing colour palettes described in this chapter will help you to produce more informative, accessible, and beautiful figures.

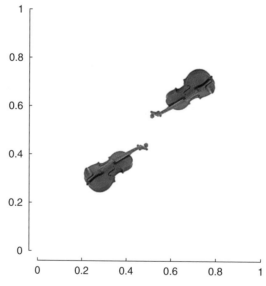

Figure 18.21 Demonstration of using the *rasterImage* function to add a square image of a violin to a plot at two rotations.

Practice questions

1. Plotting an axis in log units can be problematic because:
   A) The units are harder to interpret
   B) The tick marks are spaced irregularly
   C) Skewed distributions appear more normal
   D) The base of the logarithm affects the appearance

2. Starting an axis at a number other than zero is sometimes (unethically) used to:
   A) Distort the distribution of the data
   B) Allow logarithmic units to be used
   C) Minimize the apparent differences between conditions
   D) Exaggerate the apparent differences between conditions

3. Standard errors will always be:
   A) Smaller than standard deviations but larger than 95% confidence intervals
   B) Smaller than standard deviations and smaller than 95% confidence intervals
   C) Larger than standard deviations but smaller than 95% confidence intervals
   D) Larger than standard deviations and larger than 95% confidence intervals

4. Adding individual data points to a plot:
   A) Allows us to check visually for outliers
   B) Makes error bars unnecessary
   C) Makes the means unnecessary
   D) Is not possible for continuous data like time courses

5. A person who cannot discriminate between green and blue, but can discriminate between red and green is most likely to have:
   A) Achromatopsia
   B) Deuteranopia
   C) Protanopia
   D) Tritanopia

6. The parula palette:
   A) Changes smoothly in hue but has luminance peaks
   B) Has plateaus of hue but changes smoothly in luminance
   C) Changes smoothly in hue and luminance
   D) Has peaks of luminance and plateaus of hue

7. Perceptually uniform palettes:
   A) Aim to introduce spurious features that are not present in the data
   B) Are created by linear changes in the values of the red, green, and blue channels
   C) Aim to produce smooth changes in perceived luminance and hue
   D) Look identical even to individuals who are colour-blind

8. Transparency can be set through a channel referred to as:
   A) Gamma
   B) Alpha

    C) Beta

    D) Hexadecimal

9. A vector image:

    A) Stores the value of every pixel

    B) Stores a description of lines, text, and shapes that make up a figure

    C) Has a fixed resolution (number of pixels)

    D) Will tend to be larger than a high-quality raster image

10. Violin plots show:

    A) The raw data points

    B) The raw data and a distribution

    C) Two mirrored distributions

    D) The mean and an error bar only

Answers to all questions are provided in the answers to practice questions at the end of the book.

Reproducible data analysis

As we have mentioned several times throughout the book, many researchers have become more aware in recent years that there are substantial issues with contemporary scientific practice. In particular, modern research can be extremely complex, involving analysis pipelines requiring many stages to go from the raw data to the graphs and summary statistics presented in a final publication. The reader of a paper must then take it on trust that the authors have conducted the analysis correctly and competently, and that the reporting is accurate and unbiased.

In the interests of transparency, many advocates of open science have begun to make raw data and analysis scripts publicly available, with the intention that others could reproduce their analysis independently (note that reproducibility is distinct from replication, which is where other researchers run an experiment themselves to see if they get the same result). This may seem like a scary prospect—most scientists are not trained programmers, and are often concerned that there might be errors in their code that would invalidate some analysis, or even that the code looks messy and inelegant, even if it works.

However, these worries aside, sharing data and code is a critical step in improving scientific practice. It gives the readers and reviewers of a paper confidence that the analysis is as described in the paper, and that the data exist and have not been manipulated inappropriately. It should ideally be possible to independently reproduce the full analysis pipeline, by downloading and running it yourself to confirm the findings. Grant agencies and journals increasingly require such materials to be made available, as they realize the importance of openness for increasing the impact of research, and improving the quality and societal impact of science.

Open data and code allow other researchers to build on your work, perhaps reanalysing the data for another purpose, or using parts of the code in their own analysis. Existing data are valuable for conducting meta-analyses (see Chapter 6), and to inform power analyses in planning new studies (see Chapter 5). Open practices should generally act to speed up scientific progress, and even suggest new projects that were not previously viable. Over the past few years I have used open data in my own research, and also had others use data that I have made available. Both experiences have been very positive, and I have no regrets about sharing data publicly. In particular I have been contacted by researchers who lack laboratory facilities to collect their own data, but are able to make progress and test hypotheses by reusing existing data sets. Such secondary data analysis is also likely to increase citations to the original work.

The aim of this chapter is to discuss some details associated with making data and scripts open and accessible. Some of these points will transcend specific implementation issues, but we will also discuss how to interface with platforms that are widely used today, in particular *GitHub* and the *Open Science Framework* (OSF). In general, making code and data available will improve the integrity and progress of science, and is worthwhile for researchers in all disciplines and at all career stages.

Version control of analysis scripts

Computer scientists and professional programmers have developed robust systems for version control of code. The most widely used is a system called *git*. Essentially, the point of *git* is to keep track of every time you change some code. This means you can roll back any changes that didn't work, and for collaborative work you can see who wrote each piece of code. A detailed explanation of how to use *git* is beyond the scope of this book, and many other (free) resources are available (see e.g. Chacon and Straub (2014), and *Happy Git With R*). However we will briefly cover integration of *RStudio* and online *git* repositories such as *GitHub* because they are an important resource for conducting open, reproducible analyses.

The first stage is to install *git* on your computer. It is a stand-alone set of tools that can be accessed through the command line, as well as integrating with *RStudio* (and most other contemporary programming environments). *Git* can be obtained in several different ways—the simplest is probably to download it from the *git* website at: **https://git-scm.com/downloads**. You will then need to tell *RStudio* where your local *git* installation lives, which you can do through the *Git/SVN* section of the *Preferences* dialogue window in *RStudio*. The location of *git* will be different on different platforms (in Windows it is usually in Program Files, and on the Mac it should be somewhere like /usr/local/git/bin/git), but some searching online will help you to find it on your own machine.

If the project you want to work on already exists as a *GitHub* repository, you can access it by *cloning* the repository to your own computer (if you want to create a totally new project, the easiest way to do this is through the *GitHub* website). Cloning creates a local copy of all the files in the repository. I have set up a public dummy example to demonstrate how this works. From within *RStudio*, choose *New Project...* from the *File* menu. Select *Version Control*, followed by *Git*. You will then see a dialogue box asking for details of the repository. The one I've created is at:

https://github.com/bakerdh/gitdemo

You should enter this URL into the dialogue box. You will also need to choose a name for the project, and a directory to save it to on your computer. Then, when you click *Create Project*, all of the files from the repository will automatically be downloaded and stored as a local copy on your computer.

At the moment the project really just has a single file called *gitdemo.R*, which doesn't do very much (it prints out a message to the Console). You can make changes to this file on your computer, and then upload the changes to the repository. There are three stages to this process, which are all accessed through the *Git* tab in the top right corner of the main *RStudio* window (where the *Environment* usually appears).

First you can click the *Diff* button to *stage* the changes—this shows you a colour-coded visual representation of everything that you have changed from the latest version you downloaded, in a separate window headed *Review Changes*. If you are happy with the changes you can *Commit* them. This means you are saying that you are ready to upload the changes to the online repository so that others can see them. You should also add a commit message, where you can briefly detail the reasons for the changes. Finally, you can *push* the changes to the repository using the upward pointing green arrow—this will send everything you have altered to the *GitHub* server and incorporate it into the main project.

If you are working on a group project, you can also periodically *pull* changes made by your collaborators using the downward pointing blue arrow. This will update your local copy of the project. But this is where things can get tricky—if multiple people are working on the same files, there might be conflicts, where two people have made incompatible changes. There are tools within *git* to cope with this, including making separate *branches* of the code that you can work on individually, before merging them back with the master branch. Dedicated *git* references go into more detail about how this works (Chacon and Straub 2014).

GitHub repositories are designed to contain computer code and other small files, so it is not generally a good idea to store large data files in them (the amount of storage space is usually limited for this reason). Happily, as we will outline in 'Identifying a robust data repository' later in the chapter, other repositories exist that can store your data, and you can include code to automatically download the data to your computer. Furthermore, the OSF website integrates well with GitHub, in that you can associate a GitHub repository with an OSF project. The files contained in the repository will appear as if they are files within the OSF project, meaning that data and code are accessible from a single location.

There are two additional advantages to maintaining a GitHub repository for your programming projects. First, your code exists independently of your computer, and is automatically backed up every time you push your changes to the repository. This means that if your computer breaks or gets lost or stolen, you will not have lost your work. (I have been extremely paranoid about accidentally deleting computer code ever since I was 10 years old and tried to write a BASIC program to remove backup files, but accidentally deleted everything on the disc.) The other advantage is that, if your repository is public, other people can see how great you are at programming. Coding is now an essential skill in many areas of science, and potential employers (particularly for those at an early career stage) will often want to see evidence of your programming abilities. Giving them a link to a well-organized GitHub page might be the thing that lands you a great job!

Writing code for others to read

As well as sharing code, it is well worth bearing a few things in mind to help others to read and understand your code. This may seem obvious, but it is surprising how rarely people do this (myself included). In addition to helping others to understand your code, these suggestions might also help future versions of yourself (by which I mean you, but in the future; I'm not suggesting you're going to be cloned) to remember exactly what your past code is doing. Coding decisions that made sense when you were immersed in a project can often be impenetrable a decade later.

First of all, it is important to give your data objects, functions, and files meaningful names. When you are in the process of writing code, it is easy to come up with arbitrary names for data objects (*a* is a perennial favourite, as is *data*, and I am aware that I have used both liberally throughout this book). But a much clearer approach is to have data object names briefly describe what they contain: *rawdata* and *cleandata* is much better than *data1* and *data2*. It can also be useful to come up with a naming convention that distinguishes data objects from functions. For example, I sometimes begin function names with d_ (d for Daniel), especially when they refer to generic functions from a 'lab toolbox' that will be used in multiple scripts.

For longer object names, many programmers favour *camel case*, where the first letters of each word are capitalized. So *rawdata* would become *RawData* and *cleandata* would be *CleanData*, which can make the object names easier to parse and avoid some ambiguities. A variant of camel case is to keep the first letter of the object name in lower case (e.g. *rawData*)—this is called either lower camel case or Dromedary case (a dromedary is a type of camel with one hump). Remember that *R* is case-sensitive, so the object names *RawData* and *rawdata* refer to different data objects—it would be very confusing to use both of these in the same script!

A second recommendation is to provide clear and detailed comments about what your code is doing. In *R*, the hash symbol (#) and anything following it are ignored by *R*, allowing you to include comments, as follows:

```
randomSum <- 0 # initialize a data object to store a number in

# loop ten times
for (n in 1:10){

# add a random number to our data object each time around the loop
randomSum <- randomSum + rnorm(1)

}
```

You can use multiple hash symbols to introduce hierarchical structure into your code, for example by using three hash symbols to indicate a new section of code. I am pretty bad at commenting my own code, so this is something I need to work on myself! But it is definitely helpful for anyone trying to understand what you have done, or attempting to modify your code for their own needs. Beginning your script with some comment lines giving an overview of what it aims to do is good practice, especially in the common situation where you might have many versions of the same piece of code which are all slightly different.

Finally, there is a balance to be struck between writing dense code with multiple nested functions on a single line, and code that is more spread out and easier to read. For example, the following section of code contains multiple nested commands and a logic statement, and is therefore rather difficult to parse:

```
if (t.test(rnorm(100,mean=2,sd=2))$p.value<0.05){print('Test
is significant')}
```

On the other hand, we can rewrite (or refactor) the code by splitting things up and de-nesting the function calls as follows:

```
a <- rnorm(100, mean=2, sd=2)
tout <- t.test(a)
if (tout$p.value<0.05){
 print('Test is significant')}
```

(In both cases, the code generates some random numbers, runs a t-test on them, and spits out some text if it is significant.) Which of these approaches you take will depend on what sort of code you are trying to write. If your aim is to use as few lines of code as possible, and perhaps save on memory allocation for additional data objects, the first approach might occasionally have some advantages. If the aim is to be understood by others, the second approach is much clearer.

If you write dense code without comment lines and meaningful object names, it will seem like a chore to go back and add them later. Indeed, many programmers will be reluctant to do this, in case it stops the code from working! However, if you bear these suggestions in mind when you are developing the code, then you can embed readability into your code as you create it. This often encourages you to think more deeply about the programming decisions you are making, and about the structure of your code, which means you are more likely to write better, more legible code.

Using *R Markdown* to combine writing and analysis

A very useful tool that is worth knowing about is *R Markdown*, which is part of the *RStudio* ecosystem. The idea of a markdown script is that you can construct a document that contains both written text and computer code. As well as formatting both types of information into a single output (such as a pdf, Word document, or web page), the code is also executed, which means you can perform data analysis and create figures entirely from within your document. This makes it possible to produce a complete piece of research, including all analyses and calculations, within a single file. This has huge advantages for reproducibility, as your readers can download both the finished file and the underlying markdown code. Indeed, the first draft of this book was written in markdown. Additional features include being able to incorporate code in other programming languages besides *R* (including *Python, Julia,* and *C++*), and good typesetting of equations (based on *LaTeX* syntax). It is not an exaggeration to say that markdown has changed the way that I (and many others) produce written outputs. Some scientific journals are also investigating how to embed working code into published articles to create computationally reproducible articles, so this approach is likely to become more widespread in the future. A good reference to learn more is *R Markdown: The Definitive Guide* (Xie, Allaire, and Grolemund 2019).

Automated package installation

One of the big strengths of *R* is the way that packages are handled (see 'Packages' in Chapter 2). This is much more straightforward than in any other programming language I have used. However there is a minor issue when writing code that will be used by other people, or even just run on more than one of your own computers. The first time you need to use a package on a given system, you need to install it. For example:

```
install.packages('zip')
```

But once the package has been installed, repeating this line of code is unnecessary. In fact it will actually reinstall the same package(s), which can substantially slow down your script, especially if there are many packages to install. As an alternative, the following code will check the currently installed packages, and only install those that are missing:

```
packagelist <- c('zip')   # a list of packages we want to install
# work out which packages are not currently installed
missingpackages <- packagelist[!packagelist %in% installed.
packages()[,1]]
# install the missing packages only
if (length(missingpackages)>0){install.packages(missingpackages)}
```

We can do a similar trick for activating the packages, so this is only done for packages that are not currently active:

```
# work out which packages are not currently activated
toactivate <- packagelist[which(!packagelist %in% (.packages()))]
# silently activate the inactive packages using the library function
invisible(lapply(toactivate,library,character.only=TRUE))
```

These five lines of code will avoid repeatedly installing and activating the same packages, and is a reasonably compact code snippet to include at the start of a script, which will make your code more effortlessly accessible to others.

Identifying a robust data repository

It is worth noting that until relatively recently (i.e. since the turn of the millennium), suitable infrastructure to facilitate data sharing was not generally available. If researchers wished to share their data, they would either indicate in a paper that they would provide it upon request (usually by email), or they might post the data files to a personal website. There are obvious problems with both of these solutions—email addresses change, as do websites, and both would be lost in the event of the author's death.

In recent years, more reliable platforms for sharing data have been developed. There are two main varieties of these. In many subject areas, there are repositories that are intended for sharing particular types of data. For example, the websites *Neurovault* and *OpenNeuro* store neuroimaging data, the *Crystallography Open Database* is a well-organized collection of crystal structure data, and the *Electron Microscopy Public Image Archive* (*EMPIAR*) shares electron microscopy images.

The other option is to use a more generic data archive website. Again there are several of these, such as *Figshare*, *DataDryad*, and the *Open Science Framework* (OSF), all of which will archive data from any research field, and make it publicly available. A key feature to look for is that the data are given a stable DOI (a digital online identifier—basically a number that will always link to that data set), as this can be included in any publications associated with the data set, and also allow the data set to be cited directly. It is also worth checking that the platform has contingency plans for the future that will ensure the longevity of the data. For example, the OSF website has a preservation fund that is sufficient to maintain hosting of its archive for over 50 years.

OSF is currently my preferred repository, because it is free, integrates well with other systems (such as *GitHub*, which we discussed in 'Version control of analysis scripts'), and has many additional features such as hosting of preregistration documents and preprints. This means it can be used as a repository for an entire project, hosting the data files, analysis scripts, and other information. An ideal to strive for is that another researcher could download all of your materials, and fully reproduce your entire analysis pipeline, through from processing the raw data to creating the figures and statistics reported in a paper. The structure of the OSF website makes this goal achievable. An example repository from a recent paper of mine (Baker et al. 2021) is available at: **https://osf.io/ebhnk/**.

Automatically uploading to and downloading from OSF

One of the main reasons that I am a particular fan of the OSF repository is that it provides an *R* package to automate uploading and downloading data from the site. For uploading data, this can save a lot of time with large data sets, and is far quicker than the standard drag-and-drop web interface. For downloading data, it raises the possibility of other users requiring only a single analysis script, which would then automatically download and analyse all the data associated with a project. Both of these are big advantages for creating efficient and reproducible workflows.

The *osfr* package is hosted on GitHub, rather than the CRAN repository. We need to use the *remotes* package to download and install it as follows:

```
install.packages('remotes')
library(remotes)
remotes::install_github("centerforopenscience/osfr")
library(osfr)
```

I have created a dummy OSF project to demonstrate how uploading and downloading works. However, in order to upload data (to a repository you own) you would first need to authorize *R* to interact with your OSF account, using an access token. The instructions for how to create a token are in the settings section of the OSF website. Once you have a token, this is entered as follows:

```
osf_auth(token = 'MY_TOKEN_HERE')
```

Of course your token will only allow you to upload to your own projects, and importantly it should not be shared with anyone else. This means you need to remember to remove the token if you are posting your analysis code somewhere publicly (just as I have done above).

We can retrieve details of the dummy project's 'node' (**https://osf.io/thm3j/**) by entering the unique five-character part of the URL (*thm3j* for this example) into the *osf_retrieve_node* function as follows:

```
osfproject <- osf_retrieve_node("thm3j")
osfproject
## # A tibble: 1 x 3
##   name            id  meta
##   <chr>          <chr> <list>
## 1 Upload & Download examples thm3j <named list [3]>
```

The *osfproject* data object then contains metadata about the project, and we can use it to list the files contained in the project with the *osf_ls_files* function:

```
filelist <- osf_ls_files(osfproject)
filelist
## # A tibble: 5 x 3
```

```
##  name     id                  meta
##  <chr>    <chr>               <list>
## 1 File1.csv 5ef0d99265982801b4cf0c9f <named list [3]>
## 2 File2.csv 5ef0d99665982801abcf2ad5 <named list [3]>
## 3 File3.csv 5ef0d99a145b1a01cc52cc8f <named list [3]>
## 4 File4.csv 5ef0d99e65982801a8cf2b7b <named list [3]>
## 5 File5.csv 5ef0d9a2145b1a01cb52c874 <named list [3]>
```

From this table, we see that the project contains five *csv* files. I originally uploaded these files using the following code (which of course could also be done in a loop):

```
osf_upload(osfproject,'data/File1.csv')
osf_upload(osfproject,'data/File2.csv')
osf_upload(osfproject,'data/File3.csv')
osf_upload(osfproject,'data/File4.csv')
osf_upload(osfproject,'data/File5.csv')
```

The OSF website has automatically generated ID numbers for each file. We can then download the files to the *R* working directory using the *osf_download* function:

```
osf_download(filelist) # download all files in the list
```

Note that we are downloading all the files at once here, which might be a dangerous strategy with large data sets! If we want to download only a single file, we can index different rows of the *filelist* data object as follows:

```
osf_download(filelist[3,]) # download the third file in the list only
```

The *osf_download* function takes additional arguments that can be used to specify the destination path to save the files to on your local computer.

If we want to find a particular file, we can use *R*'s native string matching functions as follows:

```
a <- which(filelist[,1]=='File3.csv')
osf_download(filelist[a,])
```

The OSF repository allows additional nodes (also called components) to be nested inside the main one, in much the same way that computer file systems can have multiple subdirectories. A list of subnodes within a given node can be generated using the *osf_ls_nodes* function, which is called in a similar way to the *osf_ls_files* function described above. The table returned by this function is a list of additional nodes, which can also be indexed using either of these listing functions, and the enclosed files uploaded and downloaded. The package also contains functions to generate components and folders (*osf_create_component* and *osf_mkdir*). By combining these functions, we can generate online data structures of arbitrary complexity entirely from within an *R* script.

Future-proofing with open data formats

One common issue with data sharing is that many data acquisition devices have a proprietary native file format. Sometimes this requires special software to access the data, and this software may not be free to obtain. This can result in a barrier to data sharing, as not everyone who wants to access the data will have the resources necessary to obtain the software. It is also an issue for future-proofing data access. Proprietary data formats tied to a particular company may no longer exist in 20 years' time, especially if the company has ceased trading, or has replaced its software with something else that is not backwards-compatible.

A good example from my own research area is electroencephalography (EEG) data. There are currently dozens of EEG systems available, and each of them has a different proprietary format for saving the resulting data. For example the *EEGlab* analysis software (Delorme and Makeig 2004) includes extensions for importing over 30 different file formats. If I were to make available the raw EEG data files from an experiment, it might become impossible (or at least very difficult) to open them in the future.

There are two main solutions to this problem. One option is to convert the data to a standard format that is widely used in the field. For MRI data, the NIFTI (Neuroimaging Informatics Technology Initiative) file format is a great example—it is an open standard that has been widely adopted and is compatible with all current analysis software. Crucially, the file format is described in detail and is actively maintained by a working group that is not tied to a particular commercial organization. Similarly, the Brain Imaging Data Structure (BIDS) is a standard for organizing and describing neuroimaging data from multiple modalities (Gorgolewski et al. 2016). In genetics and bioinformatics, the FASTA format is widely used to store DNA data, and has similar advantages.

The other alternative is to convert your data into a generic data format that is not specific to a particular scientific field or data type. The most basic example is to use a text file format, such as the *csv* (comma-separated values) format. Each individual value in a *csv* file is separated by a comma, with new lines indicated by a carriage return. This format can store both numbers and text, and the data can be easily read by any modern programming language, as well as spreadsheet packages such as Microsoft Excel, Google Sheets, or OpenOffice Calc. Finally, column headings can be included as the first row of data.

A shortcoming of text files is that they are not very efficient for storing large amounts of data. If storage space is an issue, lossless compression algorithms (such as *zip* and *gzip*) can be used to reduce file size. Alternatively, generic formats for structuring and storing large amounts of data also exist. One example is the HDF5 (Hierarchical Data Format 5) specification, which can store arbitrarily large data structures. This is an open source format, developed by a not-for-profit group dedicated to ensuring accessibility of data stored in the format. The *hdf5r* package contains tools for reading and writing in this format within *R*, and there are similar libraries for other programming languages.

For several reasons, data formats associated with a specific programming language or software package are not typically very open. For example, the widely used Matlab language is owned by a company (The MathWorks Inc.), which also owns the specification for the *.mat* file format. Although toolboxes currently exist to import such data files into other programming environments (such as the *R.matlab* toolbox in *R*), there is no guarantee that this will always be the case. Similar arguments apply to Microsoft's *xls* and *xlsx* spreadsheet formats. Overall, making your data available in an open format is a better choice for ensuring it is accessible to others, both now and in the future.

Providing informative metadata

In addition to being able to access and open data files, it is important for potential users to understand what they contain. So including useful metadata is critical. Sometimes this can be incorporated into the data files themselves, for example in the HDF5 format. But it can also make sense to include a separate metadata file describing the data format, the experimental conditions, and so on. At its simplest, this could just be a plain text file (like a readme file) that explains the structure of the data.

I have recently begun sharing EEG data as compressed *csv* text files, and include a metadata file that describes the full data set, the experimental conditions, and how these map to trigger codes, the participants, and the equipment parameters (e.g. electrode locations). An example of the first few rows and columns might look like this:

```
hdata <- read.csv('data/headerfile.csv')
head(hdata[,1:4],n=12)
##                         Details1       Details2 Trigger      Description
## 1      SSVEP masking experiment           2014      10    Target 0 Mask 0
## 2                   Sample rate         1000Hz      20    Target 2 Mask 0
## 3             Stimulus duration            11s      30    Target 4 Mask 0
## 4              Target frequency            7Hz      40    Target 8 Mask 0
## 5                Mask frequency            5Hz      50   Target 16 Mask 0
## 6      Repetitions per condition              8      60   Target 32 Mask 0
## 7               Trials per block             40      70   Target 64 Mask 0
## 8             Total participants            100      80    Target 0 Mask 32
## 9            Number of electrodes            64      90    Target 2 Mask 32
## 10               Montage layout      5 percent     100    Target 4 Mask 32
## 11                   EEG system   ANT Neuroscan     110    Target 8 Mask 32
## 12  Original file format ANT EEprobe .cnt files            120   Target 16 Mask 32
```

Some of this information could be read in by a script, and used to automate parts of the analysis. The aim of including metadata should be that another researcher could use it, along with the description of the experiment in an associated publication, in order to work out how to reanalyse your data. Two added bonuses: if you manage to do this you are much less likely to be contacted by other researchers asking for help and explanations, and also your future self will have a much easier job if you ever need to go back to look at your own data again.

Practice questions

1. Which of the following is **not** an advantage of sharing analysis code?
   A) Other researchers can see what you did without having to contact you directly
   B) If you share code, it allows you to access other people's shared code
   C) If there are mistakes in the code, other researchers might find them for you
   D) Future work can more easily build on the work you have done

2. One problem with analysis packages that involve graphical interfaces instead of writing code is that:
   A) It is often hard to reproduce exactly the options that were selected for the analysis
   B) There is only one possible way to analyse a given data set
   C) The calculations are less accurate than they would be in a programming language
   D) There is no way to save the results of the analysis

3. An important feature to look for in a data repository is:
   A) That it is maintained by a commercial company
   B) That it stores only data from the particular type of experiment you have run
   C) That it will remove your data if you die
   D) A stable digital online identifier is created for each data set

4. An OSF personal access token will allow you to use *R* to:
   A) Upload to your own projects
   B) Download from public projects
   C) Upload to public projects
   D) Access all projects on the site

5. Which of the following are both open data formats?
   A) HDF5 and *xlsx*
   B) *csv* and *.mat*
   C) *csv* and HDF5
   D) *.mat* and *xlsx*

6. When using *git*, a *pull* request is where you:
   A) Upload your changes to the repository
   B) Download other programmers' changes from the repository
   C) Create a new branch of the code
   D) Integrate any changes into the main branch

7. In *git*, the order of the three stages to upload some changes is:
   A) Clone, push, and pull
   B) Pull, push, and commit
   C) Commit, pull, and stage
   D) Stage, commit, and push

8. Which of the following data object names is in lower camel case?
   A) AllData
   B) allData
   C) alldata
   D) Alldata

9. In R, comments are indicated by:

A) #

B) %

C) '' (the comment is enclosed in quote marks)

D) *

10. What is the difference between reproducibility and replication?

A) Replication is where other researchers try and produce the same result as you but using different methods, whereas with reproducibility they use the same methods

B) Reproducibility is where other researchers try and produce the same result as you but using different methods, whereas with replication they use the same methods

C) Reproducibility means that others can recreate your analysis pipeline (using your data), whereas replication is where they try to repeat your experiment

D) Replication means that others can recreate your analysis pipeline (using your data), whereas reproducibility is where they try to repeat your experiment

Answers to all questions are provided in the answers to practice questions at the end of the book.

A parting note

I hope that you have found at least some parts of this book useful. As discussed in Chapter 1, the aim was to provide enough information to grasp and implement the basics of several techniques. Hopefully this will serve as a starting point for your own further reading and investigation of the methods. Inevitably there are many things I have not included here, and I welcome any feedback and suggestions for new material in future editions. Remember that the CRAN repository contains many thousands of packages to investigate, and that the code to produce all figures in this book is available in a *GitHub* repository at: **https://github.com/ bakerdh/ARMbookOUP.**

As your abilities using *R* increase, you will certainly run into problems, many of which will take time and patience to solve. As mentioned in Chapter 1, the *R* community is generally very supportive and inclusive, and there is a wealth of useful information available online. This includes discussion boards, blog posts, and tutorials created by other users, and formal groups like *RLadies* and the local *R* user groups coordinated by the *R* consortium. You could also set up an *R* interest group in your own town, workplace, or university department to build a peer group of *R* users who can help each other develop their skills.

In several places I have talked about the *Open Science* movement, for example discussing statistical power in Chapter 5, publication bias in Chapter 6, and reproducibility and data sharing in Chapter 19. My hope is that open research practices will continue to become embedded in the way science is conducted across all fields and disciplines. Future readers of this book might find it endearingly quaint that I have talked explicitly about things like sharing data and code, if these have become second nature to all scientists. However, at the time of writing open practices are only just gaining traction, and are still applied to only a minority of projects.

Ultimately the scientific enterprise is about finding out how the world works. But real data are always noisy, and rarely lead to a clear interpretation in their raw state. To this end, the analysis methods described in this book are tools that help us to test hypotheses and better understand empirical data. Yet the choice of statistical tool, and the way we interpret its outcome, will often be somewhat subjective. My aim is that by understanding the fundamentals of a range of techniques, researchers can make more informed decisions about how to conduct their analyses. Mastery of advanced methods allows for creative flair in the presentation of results, and in some situations may lead to new knowledge that would be missed by more basic analyses. Although it may not feel like it to a beginner, it can be very satisfying to immerse oneself in data analysis ('wallowing in your data' as a former colleague of mine used to say). I hope that this book helps others on their way to experiencing data analysis as a pleasure rather than a chore, and leads to many new and exciting discoveries.

Answers to practice questions

Chapter 2

1. B—*R* is based on the *S* language, which was first released in 1976.

2. A—the *Environment* tab is in the upper right corner, and contains variables currently held in memory.

3. B—although scripts are useful, the syntax of different programming languages means that you usually cannot run a script in a language other than the one it was written for.

4. D—variable names cannot begin with a number, so 9thNumber is not permitted.

5. C—matrices and data frames have both rows and columns, much like an Excel spreadsheet.

6. B—square brackets are used to index variables, which distinguishes them from function calls (which use normal brackets).

7. D—the packages installed on your computer are visible in the Packages tab in the lower right panel of *RStudio*.

8. B—the loop will repeat five times, with the variable *n* having the values 31, 32, 33, 34, and 35.

9. D—the loop will repeat three times, so the end value of *a* will be 0 + 4 + 4 + 4 = 12. This means that the first and fourth *if* statements will be true: a>10 and a==12. So the code will print out *Bananas*, followed by *Oranges*.

10. D—the transpose function in *R* is called *t*. You can work this out by trying to call the *help* for each of the four options, or by searching online.

Chapter 3

1. B—the data layouts are called *wide* and *long*—see the chapter for examples.

2. C—the error bars on a boxplot usually show the inner fence. Data points falling outside of the whiskers are classified as outliers using Tukey's method.

3. D—the interquartile range gives the points between which 50% of the data points lie. This is important for some methods of identifying outliers, but it is not used to replace outlying values.

4. D—the Mahalanobis distance is suitable for identifying outliers with multivariate data. The other three methods are used for univariate data.

5. C—the inner fence is 1.5 times the interquartile range beyond Q1 and Q3, which works out as 2.698 standard deviations from the mean.

6. A—a 20% trimmed mean would exclude the highest 20% and lowest 20% of values, leaving only the central 60%.

7. B—the Kolmogorov-Smirnov test is a test of the normality assumption, rather than an alternative to parametric methods. The other responses are all plausible alternatives to using parametric tests.

8. C—the *scale* function performs normalization of the mean and standard deviation. If we want to avoid subtracting the mean, we set the 'centre' argument to FALSE, as in the third answer.

9. D—all of the previous options are used to assess deviations from normality. The Q-Q plot shows the quantiles of the data and reference distributions, and the two tests give a quantitive comparison.

10. A—logarithmic transforms will squish large values closer together, so the skewed tail of the distribution will shrink.

Chapter 4

1. A—Gossett was employed by the Guinness corporation in Dublin as their head brewer, to analyse data on batches of beer.

2. B—the model formula uses the tilde (~) to indicate relationships, and is always of the form DV ~ IV. Since we want to know how age predicts brain volume, we need brainvolume to come first in the formula.

3. C—you can confirm that the other three options are not *R* functions by trying to call the *Help* function for them—they do not exist.

4. D—the residuals are the left over variance that cannot be explained by the model. A good way to think about this is the error between the model's predictions and the data points.

5. C—a null regression model is flat, so it has a slope of zero. See Figure 4.2 for an example.

6. D—the slope of a fitted regression line will depend on the data we are fitting, and the strength of any relationship between the two variables.

7. B—the degrees of freedom are always one less than the number of groups. A good way to think about this is the number of straight lines required to join successive pairs of points (see, e.g., Figure 4.5).

8. C—factors are categorical variables, used to define groups, e.g. in ANOVA.

9. B—we cannot use the *t.test* function because we have more than two levels, but we could use either the *aov* or *lm* functions, as illustrated in the chapter.

10. A—the asterisk symbol is used to indicate factorial combination, and to generate interaction terms. A plus symbol would request additive combination, as in multiple regression.

Chapter 5

1. B—the power is the proportion of significant tests, so 2000/10000 = 0.2.

2. C—Cohen's *d* is the difference in means divided by the standard deviation, so (15 – 12)/20 = 0.15.

3. C—as we collect more data, our precision in estimating the true effect size is less subject to noise from random sampling error.

4. D—power is the probability of producing a significant effect, so a low-powered study is unlikely to do this.

5. A—when power is low, only studies with large effect sizes will be significant (see Figure 5.2), which overestimates the true effect.

6. D—enter *pwr.t.test(d = 0.8, sig.level = 0.05, power = 0.8, type='one.sample')*. The answer is 14.3, which rounds up to 15.

7. A—enter *pwr.r.test(r=0.3, n=24, sig.level=0.05)*. The answer is 0.30.

8. B—enter *pwr.anova.test(f=0.33,k=8,n=30,sig.level=0.01)*. The answer is 0.91.

9. C—enter *pwr.chisq.test(N = 12, df = 10, power = 0.5, sig.level=0.05)*. The answer is 0.88.

10. A—enter *pwr.f2.test(u=2, v=12, sig.level=0.05, power=0.8)*. The answer is 0.83.

Chapter 6

1. D—enter *mes(0.34,0.3,0.1,0.1,24,24)*; the value of *d* is 0.4.

2. B—enter *res(0.7,n=10)*; the value of *d* is 1.96.

3. A—enter *pes(0.01,30,30)*; the value of the odds ratio is 3.48.

4. D—enter *fes(13.6,17,17)*; the value of *g* is 1.24.

5. C—enter *propes(0.7,0.6,3,3)*; the value of the log OR is 0.44.

6. C—enter *meta.summaries(c(0.1,0.6,-0.2,0.9,1.1), c(0.2,0.3,0.1,0.4,0.5),method='random')*; the summary effect size is 0.371.

7. B—enter *meta.summaries(c(0.1,0.6,-0.2,0.9,1.1),c(0.2,0.3,0.1,0.4,0.5),method='fixed')*; the summary effect size is −0.00552.

8. A—assign the output of the *meta.summaries* function to a variable (e.g. called *metaoutput*). Then enter: *metaplot(effectsizes,standarderrors,summn=metaoutput$summary,sumse=metaoutput$se.summary,sumnn=metaoutput$se.summary^-2,xlab='Effect size (d)')*. The diamond overlaps the vertical dashed line, so there is no true effect.

9. A—small sample studies have lower statistical power, so in general they are less likely to produce significant effects.

10. D—the odds for the treatment group is 10/(500 – 10) = 0.02040816. The odds for the control group is 5/(400 – 5) = 0.01265823. The ratio of the two numbers is 1.61.

Chapter 7

1. B—fixed effects is the term that encompasses traditional independent variables.

2. C—random effects involve group differences on the dependent variable.

3. D—mixed-effects models are no more guaranteed to produce significant results than any other technique, as this depends on the data.

4. A—mixed-effects models are able to estimate model parameters (e.g. slope and intercept terms), despite some missing data.

5. A—the term defining the random effects (in the brackets) features a 0 for the intercept, meaning that only the slopes are random.

6. B—the term defining the random effects (in the brackets) both features a 1 requesting random intercepts, and the IV term is included, specifying random slopes.

7. C—the quantile-quantile (Q-Q) plot is used to check the residuals. It plots the data quantiles against the model prediction quantiles.

8. A—the conditional R^2 value gives the overall proportion of the variance explained by the model, including both fixed and random effects.

9. B—the AIC score incorporates the inverse likelihood and also the number of free parameters as a penalty term. Smaller scores indicate better fits. For the other three statistics, larger values indicate better fits.

10. D—celebrities will have different amounts of fame and recognizability, so image should be a random effect. Although country is a grouping variable that might be treated as a random effect, it has only two levels and is likely to be the independent variable of interest.

Chapter 8

1. D—pseudo-random numbers are generated by an algorithm. The computer's clock is entirely predictable, but can be used to seed a random number generator. The other two options are inherently random.

2. A—regardless of the shape of the individual distributions, their sum should always be normal (see the example in Figure 8.3).

3. B—the seed is the name for the value used by the algorithm to generate a sequence of random numbers.

4. C—bootstrapping is primarily used to estimate confidence intervals, though this can be applied to any statistic including the mean, median, or t-statistic.

5. C—it is very difficult to distinguish between truly random and pseudo-random numbers, and it is unlikely that this could be done with stochastic simulations.

6. A—enter *median(rgamma(100000,shape=2,scale=2))*. The output will be approximately 3.35.

7. D—we use the *quantile* function to request points from an empirical distribution that correspond to particular probabilities.

8. B—the 95% confidence intervals are taken at 2.5% and 97.5% on a distribution, which means that 95% of the values lie between those points.

9. A—this is a permutation of the original set, because each of the five numbers appears once. The other three examples include either duplicate numbers (implying resampling with replacement), or numbers that are not in the original set.

10. C—enter *hist(rpois(10000,lambda=2))*. Poisson distributions have positive skew as they are bounded at 0.

Chapter 9

1. C—there is always one more dimension than the number of free parameters (which represents the error between model and data).

2. B—for a three-dimensional space, the x- and y-coordinates represent the parameter values, and the height represents the error.

3. A—circulation is not something associated with simplex algorithms, but the other three options are geometric operations that change the shape and location of the simplex.

4. D—local minima are low regions of the error space that are not as low as the global minimum.

5. A—because the model code needs to be repeated many times, making it more efficient can speed up the fitting process.

6. C—since negative rates are not possible, this would be a sensible constraint.

7. D—*nelder_mead* can minimize arbitrary functions that we define, but the function must return the error between the model and data.

8. B—the AIC combines the error of the fit with a penalty term that is a multiple of the number of free parameters, so it takes the degrees of freedom into account when comparing models.

9. B—a small RMS error indicates a good fit. The number of free parameters will depend on a balance between model simplicity and accuracy, so there are no firm rules about how many parameters there should be.

10. A—functional models try to say something about the processes involved, as compared with descriptive models that aim only to summarize the data.

Chapter 10

1. C—the inverse transform converts from the Fourier spectrum back into the temporal/spatial domain.

2. A—the Nyquist limit is the highest frequency that can be represented, and is always half the sample rate.

3. C—the signal duration determines the frequency resolution, with the minimum frequency step being determined by 1/duration. For example, a 1-second signal has a frequency resolution of 1 Hz.

4. D—phase is an angular term measured in degrees (or alternatively, radians).

5. B—low pass filters allow through low frequencies, but block high frequencies.

6. C—these arguments calculate the Fourier transform, and then take the absolute values, which discards the phase information.

7. D—convolving two signals in the temporal domain is the same as taking the Fourier transform of both signals and multiplying them together.

8. A—the convention is to plot the DC component in the centre, with spatial frequency increasing as a function of distance from the centre, so the highest frequencies appear in the corners of the spectrum.

9. B—the phase information is extracted from the complex spectrum using the *angle* function.

10. B—assuming that the filter has already been Fourier-transformed, this line of code calculates the Fourier transform, applies the filter, then transforms back and takes the real values of the output.

Chapter 11

1. D—multivariate tests involve multiple dependent variables, regardless of the number of independent variables.

2. A—Hotelling's T^2 does not make assumptions about the relative variances of the DVs, so their units do not have to be the same. However there should be at least two dependent variables, otherwise you could just use a standard t-test.

3. B—the T^2 statistic includes terms for the means and variances of each dependent variable, as well as the covariance between them.

4. B—the F-distribution is used to determine significance, following a transform of the test statistic based on the sample size and number of dependent variables.

5. C—both N and m are important in calculating the degrees of freedom, which are used to assess statistical significance.

6. A—if the variances are not equal or the variables are correlated, the Type I error rate (false positive rate) will be inflated.

7. B—the eigenvalues are the lengths, the eigenvectors are the lines themselves, and the condition index is the square root of the ratio of eigenvalues.

8. C—the pairwise Mahalanobis distance takes into account the covariance matrices of both groups, whereas the standard Mahalanobis distance just considers one group's covariance when comparing it to a point.

9. D—the two-sample Hotelling's T^2 test requires pooling the covariances across both samples, taking into account the sample size (see equation 11.4).

10. A—surprisingly, the T^2 statistic discounts any correlation between the dependent variables, so it has no effect on statistical power.

Chapter 12

1. C—latent variables are by definition internal variables that are unobservable. However we can infer their existence from other observations.

2. B—goodness of fit is calculated by comparing the model covariance matrix with the empirical covariance matrix.

3. B—measured variables are always squares or rectangles, whereas latent variables are ovals or circles.

4. D—double-headed arrows indicate the residual variance that the other model components cannot explain.

5. C—using the equation $N = p(p+1)/2$ for $p = 5$ gives $5*(5+1)/2 = 15$.

6. A—correlations are indicated by the double tilde, whereas a latent variable is =~.

7. A—the number of data points is $N = p(p+1)/2$, which for $p = 4$ gives $4*(4+1)/2 = 10$. The number of free parameters is the four measured variables, plus the latent variable, plus the two covariances, giving seven free parameters. This means that the model is over-identified (more data points than parameters).

8. C—the root mean square error tends to zero as the model fit improves, because it is calculated by taking the square root of the mean squared error between the model and the empirical data. A perfect fit would have no residual error, so the RMSEA would be zero.

9. D—the Wald test indicates which parameters can be removed.

10. A—this is a rule of thumb, but in general correlation coefficients are hard to estimate accurately with $N<200$.

Chapter 13

1. D—Euclidean distances are the shortest straight lines between the points.

2. A—the residual lines are used to calculate the goodness of fit, as shown in Figure 13.2.

3. B—the Akaike information criterion (AIC) includes a penalty term for the number of free parameters (clusters).

4. C—rescaling ensures that each dependent variable is approximately equal in its contribution to the distances between points.

5. C—I made up inverse k-means clustering, so as far as I know it isn't a real thing!

6. A—MDS is a visualization technique that collapses multidimensional data to two or three dimensions for plotting.

7. B—Shepard diagrams plot the pairwise distances in the original space against those in the rescaled space. If the values are highly correlated this suggests that the algorithm has preserved the ordering of distances.

8. D—metric algorithms care about the distances between points, whereas non-metric algorithms just work on the rank-ordered distances.

9. D—the distance matrix does not depend on the number of dimensions (m), just the number of observations (N). There will be an entry in the matrix for each pairwise combination of points, so it will have size $N \times N$.

10. D—although technically possible, it does not make sense to produce a solution with more dimensions than in the original data set.

Chapter 14

1. B—the guess rate is $1/n$, which for four categories is 1/4 or 25%.

2. A—Cohen's d is a measure of effect size, calculated as the difference between means scaled by the standard deviation. As it increases, the group differences get bigger and so classification becomes easier (and therefore more accurate).

3. B—linear classifiers use straight lines to partition data, but with many dimensions this will become a hyperplane.

4. D—there are versions of MVPA that use correlation instead of classification to identify consistent patterns in data.

5. C—normalization involves subtracting the mean and scaling by the variances. This will often speed up classification and make it more accurate.

6. C—neural networks have hidden (sometimes called 'deep') layers that convert the inputs to outputs using various mathematical operations.

7. B—overfitting inflates accuracy, but it will not necessarily be perfect because of overlap between the data across categories (see Figure 14.1).

8. A—as with ANOVA, we can follow up the omnibus test (involving all three categories) using pairwise classifications between different groups.

9. B—support vectors are calculated using the closest few points to the decision boundary, but not those that are further away.

10. D—the structure of biological visual and auditory systems are the inspiration for deep neural networks. Note that this structure does often involve banks of filters, but these are not the primary motivation.

Chapter 15

1. A—the familywise error rate is calculated as $1-(1-\alpha)^m$, which here is $1-(1-0.01)^6 = 0.059$.

2. B—this is calculated as $\frac{\alpha}{m} = \frac{0.05}{20} = 0.0025$.

3. C—this is calculated as $\frac{\alpha}{m-(k-1)} = \frac{0.05}{12-(9-1)} = 0.0125$.

4. D—adjusting the alpha level reduces power. However Sidak correction fixes the Type I error rate so that it does not change.

5. A—the FDR is the proportion of discoveries that are false. Since Type I errors are false positives, and only significant tests are counted as discoveries, option A is the best description.

6. B—this is calculated as $\frac{q^*k}{m} = \frac{0.05 \times 50}{88}$, which is option B.

7. C—cluster correction makes sense when there are some correlations between conditions, so it is only well suited to repeated measures-type designs. If only one outcome measure is contributed by each participant, cluster correction would probably not be suitable.

8. D—the summed test-statistic (i.e. t-value) is used to calculate cluster statistics.

9. A—the cluster correction method avoids cluster shrinkage due to adjusting the α level.

10. B—null distributions are generated by randomly permuting the conditions for each participant. This is done many times for the largest cluster, to build an empirical null distribution that can be used to determine significance for each cluster.

Chapter 16

1. D—if the signal is absent and we respond 'no', we have 'correctly rejected' the possibility that the signal was present.

2. B—if the signal is present and we respond 'no', we have missed the signal, so this is called a miss.

3. A—this is calculated as hits/(hits+misses) = 28/(28+15) = 0.65.

4. C—although the statistic is closely related to Cohen's d, and is calculated using z-scores, the sensitivity index is known as d'.

5. B—d' is the difference between z-scores for hits and false alarms.

6. B—find the value of 0.9 on the y-axis of the graph, and trace along until this intercepts the curve. Then trace down to find the appropriate value on the x-axis—it is between 1 and 2, so the closest answer is x = 1.3.

7. D—the ROC curve plots hits against false alarms for a range of criteria.

8. B—for a design to be true 2AFC there must always be two stimuli to choose between. Yes/no paradigms have two responses, but only a single stimulus.

9. C—to calculate d' we need the hit rate (0.95) and the false alarm rate. The false alarm rate is 1 − correct rejections, so it is 1 − 0.8 = 0.2. Then enter qnorm(0.95) – qnorm(0.2) to get the answer 2.49.

10. A—enter qnorm(0.76) * sqrt(2) to get the answer 0.9988627, which rounds to 1.

Chapter 17

1. C—a significant p-value means we are very unlikely to have observed the data if the null hypothesis were true.

2. C—frequentist methods have a fixed Type I error rate, determined by the α level (often 0.05).

3. B—the prior describes our existing knowledge about a situation.

4. A—Bayes' theorem describes how conditional probabilities are combined and calculated, for example $P(A|B)$—the probability of A given B.

5. D—in the subscript 0 indicates the null hypothesis, and 1 indicates the alternative hypothesis. So BF_{01} is the ratio of probabilities of null over experimental.

6. A—from Jeffreys's (1961) approximation, Bayes factor (BF_{10}) values above 3 indicate some evidence in support of the alternative hypothesis (see Table 17.1).

7. B—Bayes factors decrease as a function of sample size when there is no true effect (see Figure 17.6).

8. A—Bayesian tests can provide evidence in support of the null hypothesis, when an experiment has appropriate power.

9. C—calculated as 0.67/0.02, the ratio is 33.5.

10. D—from equation 17.1: $P(A|B) = \dfrac{P(B|A)P(A)}{P(B)}$, which here will be

$P(Nov|moustache) = \dfrac{P(moustache|Nov)P(Nov)}{P(moustache)}$.

Plugging in the numbers:

$P(Nov|moustache) = \dfrac{0.3 \times \dfrac{30}{365}}{0.05} = 0.49$.

Chapter 18

1. A—log units can be quite hard to interpret. The tick marks will usually be in equal log steps, and only positively skewed distributions appear more normal. The choice of base number does not affect the appearance, but does affect the units.

2. D—starting an axis closer to the smallest value can make the differences between conditions appear larger without changing the data.

3. B—standard errors are smaller than standard deviations because $SE = SD/\sqrt{(N)}$. Standard errors will be smaller than 95% confidence intervals because $CI = 1.96 \times SE$.

4. A—individual data help to reassure the reader that there are no problems with outliers, or when there are the outliers become clear and explicit.

5. D—tritanopia occurs when the blue photoreceptors are absent, making it hard to distinguish between blue and green hues.

6. C—the parula palette is designed to change smoothly in both hue and luminance.

7. C—the idea of a perceptually uniform palette is that the perceived colour changes smoothly, even if this involves non-linear changes to the output of the three colour channels.

8. B—the alpha channel codes transparency information.

9. B—vector files contain a description of the components that make up a figure. They are typically smaller than their raster equivalents, and do not have a fixed resolution or an explicit representation of pixels.

10. C—violin plots show a kernel density function mirrored about its midpoint.

Chapter 19

1. B—if code is shared publicly, it can be accessed by anyone, so there is no need to have shared your own code. Note that answer C might seem like a disadvantage in some ways, but really it is a good thing to correct mistakes, even if it requires corrections to a publication.

2. A—with some complex analyses, graphical interfaces often have so many options that it can be difficult to precisely reproduce the original choices (even if you are the one who made them!)

3. D—digital online identifiers are important to ensure that it will always be possible to find the data even if website URLs change in the future.

4. A—the access token gives R the permission to access your own OSF projects so that you can upload and download data. Public projects are already accessible without requiring the token, and there is no permission setting that will allow you to access private projects belonging to other users.

5. C—HDF5 and csv files are both open data formats, whereas .mat and xlsx are proprietary.

6. B—A pull request downloads the current version, including any changes made by other programmers working on the project.

7. D—we first stage the changes to check differences from the main branch, then we commit those changes, and finally push them to the repository.

8. B—in lower camel case the first letter is always lower case, but subsequent words start with a capital letter.

9. A—the hash symbol (#) is used to indicate comments in R (other programming languages use different characters—for example Matlab uses a % symbol).

10. C—reproducibility is about repeating an analysis on a given data set, whereas replication is about repeating an experiment.

Alphabetical list of key *R* packages used in this book

BayesFactor—package for running Bayesian versions of various statistical tests, used in Chapter 17. Key functions include *ttestBF*, *anovaBF*, and *lmBF*.

caret—package containing functions for **C**lassification **A**nd **RE**gression **T**raining, used in Chapter 14. Key functions include *train* and *predict*.

compute.es—tools for calculating and converting effect sizes, used in Chapter 6. Key functions include *des*, *mes*, *pes*, and *tes*.

FourierStats—functions to conduct Hotelling's T^2 and T^2_{circ} tests, used in Chapter 11. Key functions include *tsqh.test*, *tsqc.test*, *CI.test*, and *pairwisemahal*.

grImport—tools to import and manipulate vector graphics files, used in Chapter 18. The key functions are *PostScriptTrace*, *readPicture*, and *grid.picture*.

jpeg—package for loading in JPEG images, used in Chapters 10 and 18. The package contains two functions, *readJPEG* and *writeJPEG*.

lavaan—**La**tent **Va**riable **An**alysis package for running structural equation models, used in Chapter 12. The key functions are *cfa*, *lavTestScore*, and *lavTestWald*.

MAd—meta-analysis with mean differences, including tools for calculating effect sizes, used in Chapter 6. Key functions include *r_to_d*, *t_to_d*, and *or_to_d*.

MASS—helper package from the textbook *Modern Applied Statistics with S* (Venables, Ripley, and Venables 2002). Contains the *isoMDS* function for non-metric multidimensional scaling used in Chapter 13, as well as the *Shepard* function to produce Shepard plots.

pals—tools for creating and evaluating colour maps and palettes for plotting, used in Chapter 18. The *kovesi* palettes are very useful, the *pal.safe* function simulates colour blindness, and the *pal.test* function can be used to evaluate a palette.

pracma—package containing many practical mathematical functions. This includes the *nelder_mead* function to implement the Nelder and Mead (1965) downhill simplex algorithm, described in Chapter 9.

PRISMAstatement—functions for creating PRISMA diagrams for use in meta-analysis, used in Chapter 6. The *prisma_graph* function generates a diagram.

psyphy—functions for analysing psychophysical data, including *d′* calculation with the *dprime.mAFC* function, as used in Chapter 16.

pwr—package of power analysis functions used in Chapter 5. Key functions include *pwr.t.test*, *pwr.r.test*, and *pwr.anova.test*.

quickpsy—package for fitting psychometric functions, used in Chapter 16. The main function is also called *quickpsy*.

remotes—tools for installing packages from repositories such as GitHub, as demonstrated in Chapter 19. The key function is *install_github*.

rmeta—meta-analysis tools used in Chapter 6. Key functions include *meta.summaries*, *metaplot*, and *funnelplot*.

semPlot—package used to produce graphical representations of structural equation models, used in Chapter 12. The key function is *semPaths*.

signal—package for signal processing. The *fir1* function was used to construct filters in Chapter 10.

References

Akaike, H. 1974. 'A New Look at the Statistical Model Identification'. *IEEE Transactions on Automatic Control* 19(6): 716–23.

Allefeld, Carsten and John-Dylan Haynes. 2014. 'Searchlight-Based Multi-Voxel Pattern Analysis of fMRI by Cross-Validated Manova'. *Neuroimage* 89: 345–57. https://doi.org/10.1016/j.neuroimage.2013.11.043.

Allen, Micah, Davide Poggiali, Kirstie Whitaker, Tom Rhys Marshall, and Rogier A Kievit. 2019. 'Raincloud Plots: A Multi-Platform Tool for Robust Data Visualization'. *Wellcome Open Research* 4: 63. https://doi.org/10.12688/wellcomeopenres.15191.2.

Allison, Paul D. 2003. 'Missing Data Techniques for Structural Equation Modeling'. *Journal of Abnormal Psychology* 112(4): 545–57. https://doi.org/10.1037/0021-843X.112.4.545.

Amrhein, Valentin, Sander Greenland, and Blake McShane. 2019. 'Scientists Rise up Against Statistical Significance'. *Nature* 567(7748): 305–7. https://doi.org/10.1038/d41586-019-00857-9.

Anscombe, F. J. 1973. 'Graphs in Statistical Analysis'. *The American Statistician* 27(1): 17–21. https://doi.org/10.1080/00031305.1973.10478966.

Baayen, R. H., D. J. Davidson, and D. M. Bates. 2008. 'Mixed-Effects Modeling with Crossed Random Effects for Subjects and Items'. *Journal of Memory and Language* 59(4): 390–412. https://doi.org/https://doi.org/10.1016/j.jml.2007.12.005.

Baker, D. H. 2021. 'Statistical Analysis of Periodic Data in Neuroscience'. *Neurons, Behavior, Data Analysis and Theory*. doi: 10.51628/001c.27680. http://arxiv.org/abs/2101.04408.

Baker, D. H., F. A. Lygo, T. S. Meese, and M. A. Georgeson. 2018. 'Binocular Summation Revisited: Beyond √2. *Psychological Bulletin* 144(11): 1186–99. https://doi.org/10.1037/bul0000163.

Baker, D. H. and B. Richard. 2019. 'Dynamic Properties of Internal Noise Probed by Modulating Binocular Rivalry'. *PLoS Computational Biology* 15(6): e1007071. https://doi.org/10.1371/journal.pcbi.1007071.

Baker, D. H., G. Vilidaite, F. A. Lygo, A. K. Smith, T. R. Flack, A. D. Gouws, and T. J. Andrews. 2021. 'Power Contours: Optimising Sample Size and Precision in Experimental Psychology and Human Neuroscience'. *Psychological Methods* 26(3): 295–314. https://doi.org/10.1037/met0000337.

Bates, D. M., Martin Mächler, Ben Bolker, and Steve Walker. 2015. 'Fitting Linear Mixed-Effects Models Using Lme4'. *Journal of Statistical Software* 67(1): 1–48. https://doi.org/10.18637/jss.v067.i01.

Begley, C. Glenn and Lee M. Ellis. 2012. 'Drug Development: Raise Standards for Preclinical Cancer Research'. *Nature* 483(7391): 531–3. https://doi.org/10.1038/483531a.

Belia, Sarah, Fiona Fidler, Jennifer Williams, and Geoff Cumming. 2005. 'Researchers Misunderstand Confidence Intervals and Standard Error Bars'. *Psychological Methods* 10(4): 389–96. https://doi.org/10.1037/1082-989X.10.4.389.

Benedek, Mathias, Barbara Wilfling, Reingard Lukas-Wolfbauer, Björn H. Katzur, and Christian Kaernbach. 2010. 'Objective and Continuous Measurement of Piloerection'. *Psychophysiology* 47(5): 989–93. https://doi.org/10.1111/j.1469-8986.2010.01003.x.

Benjamin, Daniel J., James O. Berger, Magnus Johannesson, Brian A. Nosek, E.-J. Wagenmakers, Richard Berk, Kenneth A. Bollen, et al. 2017. 'Redefine Statistical Significance'. *Nature Human Behaviour* 2(1): 6–10. https://doi.org/10.1038/s41562-017-0189-z.

Benjamini, Y. and Y. Hochberg. 1995. 'Controlling the False Discovery Rate: A Practical and Powerful Approach to Multiple Testing'. *Journal of the Royal Statistical Society: Series B (Methodological)* 57(1): 289–300. http://www.jstor.org/stable/2346101.

Benjamini, Yoav and Daniel Yekutieli. 2001. 'The Control of the False Discovery Rate in Multiple Testing Under Dependency'. *The Annals of Statistics* 29(4): 1165–88. https://doi.org/10.1214/aos/1013699998.

Biederman, I. and M. M. Shiffrar. 1987. 'Sexing Day-Old Chicks: A Case Study and Expert Systems Analysis of a Difficult Perceptual-Learning Task'. *Journal of Experimental Psychology: Learning, Memory and Cognition* 13(4): 640–5.

Bijmolt, Tammo H. A., Michel Wedel, and Wayne S. DeSarbo. 2021. 'Adaptive Multidimensional Scaling: Brand Positioning Based on Decision Sets and Dissimilarity Judgments'. *Customer Needs and Solutions* 8(7): 1–15. https://doi.org/10.1007/s40547-020-00112-7.

Bishop, Christopher M. 2006. *Pattern Recognition and Machine Learning*. New York: Springer. http://www.loc.gov/catdir/enhancements/fy0818/2006922522-d.html.

Bland, M. 2000. *An Introduction to Medical Statistics*. Oxford: Oxford University Press.

Blohm, Gunnar, Konrad P. Kording, and Paul R. Schrater. 2020. 'A How-to-Model Guide for Neuroscience'.

eNeuro 7(1): eneuro.0352-19.2019. https://doi.org/10.1523/ENEURO.0352-19.2019.

Bollen, Kenneth A., James B. Kirby, Patrick J. Curran, Pamela M. Paxton, and Feinian Chen. 2007. 'Latent Variable Models Under Misspecification: Two-Stage Least Squares (2SLS) and Maximum Likelihood (ML) Estimators'. *Sociological Methods & Research* 36(1): 48–86. https://doi.org/10.1177/0049124107301947.

Borland, David and M. Russell Taylor 2nd. 2007. 'Rainbow Color Map (Still) Considered Harmful'. *IEEE Computer Graphics and Applications* 27(2): 14–17. https://doi.org/10.1109/mcg.2007.323435.

Brodeur, Mathieu B., Emmanuelle Dionne-Dostie, Tina Montreuil, and Martin Lepage. 2010. 'The Bank of Standardized Stimuli (BOSS), a New Set of 480 Normative Photos of Objects to be Used as Visual Stimuli in Cognitive Research'. *PLoS One* 5(5): e10773. https://doi.org/10.1371/journal.pone.0010773.

Button, K. S., J. P. A. Ioannidis, C. Mokrysz, B. A. Nosek, J. Flint, E. S. J. Robinson, and M. R. Munafò. 2013. 'Power Failure: Why Small Sample Size Undermines the Reliability of Neuroscience'. *Nature Reviews Neuroscience* 14(5): 365–76. https://doi.org/10.1038/nrn3475.

Bürkner, Paul-Christian. 2017. 'brms: An R Package for Bayesian Multilevel Models Using Stan'. *Journal of Statistical Software* 80(1): 1–28. https://doi.org/10.18637/jss.v080.i01.

Bürkner, Paul-Christian. 2018. 'Advanced Bayesian Multilevel Modeling with the R Package brms'. *The R Journal* 10(1): 395–411. https://doi.org/10.32614/RJ-2018-017.

Cangur, Sengul and Ilker Ercan. 2015. 'Comparison of Model Fit Indices Used in Structural Equation Modeling Under Multivariate Normality'. *Journal of Modern Applied Statistical Methods* 14(1): 152–67. https://doi.org/10.22237/jmasm/1430453580.

Chacon, S. and B. Straub. 2014. *Pro Git*. 2nd edn. New York: Apress. https://git-scm.com/book/en/v2.

Cham, Heining, Evgeniya Reshetnyak, Barry Rosenfeld, and William Breitbart. 2017. 'Full Information Maximum Likelihood Estimation for Latent Variable Interactions with Incomplete Indicators'. *Multivariate Behavioral Research* 52(1): 12–30. https://doi.org/10.1080/00273171.2016.1245600.

Chauvenet, W. 1863. *A Manual of Spherical and Practical Astronomy*, Vol. II. Philadelphia: J. B. Lippincott & Co.

Cleveland, William S. and Robert McGill. 1983. 'A Color-Caused Optical Illusion on a Statistical Graph'. *The American Statistician* 37(2): 101–5. http://www.jstor.org/stable/2685868.

Coggan, David D., Daniel H. Baker, and Timothy J. Andrews. 2016. 'The Role of Visual and Semantic Properties in the Emergence of Category-Specific Patterns of Neural Response in the Human Brain'. *eNeuro* 3(4): eneuro.0158-16.2016. https://doi.org/10.1523/ENEURO.0158-16.2016.

Coggan, David D., Afrodite Giannakopoulou, Sanah Ali, Burcu Goz, David M. Watson, Tom Hartley, Daniel H. Baker, and Timothy J. Andrews. 2019. 'A Data-Driven Approach to Stimulus Selection Reveals an Image-Based Representation of Objects in High-Level Visual Areas'. *Human Brain Mapping* 40(16): 4716–31. https://doi.org/10.1002/hbm.24732.

Cohen, J. 1988. *Statistical Power Analysis for the Behavioral Sciences*. Hillsdale, NJ: Lawrence Erlbaum.

Colegrave, N. and G. D. Ruxton. 2020. *Power Analysis: An Introduction for the Life Sciences*. Oxford Biology Primers. Oxford: Oxford University Press.

Cramer, Angélique O. J., Don van Ravenzwaaij, Dora Matzke, Helen Steingroever, Ruud Wetzels, Raoul P. P. P. Grasman, Lourens J. Waldorp, and Eric-Jan Wagenmakers. 2016. 'Hidden Multiplicity in Exploratory Multiway ANOVA: Prevalence and Remedies'. *Psychonomic Bulletin & Review* 23(2): 640–7. https://doi.org/10.3758/s13423-015-0913-5.

Crameri, Fabio, Grace E. Shephard, and Philip J. Heron. 2020. 'The Misuse of Colour in Science Communication'. *Nature Communications* 11(1): 5444. https://doi.org/10.1038/s41467-020-19160-7.

Crowley, P., I. Chalmers, and M. J. Keirse. 1990. 'The Effects of Corticosteroid Administration Before Preterm Delivery: An Overview of the Evidence from Controlled Trials'. *British Journal of Obstetrics & Gynaecology* 97(1): 11–25. https://doi.org/10.1111/j.1471-0528.1990.tb01711.x.

Curry, Haskell B. 1944. 'The Method of Steepest Descent for Non-Linear Minimization Problems'. *Quarterly of Applied Mathematics* 2(3): 258–61. https://doi.org/10.1090/qam/10667.

De Fauw, Jeffrey, Joseph R. Ledsam, Bernardino Romera-Paredes, Stanislav Nikolov, Nenad Tomasev, Sam Blackwell, Harry Askham, et al. 2018. 'Clinically Applicable Deep Learning for Diagnosis and Referral in Retinal Disease'. *Nature Medicine* 24(9): 1342–50. https://doi.org/10.1038/s41591-018-0107-6.

Delorme, Arnaud and Scott Makeig. 2004. 'EEGLAB: An Open Source Toolbox for Analysis of Single-Trial EEG Dynamics Including Independent Component Analysis'. *Journal of Neuroscience Methods* 134(1): 9–21. https://doi.org/10.1016/j.jneumeth.2003.10.009.

Dienes, Zoltan and Neil Mclatchie. 2018. 'Four Reasons to Prefer Bayesian Analyses over Significance Testing'. *Psychonomic Bulletin & Review* 25(1): 207–18. https://doi.org/10.3758/s13423-017-1266-z.

Dong, Ensheng, Hongru Du, and Lauren Gardner. 2020. 'An Interactive Web-Based Dashboard to Track Covid-19 in Real Time'. *Lancet Infectious Disease* 20(5): 533–4. https://doi.org/10.1016/S1473-3099(20)30120-1.

Efron, Bradley and Robert Tibshirani. 1993. *An Introduction to the Bootstrap*, Vol. 57. New York: Chapman & Hall. http://www.loc.gov/catdir/enhancements/fy0730/93004489-d.html.

Eklund, Anders, Thomas E. Nichols, and Hans Knutsson. 2016. 'Cluster Failure: Why fMRI Inferences for Spatial Extent Have Inflated False-Positive Rates'. *Proceedings of the National Academy of Sciences USA* 113(28): 7900–5. https://doi.org/10.1073/pnas.1602413113.

Evans, Karla K., Tamara Miner Haygood, Julie Cooper, Anne-Marie Culpan, and Jeremy M. Wolfe. 2016. 'A Half-Second Glimpse Often Lets Radiologists Identify Breast Cancer Cases Even When Viewing the Mammogram of the Opposite Breast'. *Proceedings of the National Academy of Sciences USA* 113(37): 10292–7. https://doi.org/10.1073/pnas.1606187113.

Fahlman, A., M. Brodsky, R. Wells, K. McHugh, J. Allen, A. Barleycorn, J. C. Sweeney, D. Fauquier, and M. Moore. 2018. 'Field Energetics and Lung Function in Wild Bottlenose Dolphins, *Tursiops Truncatus*, in Sarasota Bay Florida'. *Royal Society Open Science* 5(1): 171280. https://doi.org/10.1098/rsos.171280.

Faul, F., E. Erdfelder, A. Buchner, and A.-G. Lang. 2009. 'Statistical Power Analyses Using G*Power 3.1: Tests for Correlation and Regression Analyses'. *Behavior Research Methods* 41(4): 1149–60. https://doi.org/10.3758/BRM.41.4.1149.

Faul, F., E. Erdfelder, A.-G. Lang, and A. Buchner. 2007. 'G*Power 3: A Flexible Statistical Power Analysis Program for the Social, Behavioral, and Biomedical Sciences'. *Behavior Research Methods* 39(2): 175–91. https://doi.org/10.3758/BF03193146.

Fieberg, John R., Kelsey Vitense, and Douglas H. Johnson. 2020. 'Resampling-Based Methods for Biologists'. *PeerJ* 8: e9089. https://doi.org/10.7717/peerj.9089.

Field, A., J. Miles, and Z. Field. 2012. *Discovering Statistics Using R*. London: Sage.

Field, Andy P. and Raphael Gillett. 2010. 'How to Do a Meta-Analysis'. *British Journal of Mathematical and Statistical Psychology* 63(Pt 3): 665–94. https://doi.org/10.1348/000711010X502733.

Field, D. J. 1987. 'Relations Between the Statistics of Natural Images and the Response Properties of Cortical Cells'. *Journal of the Optical Society of America A* 4(12): 2379–94. https://doi.org/10.1364/josaa.4.002379.

Fischer, H. (2011) *A History of the Central Limit Theorem: From Classical to Modern Probability Theory*. New York: Springer.

Fisher, Ronald A. 1926. 'The Arrangement of Field Experiments'. *Journal of the Ministry of Agriculture* 33: 503–15. https://doi.org/10.23637/ROTHAMSTED.8V61Q.

Forgy, E.W. 1965. 'Cluster Analysis of Multivariate Data: Efficiency Vs Interpretability of Classifications'. *Biometrics* 21(3): 768–9.

Fox, J. and S. Weisberg. 2018. *An R Companion to Applied Regression*. 3rd edn. California: SAGE Publications.

Friston, Karl. 2012. 'Ten Ironic Rules for Non-Statistical Reviewers'. *Neuroimage* 61(4): 1300–10. https://doi.org/10.1016/j.neuroimage.2012.04.018.

Friston, Karl J., Thomas Parr, Peter Zeidman, Adeel Razi, Guillaume Flandin, Jean Daunizeau, Ollie J. Hulme, et al. 2020. 'Dynamic Causal Modelling of Covid-19'. *Wellcome Open Research* 5: 89. https://doi.org/10.12688/wellcomeopenres.15881.2.

Galecki, A. and T. Burzykowski. 2013. *Linear Mixed-Effects Models Using R*. New York: Springer-Verlag.

Glasziou, P. P. and D. E. Mackerras. 1993. 'Vitamin A Supplementation in Infectious Diseases: A Meta-Analysis'. *British Medical Journal* 306(6874): 366–70. https://doi.org/10.1136/bmj.306.6874.366.

Gorgolewski, Krzysztof J., Tibor Auer, Vince D. Calhoun, R. Cameron Craddock, Samir Das, Eugene P. Duff, Guillaume Flandin, et al. 2016. 'The Brain Imaging Data Structure, a Format for Organizing and Describing Outputs of Neuroimaging Experiments'. *Scientific Data* 3: 160044. https://doi.org/10.1038/sdata.2016.44.

Green, D. M. and John A. Swets. 1966. *Signal Detection Theory and Psychophysics*. New York: Wiley.

Gronau, Quentin F., Alexander Ly, and Eric-Jan Wagenmakers. 2019. 'Informed Bayesian T-Tests'. *The American Statistician* 74(2): 137–43. https://doi.org/10.1080/00031305.2018.1562983.

Grootswagers, Tijl, Susan G. Wardle, and Thomas A. Carlson. 2017. 'Decoding Dynamic Brain Patterns from Evoked Responses: A Tutorial on Multivariate Pattern Analysis Applied to Time Series Neuroimaging Data'. *Journal of Cognitive Neuroscience* 29(4): 677–97. https://doi.org/10.1162/jocn_a_01068.

Harbord, Roger M., Matthias Egger, and Jonathan A. C. Sterne. 2006. 'A Modified Test for Small-Study Effects in Meta-Analyses of Controlled Trials with Binary Endpoints'. *Statistics in Medicine* 25(20): 3443–57. https://doi.org/10.1002/sim.2380.

Hartigan, J. A. and M. A. Wong. 1979. 'Algorithm AS 136: A K-Means Clustering Algorithm'. *Journal of the Royal Statistical Society. Series C (Applied Statistics)* 28(1): 100–108. http://www.jstor.org/stable/2346830.

Haxby, J. V., M. I. Gobbini, M. L. Furey, A. Ishai, J. L. Schouten, and P. Pietrini. 2001. 'Distributed and Overlapping Representations of Faces and Objects in Ventral Temporal Cortex'. *Science* 293(5539): 2425–30. https://doi.org/10.1126/science.1063736.

Higgins, J.P.T. and S. Green. 2011. *Cochrane Handbook for Systematic Reviews of Interventions Version 5.1.0*.

The Cochrane Collaboration. http://handbook.
cochrane.org.

Hill, Andrew A., Peter LaPan, Yizheng Li, and Steve
Haney. 2007. 'Impact of Image Segmentation on High-
Content Screening Data Quality for Sk-Br-3 Cells'. *BMC
Bioinformatics* 8: 340. https://doi.org/10.1186/1471-
2105-8-340.

Hoenig, J. M. and D. M. Heisey. 2001. 'The Abuse of
Power'. *The American Statistician* 55(1): 19–24. https://
doi.org/10.1198/000313001300339897.

Holland, John H. 1992. *Adaptation in Natural and
Artificial Systems: An Introductory Analysis with
Applications to Biology, Control, and Artificial
Intelligence*. 1st MIT Press edn. Cambridge, MA: MIT
Press.

Holm, S. 1979. 'A Simple Sequentially Rejective Multiple
Test Procedure'. *Scandinavian Journal of Statistics* 6(2):
65–70. http://www.jstor.org/stable/4615733.

Holzinger, K. and F. Swineford. 1939. *A Study in
Factor Analysis: The Stability of a Bifactor Solution*.
Supplementary Educational Monograph 48. Chicago:
University of Chicago Press.

Hotelling, H. 1931. 'The Generalization of Student's
Ratio'. *The Annals of Mathematical Statistics* 2(3):
360–78. http://www.jstor.org/stable/2957535.

Huber, P.J. 2004. *Robust Statistics*. New York: Wiley.

Ioannidis, J. P. A. 2008. 'Why Most Discovered True
Associations are Inflated'. *Epidemiology* 19(5): 640–8.
https://doi.org/10.1097/EDE.0b013e31818131e7.

Isik, Leyla, Ethan M. Meyers, Joel Z. Leibo, and Tomaso
Poggio. 2014. 'The Dynamics of Invariant Object
Recognition in the Human Visual System'. *Journal
of Neurophysiology* 111(1): 91–102. https://doi.
org/10.1152/jn.00394.2013.

Jain, Neha and Vandana Ahuja. 2014. 'Segmenting
Online Consumers Using *K*-Means Cluster Analysis'.
*International Journal of Logistics Economics and
Globalisation* 6(2): 161–78. https://doi.org/10.1504/
ijleg.2014.068274.

Jeffreys, H. 1961. *Theory of Probability*. 3rd edn. Oxford:
Oxford University Press.

Karaboga, D. 2005. *An Idea Based on Honey Bee Swarm
for Numberical Optimization*. Technical Report TR06.
Turkey: Erciyes University.

Kass, R. E. and A. E. Raftery. 1995. 'Bayes Factors'. *Journal
of the American Statistical Association* 90(430): 773–95.

Kennedy, J. and R. Eberhart. 1995. 'Particle Swarm
Optimization'. *Proceedings of ICNN'95* 4: 1942–8.
https://doi.org/10.1109/ICNN.1995.488968.

Kingdom, F. A. A. and N. Prins. 2010. *Psychophysics:
A Practical Introduction*. London: Elsevier.

Kline, Rex B. 2015. *Principles and Practice of Structural
Equation Modelling*. 4th edn. New York: Guilford.

Kolmogorov, A. N. 1992. *Selected Works II: Probability
Theory and Mathematical Statistics*. Edited by A. N.
Shiryaev. Vol. 26. Dordrecht: Springer Netherlands.

Kovesi, P. 2015. 'Good Colour Maps: How to Design
Them'. *arXiv*, 1509: 03700. https://arxiv.org/
abs/1509.03700.

Kruschke, John K. 2014. *Doing Bayesian Data Analysis:
A Tutorial with R, JAGS, and Stan*. 2nd edn. London:
Elsevier, Academic Press.

Kuhn, Max. 2008. 'Building Predictive Models in *R* Using
the caret Package'. *Journal of Statistical Software* 28(5):
1–26. https://doi.org/10.18637/jss.v028.i05.

Kuznetsova, Alexandra, Per B. Brockhoff, and Rune H. B.
Christensen. 2017. 'LmerTest Package: Tests in Linear
Mixed Effects Models'. *Journal of Statistical Software*
82(13): 1–26. https://doi.org/10.18637/jss.v082.i13.

Lakens, Daniel, Federico G. Adolfi, Casper J. Albers, Farid
Anvari, Matthew A. J. Apps, Shlomo E. Argamon, Thom
Baguley, et al. 2018. 'Justify Your Alpha'. *Nature Human
Behaviour* 2(3): 168–71. https://doi.org/10.1038/
s41562-018-0311-x.

Lambert, Ben. 2018. *A Student's Guide to Bayesian
Statistics*. London: SAGE Publications.

Lenhard, W. and A. Lenhard. 2016. *Calculation of Effect
Sizes*. Dettelbach (Germany): Psychometrica. https://
doi.org/10.13140/RG.2.2.17823.92329.

Lilja, David. 2016. *Linear Regression Using R: An
Introduction to Data Modeling*. Minneapolis: University
of Minnesota Libraries Publishing. https://doi.
org/10.24926/8668/1301.

Linares, Daniel, and Joan López-Moliner. 2016.
"quickpsy: An R Package to Fit Psychometric Functions
for Multiple Groups." *The R Journal* 8(1): 122–31.
https://doi.org/10.32614/RJ-2016-008.

Liu, Yiwen, Robert Carmer, Gaonan Zhang, Prahatha
Venkatraman, Skye Ashton Brown, Chi-Pui Pang,
Mingzhi Zhang, Ping Ma, and Yuk Fai Leung. 2015.
'Statistical Analysis of Zebrafish Locomotor Response'.
PLoS One 10(10): e0139521. https://doi.org/10.1371/
journal.pone.0139521.

Lloyd, S. 1982. 'Least Squares Quantization in PCM'. *IEEE
Transactions on Information Theory* 28(2): 129–37.
https://doi.org/10.1109/TIT.1982.1056489.

Lopes, António M., José P. Andrade, and J. A. Tenreiro
Machado. 2016. 'Multidimensional Scaling Analysis
of Virus Diseases'. *Computer Methods and Programs in
Biomedicine* 131: 97–110. https://doi.org/10.1016/j.
cmpb.2016.03.029.

Luck, Steven J. and Nicholas Gaspelin. 2017. 'How to Get
Statistically Significant Effects in Any ERP Experiment
(and Why You Shouldn't)'. *Psychophysiology* 54(1):
146–57. https://doi.org/10.1111/psyp.12639.

Macmillan, N. A. and C. D. Creelman. 1991. *Detection
Theory: A User's Guide*. Cambridge: Cambridge
University Press.

MacQueen, J. 1967. 'Some Methods for Classification and Analysis of Multivariate Observations'. In *Proceedings of the Fifth Berkeley Symposium on Mathematical Statistics and Probability*, 281–97. Berkeley, CA: University of California Press.

Mahalanobis, P. C. 1936. 'On the Generalised Distance in Statistics'. *Proceedings of the National Academy of Sciences of India* 2(1): 49–55.

Maris, Eric and Robert Oostenveld. 2007. 'Nonparametric Statistical Testing of EEG- and MEG-Data'. *Journal of Neuroscience Methods* 164(1): 177–90. https://doi.org/10.1016/j.jneumeth.2007.03.024.

Matsumoto, Makoto and Takuji Nishimura. 1998. 'Mersenne Twister: A 623-Dimensionally Equidistributed Uniform Pseudo-Random Number Generator'. *ACM Transactions on Modeling and Computer Simulation* 8(1): 3–30. https://doi.org/10.1145/272991.272995.

Meteyard, Lotte and Robert A. I. Davies. 2020. 'Best Practice Guidance for Linear Mixed-Effects Models in Psychological Science'. *Journal of Memory and Language* 112: 104092. https://doi.org/10.1016/j.jml.2020.104092.

Miranda-Mendizábal, A., P. Castellvı́, O. Parés-Badell, J. Almenara, I. Alonso, M. J. Blasco, A. Cebrià, et al. 2017. 'Sexual Orientation and Suicidal Behaviour in Adolescents and Young Adults: Systematic Review and Meta-Analysis'. *British Journal of Psychiatry* 211(2): 77–87. https://doi.org/10.1192/bjp.bp.116.196345.

Moher, David, Alessandro Liberati, Jennifer Tetzlaff, and Douglas G. Altman. 2009. 'Preferred Reporting Items for Systematic Reviews and Meta-Analyses: The Prisma Statement'. *PLoS Medicine* 6(7): e1000097. https://doi.org/10.1371/journal.pmed.1000097.

Morgan, Michael, Barbara Dillenburger, Sabine Raphael, and Joshua A. Solomon. 2012. 'Observers Can Voluntarily Shift Their Psychometric Functions Without Losing Sensitivity'. *Attention Perception, & Psychophysics* 74(1): 185–93. https://doi.org/10.3758/s13414-011-0222-7.

Navarro, D. 2019. *Learning Statistics with R*. https://learningstatisticswithr.com/.

Nelder, J. A. and R. Mead. 1965. 'A Simplex Method for Function Minimization'. *The Computer Journal* 7(4): 308–13. https://doi.org/10.1093/comjnl/7.4.308.

Newman, George E. and Brian J. Scholl. 2012. 'Bar Graphs Depicting Averages Are Perceptually Misinterpreted: The Within-the-Bar Bias'. *Psychonomic Bulletin & Review* 19(4): 601–7. https://doi.org/10.3758/s13423-012-0247-5.

Oliva, Aude and Antonio Torralba. 2001. 'Modeling the Shape of the Scene: A Holistic Representation of the Spatial Envelope'. *International Journal of Computer Vision* 42(3): 145–75. https://doi.org/10.1023/a:1011139631724.

Open Science Collaboration. 2015. 'Estimating the Reproducibility of Psychological Science'. *Science* 349(6251): aac4716. https://doi.org/10.1126/science.aac4716.

Pincus, Martin. 1970. 'A Monte Carlo Method for the Approximate Solution of Certain Types of Constrained Optimization Problems'. *Operations Research* 18(6): 1225–8. https://doi.org/10.1287/opre.18.6.1225.

Pirrone, Angelo, Wen Wen, Sheng Li, Daniel H. Baker, and Elizabeth Milne. 2018. 'Autistic Traits in the Neurotypical Population Do Not Predict Increased Response Conservativeness in Perceptual Decision Making'. *Perception* 47(10–11): 1081–96. https://doi.org/10.1177/0301006618802689.

Press, W. H., S. A. Teukolsky, W. T. Vetterling, and B. P. Flannery. 1986. *Numerical Recipes: The Art of Scientific Computing*. Cambridge: Cambridge University Press.

R Core Team. 2018. *R: A Language and Environment for Statistical Computing*. Vienna: R Foundation for Statistical Computing. http://www.R-project.org/.

Rietveld, Toni and Roeland van Hout. 2007. 'Analysis of Variance for Repeated Measures Designs with Word Materials as a Nested Random or Fixed Factor'. *Behavior Research Methods* 39(4): 735–47. https://doi.org/10.3758/bf03192964.

Rosenthal, R. 1979. 'The File Drawer Problem and Tolerance for Null Results'. *Psychological Bulletin* 86(3): 638–41.

Rosseel, Yves. 2012. 'Lavaan: An R Package for Structural Equation Modeling'. *Journal of Statistical Software* 48(2): 1–36. https://doi.org/10.18637/jss.v048.i02.

Rouder, Jeffrey N., Paul L. Speckman, Dongchu Sun, Richard D. Morey, and Geoffrey Iverson. 2009. 'Bayesian *T* Tests for Accepting and Rejecting the Null Hypothesis'. *Psychonomic Bulletin & Review* 16(2): 225–37. https://doi.org/10.3758/PBR.16.2.225.

Rousselet, Guillaume A., John J. Foxe, and J. Paul Bolam. 2016. 'A Few Simple Steps to Improve the Description of Group Results in Neuroscience'. *European Journal of Neuroscience* 44(9): 2647–51. https://doi.org/10.1111/ejn.13400.

Sánchez-Meca, Julio and Fulgencio Marin-Martinez. 2008. 'Confidence Intervals for the Overall Effect Size in Random-Effects Meta-Analysis'. *Psychological Methods* 13(1): 31–48. https://doi.org/10.1037/1082-989X.13.1.31.

Satorra, A. and W. E. Saris. 1985. 'Power of the Likelihood Ratio Test in Covariance Structure Analysis'. *Psychometrika* 50(1): 83–90.

Schwarz, Gideon. 1978. 'Estimating the Dimension of a Model'. *The Annals of Statistics* 6(2): 461–4. https://doi.org/10.1214/aos/1176344136.

Schwarzkopf, D. S. and G. Rees. 2011. 'Pattern Classification Using Functional Magnetic Resonance

Imaging'. *Wiley Interdisciplinary Reviews: Cognitive Science* 2(5): 568–79. https://doi.org/10.1002/wcs.141.

Scott Chialvo, Clare H., Pablo Chialvo, Jeffrey D. Holland, Timothy J. Anderson, Jesse W. Breinholt, Akito Y. Kawahara, Xin Zhou, Shanlin Liu, and Jennifer M. Zaspel. 2018. 'A Phylogenomic Analysis of Lichen-Feeding Tiger Moths Uncovers Evolutionary Origins of Host Chemical Sequestration'. *Molecular Phylogenetics and Evolution* 121: 23–34. https://doi.org/10.1016/j.ympev.2017.12.015.

Shanks, David R., Miguel A. Vadillo, Benjamin Riedel, Ashley Clymo, Sinita Govind, Nisha Hickin, Amanda J. F. Tamman, and Lara M. C. Puhlmann. 2015. 'Romance, Risk, and Replication: Can Consumer Choices and Risk-Taking be Primed by Mating Motives?' *Journal of Experimental Psychology: General* 144(6): e142–58. https://doi.org/10.1037/xge0000116.

Shapiro, S. S. and M. B. Wilk. 1965. 'An Analysis of Variance Test for Normality (Complete Samples)'. *Biometrika* 52(3–4): 591–611. https://doi.org/10.1093/biomet/52.3-4.591.

Smirnov, N. 1948. 'Table for Estimating the Goodness of Fit of Empirical Distributions'. *The Annals of Mathematical Statistics* 19(2): 279–81. https://doi.org/10.1214/aoms/1177730256.

Spearman, C. 1904. '"General Intelligence," Objectively Determined and Measured'. *American Journal of Psychology* 15(2): 201–92. http://www.jstor.org/stable/1412107.

Storey, John D. and Robert Tibshirani. 2003. 'Statistical Significance for Genomewide Studies'. *Proceedings of the National Academy of Sciences USA* 100(16): 9440–5. https://doi.org/10.1073/pnas.1530509100.

Stroop, J. R. 1935. 'Studies of Interference in Serial Verbal Reactions'. *Journal of Experimental Psychology* 18(6): 643–62. https://doi.org/10.1037/h0054651.

Thompson, R. 1985. 'A Note on Restricted Maximum Likelihood Estimation with an Alternative Outlier Model'. *Journal of the Royal Statistical Society: Series B (Methodological)* 47: 53–5. http://www.jstor.org/stable/2345543.

Tukey, J. W. 1977. *Exploratory Data Analysis*. Reading, MA: Addison-Wesley.

Venables, W. N., Brian D. Ripley, and W. N. Venables. 2002. *Modern Applied Statistics with S*. 4th edn. New York: Springer. http://www.loc.gov/catdir/toc/fy042/2002022925.html.

Victor, J. D. and J. Mast. 1991. 'A New Statistic for Steady-State Evoked Potentials'. *Electroencephalography and Clinical Neurophysiology* 78(5): 378–88. https://doi.org/10.1016/0013-4694(91)90099-p.

Vigen, T. 2015. *Spurious Correlations*. New York: Hachette Books.

Vilidaite, G., Anthony M. Norcia, Ryan J. H. West, Christopher J. H. Elliott, Francesca Pei, Alex R. Wade, and Daniel H. Baker. 2018. 'Autism Sensory Dysfunction in an Evolutionarily Conserved System'. *Proceedings of the Royal Society B: Biological Sciences* 285(1893): 20182255. https://doi.org/10.1098/rspb.2018.2255.

Wagenmakers, Eric-Jan. 2007. 'A Practical Solution to the Pervasive Problems of P Values'. *Psychonomic Bulletin & Review* 14(5): 779–804. https://doi.org/10.3758/bf03194105.

Wajrock, Sophie, Nicolas Antille, Andreas Rytz, Nicolas Pineau, and Corinne Hager. 2008. 'Partitioning Methods Outperform Hierarchical Methods for Clustering Consumers in Preference Mapping'. *Food Quality and Preference* 19(7): 662–9. https://doi.org/https://doi.org/10.1016/j.foodqual.2008.06.002.

Walters, Charlotte L., Robin Freeman, Alanna Collen, Christian Dietz, M. Brock Fenton, Gareth Jones, Martin K. Obrist, et al. 2012. 'A Continental-Scale Tool for Acoustic Identification of European Bats'. *Journal of Applied Ecology* 49(5): 1064–74. https://doi.org/10.1111/j.1365-2664.2012.02182.x.

Wang, Y. Andre and Mijke Rhemtulla. 2021. 'Power Analysis for Parameter Estimation in Structural Equation Modeling: A Discussion and Tutorial'. *Advances in Methods and Practices in Psychological Science* 4(1): 1–17. https://doi.org/10.1177/2515245920918253.

Wang, Yilun and Michal Kosinski. 2018. 'Deep Neural Networks are More Accurate than Humans at Detecting Sexual Orientation from Facial Images'. *Journal of Personality and Social Psychology* 114(2): 246–57. https://doi.org/10.1037/pspa0000098.

Ware, C. 1988. 'Color Sequences for Univariate Maps: Theory, Experiments and Principles'. *IEEE Computer Graphics and Applications* 8(5): 41–9. https://doi.org/10.1109/38.7760.

Weiskrantz, Larry. 2007. 'Blindsight'. *Scholarpedia* 2(4): 3047. https://doi.org/10.4249/scholarpedia.3047.

Weisstein, N. 1980. 'The Joy of Fourier Analysis'. In *Visual Coding and Adaptability*, edited by S. C. Harris, 365–80. Hillsdale, NJ: Lawrence Erlbaum.

Westfall, Jacob, David A. Kenny, and Charles M. Judd. 2014. 'Statistical Power and Optimal Design in Experiments in which Samples of Participants Respond to Samples of Stimuli'. *Journal of Experimental Psychology: General* 143(5): 2020–45. https://doi.org/10.1037/xge0000014.

Wickham, H. 2016. *ggplot2: Elegant Graphics for Data Analysis*. New York: Springer-Verlag.

Wixted, John T. 2020. 'The Forgotten History of Signal Detection Theory'. *Journal of Experimental Psychology:*

Learning, Memory, and Cognition 46(2): 201–33. https://doi.org/10.1037/xlm0000732.

Wolf, Erika J., Kelly M. Harrington, Shaunna L. Clark, and Mark W. Miller. 2013. 'Sample Size Requirements for Structural Equation Models: An Evaluation of Power, Bias, and Solution Propriety'. *Educational and Psychological Measurement* 76(6): 913–34. https://doi.org/10.1177/0013164413495237.

Xie, Y., J.J. Allaire, and G. Grolemund. 2019. *R Markdown: The Definitive Guide*. London: Chapman & Hall.

Yap, B. W. and C. H. Sim. 2011. 'Comparisons of Various Types of Normality Tests'. *Journal of Statistical Computation and Simulation* 81(12): 2141–55. https://doi.org/10.1080/00949655.2010.520163.

Yarkoni, Tal, Russell A. Poldrack, Thomas E. Nichols, David C. Van Essen, and Tor D. Wager. 2011. 'Large-Scale Automated Synthesis of Human Functional Neuroimaging Data'. *Nature Methods* 8(8): 665–70. https://doi.org/10.1038/nmeth.1635.

Yen, Tian-Ming. 2016. 'Culm Height Development, Biomass Accumulation and Carbon Storage in an Initial Growth Stage for a Fast-Growing Moso Bamboo (*Phyllostachy Pubescens*)'. *Botanical Studies* 57(1): 10. https://doi.org/10.1186/s40529-016-0126-x.

Index